Traveling with the Atom
A Scientific Guide to Europe

Traveling with the Atom
A Scientific Guide to Europe and Beyond

By

Glen E. Rodgers
Allegheny College, PA
Email: grodgers@allegheny.edu

Print ISBN: 978-1-78801-528-8
EPUB ISBN: 978-1-78801-702-2

A catalogue record for this book is available from the British Library

© Glen E. Rodgers 2020

All rights reserved

Apart from fair dealing for the purposes of research for non-commercial purposes or for private study, criticism or review, as permitted under the Copyright, Designs and Patents Act 1988 and the Copyright and Related Rights Regulations 2003, this publication may not be reproduced, stored or transmitted, in any form or by any means, without the prior permission in writing of The Royal Society of Chemistry or the copyright owner, or in the case of reproduction in accordance with the terms of licences issued by the Copyright Licensing Agency in the UK, or in accordance with the terms of the licences issued by the appropriate Reproduction Rights Organization outside the UK. Enquiries concerning reproduction outside the terms stated here should be sent to The Royal Society of Chemistry at the address printed on this page.

Whilst this material has been produced with all due care, The Royal Society of Chemistry cannot be held responsible or liable for its accuracy and completeness, nor for any consequences arising from any errors or the use of the information contained in this publication. The publication of advertisements does not constitute any endorsement by The Royal Society of Chemistry or Authors of any products advertised. The views and opinions advanced by contributors do not necessarily reflect those of The Royal Society of Chemistry which shall not be liable for any resulting loss or damage arising as a result of reliance upon this material.

The Royal Society of Chemistry is a charity, registered in England and Wales, Number 207890, and a company incorporated in England by Royal Charter (Registered No. RC000524), registered office: Burlington House, Piccadilly, London W1J 0BA, UK, Telephone: +44 (0) 20 7437 8656.

Visit our website at www.rsc.org/books

Printed in the United Kingdom by CPI Group (UK) Ltd, Croydon, CR0 4YY, UK

Foreword

This book is essential reading for any physicist, chemist, science student, or indeed anyone interested in the physical sciences while they are on their travels. There are other travel guides which cover science as a whole, and travel guides which specifically cover physics and chemistry – I contributed to a travel guide for chemists a few years ago – but none as far as I am aware that covers the history of the atom, a topic which will have a wide appeal across all the sciences. *Traveling with the Atom* enables the reader to explore historical landmarks, mostly in Europe but also elsewhere, which have an association with one of the most important ideas in the history of humankind: the concept of the atom. It covers sites as wide-ranging as homesteads, graveyards, laboratories, apartments, abbeys, and castles, which are found in rural areas, working-class conurbations, and some of the most romantic cities in Europe. They include Lismore Castle in Ireland, Bowood House in England, Westminster Abbey in London, Edinburgh New Town, the Musée des Arts et Metiers in Paris, and the Dmitry Mendeleev Memorial Museum in St. Petersburg, Russia, among many others. Outside Europe, the guide covers Priestley House in Northumberland, Pennsylvania, USA, the Rutherford Museum in Montreal, Canada, and Rutherford's Den in Christchurch, New Zealand. The sites are all ranked in

---

Traveling with the Atom: A Scientific Guide to Europe and Beyond
By Glen E. Rodgers
© Glen E. Rodgers 2020
Published by the Royal Society of Chemistry, www.rsc.org

importance, so that the visitor can decide whether a particular site is worth extending their visit for a day or making a detour to see it.

But *Traveling with the Atom* is more than a mere travel guide, it uses these historic sites to inform the reader about one of the most important developments in modern science, how scientists came to understand what the atom was and the role it plays in the cosmos: the story of brilliant insights and great blunders, of carefully designed experiments and serendipitous accidents. All of which are clearly explained in this book. You might wish to plan a vacation based on the sites listed here, as Glen and Kitty Rodgers have done for many years, or you may simply take this book on every vacation in the hope of finding a site associated with the atom while you are travelling. Either way, you will be well informed and will gain the pleasure having a deeper appreciation of a locality.

I first met Glen and Kitty when they visited the Science Museum in London, where I was in charge of the chemistry collections in 1998. I immediately realised that Glen was an excellent teacher, with massive enthusiasm and knowledge of the history of the atom. Subsequently, in 2006 we met up again to visit Corpus Christi Church in London's Covent Garden where Robert Boyle had his laboratory, helped by his German assistant Gottfried Hanckwitz (aka Ambrose Godfrey). We then had a nice meal in Rule's – London's oldest restaurant – across the road. I think any book written by Glen is worth reading, especially one about the history of the atom. Hence, I am very happy to write this foreword and to warmly commend this book to any reader with even the slightest interest in that crucial building block of matter, the atom.

Peter J. T. Morris,
Former Principal Curator (Science), Science Museum, London.

Preface

As a professor of chemistry for 35 years, I was always intrigued by the history of the atomic concept. For years, I had dreamed of visiting the actual places where these "atomic scientists" had lived and worked. In 1998, I had the opportunity to design a sabbatical leave from Allegheny College entitled "Scientific/Historical Travelling: A Self-Designed Tour of England, Scotland, France, Switzerland, and Germany". The culmination of that 1998 sabbatical was a 9-week trip, the first of 12 trips involving varying degrees of traveling with the atom.

About that time, the college initiated its Allegheny College Center for Experiential Learning (ACCEL) program, through which faculty could design "travel seminars", leading students for several weeks at a time to places all over the world to study by experiencing first-hand an issue or set of issues that were often not the primary professional interest of the faculty member. In 2002, I (along with my wife, Kitty) led "Traveling with the Atom: London and Paris" with seven students. In 2003, the college partially supported an exploratory trip to Germany, Poland, and Russia to visit more "traveling-with-the-atom" sites. In 2004, with colleagues in the economics and German departments, I co-led "Traveling in the Liberal Arts Tradition: Berlin, Leipzig, Warsaw and Prague" with 19 students. Retiring early

---

Traveling with the Atom: A Scientific Guide to Europe and Beyond
By Glen E. Rodgers
© Glen E. Rodgers 2020
Published by the Royal Society of Chemistry, www.rsc.org

gave my wife and I the opportunity to continue travel focused on the history of the atom and to start putting together the book you have in front of you. We have now been scientific/historical travelers for more than twenty years and have visited England, Scotland, Ireland, Canada, France, the Netherlands, Denmark, Switzerland, Germany, Norway, Sweden, Finland, Italy, the Czech Republic, Poland, Russia, and New Zealand. For more details on these trips, the reader can consult www.travelingatom.com.

Modeled after the way my wife and I (and occasionally other fellow travelers) traveled together, *Traveling with the Atom*, although decidedly a travel book first and foremost, became a way to prepare scientists, their students, friends, and companions to best appreciate sites they were about to visit. Given the dearth of information in traditional guidebooks, we wanted travelers to think about scientific sites as they planned their travels. An important secondary goal of the book was to demonstrate to the traveler that humankind only slowly and haltingly unraveled these atomic insights. In addition to describing the actual sites, it became important to remind fellow travelers that the pathway to the modern atom is one characterized by ingenious experiments and clear-headed observations as well as serendipitous accidents and irrelevant or ill-conceived manipulations; by great insights as well as wrong-headed ideas; by persuasive arguments by humble men and women as well as pig-headed opinions driven by big egos; by logical discussions, papers, and meetings, as well as by personal attacks, stinging diatribes, and heated debates. As a result, this book relates many of the fascinating stories associated with the development of the atomic concept. For the scientifically literate, the background it provides will serve to jog the memory and remind us of what we might have recently studied in school, or perhaps what we read or studied a number of years ago. For others not so familiar with the history of the atom, it hopefully provides a framework sufficient to help understand the significance of a given site, experiment, person, or piece of equipment. The book, then, focusses on two types of landmarks – the temporal landmarks of the history of the atomic concept and the physical landmarks that have been preserved all over the European continent to commemorate this achievement.

Preface   ix

The author being a chemical educator and not someone trained in the history of chemistry, physics, and/or science in general, this book should not be viewed as an authoritative capsule history of the atom. The historical aspects are somewhat driven by the commemorations one finds on the ground in Europe and a few places farther abroad. For example, if there are great sites devoted to Dmitri Mendeleev, the book dwells a bit on him and perhaps not so much on others who devised the predecessors to the periodic table.

Travel sites are rated from one to five atoms on a somewhat arbitrary scale based on their perceived appeal to travelers interested in the history of the atomic concept. A physical address, a set of coordinates and a brief description of the site are always provided. These things change, of course, so it's important to check them out a little before setting out on a trip. www.travelingatom.com provides a "Travelers Exchange" forum where travelers can exchange information, share experiences, and ask questions of each other and the author. I invite you to contribute to this forum.

ACKNOWLEDGMENTS

What fun it has been to be in touch with folks all over Europe and beyond about the places we initially wanted to visit and, often, other places that had not occurred to us. Listing people who have encouraged us, guided us, physically met with us, even shared a meal with us, *etc.* is always a little chancy because, for sure, we do not want to omit those who have contributed so much to our travels. Nevertheless, one should give this a try. So here we go.

Dr John Reglinski, Department of Chemistry, Strathclyde University and, with his wife, Susan, gracious hosts in Scotland; Christopher Cooksey at University College London; Dr Peter Morris formerly the Senior Curator, Experimental Chemistry, at The Science Museum; Yvonne Twomey (scientific tour leader); Dr Peter Wothers, then at St. Catharine's College, Cambridge; Dr Stephen Johnson, the Museum of the History of Science; Christiane Delpy of the National Technical Museum (Paris); Magnus Mueller at the Liebig-Museum; Dr Frank A. J. L. James, then Reader in History of Science, The Royal Institution;

Dr Gordon L. Squires, then curator of the new Cavendish Laboratory Museum; Katie Eagleton (student tour guide of scientific sites in Cambridge); Dr Rosamund Cleal, curator, the Great Circle of Avebury; Ms Ginette Gablot of Parcours des Sciences in Paris; Dr Lionel Beluze, tour guide for Parcours des Sciences; Dr Thierry Leland, curator of scientific instruments at the Conservatoire National des Arts et Metiers (Paris); Dr and Mrs Phillip Wolfe, Allegheny College Professor of French (hosts and tour leaders, Paris); Dr Antonio Moskwa, Allegheny Professor of Economics and co-leader of 2003 tour; Dr Peter Ensberg, Allegheny Professor of German (co-leader, host in Germany); Karl and Anna-Elisabeth Hansel, then at the Wilhelm-Ostwald-Gesellschaft zu Grossbothen (Anna is Ostwald's great granddaughter); Dr Karl Doblhofer at Fritz Haber Institute and, with his wife, Heide, gracious hosts in Berlin and Potsdam; Martin Fuchs, Fritz Haber Institute, Berlin-Dahlem; Dr Eckart Henning, then Director at the Max Planck Institute (M-P-Gesellschaft); Dr Siegfried Richter at the Clemens Winkler Laboratory Museum; Christopher Hamilton (adviser, St. Petersburg, Russia); Dr Igor S. Dimitriev, director of D. I. Mendeleev Museum and Archives; Eleonora Dubovitzykaya at the Museum and Archive; Dr Richard Cook, President of Allegheny College (participant, first week of 2004 tour and interpreter at the Ostwald Energy House); Rupert Baker, Library Manager at The Royal Society; Dr Michael Hunter, Director of the "The Boyle Project" at Birkbeck College; Christine Reynolds, then the Assistant Keeper of the Muniments, Westminster Abbey; Mary Hoolihan, then Curator at the Lismore Heritage Center; Dr Tracy Popey (former student and valued tour guide in Germany); Dr David Sartori (former student and valued tour guide in Germany); Dr Roland Weigand, host at the Röntgen Rooms, Wurzburg; Dr Lucio Fregonese, host at Volta Cabinet, Pavia; Dr Anna Giatti, host at Fondazione Scienza e Tecnica, Florence; Dr Giorgio Strano, host at the History of Science Museum (Museo di Storia della Scienza) now the Galileo Museum in Florence; Dr Giovanni Battinelli, then historian of physics and our host at the Physics Department, University of Rome; Felicity Pors, archivist and host, Niels Bohr Institute; Alexander Knapp (friend and tour guide, London); Stuart Martin, then Secretary/Treasurer, the Priestley Society and host in Leeds; Capt. Duncan Ferguson, The Maxwell at Glenlair Trust, host at

Preface xi

the Glenlair House; David Forfar, then Chairman of The Maxwell Foundation; Dr John Arthur at the Maxwell Foundation and tour leader on "Maxwell Walking Tour"; Jenny Stuart, Lead Visitor Coordinator, Rutherford's Den, Christchurch, NZ; Dr Norman Pohl, Director of "Historicum", TU Bergakademie Freiberg; Dr Edwin Kroke, Department of Chemistry and Physics, TU Bergakademie Freiberg; Anton N. Pronin, Acting Director, D. I. Mendeleev Institute for Metrology; Elena Ginak, Director of Metrological Museum, St. Petersburg.

Students on our two traveling-with-the-atom trips had to prepare to briefly speak about two topics during the trip. Often, these were delivered on the way to a site or even on-site at an appropriate place. For example, we gathered around the statue of Marie Skladovska-Curie at the Curie Cancer Centre in Warsaw and a student spoke about Curie's activities during World War I. Another student spoke to us about Benjamin Franklin as we all stood at the base of his statue close-by to the Eiffel Tour in Paris. These talks contributed greatly to the group's understanding of the significance of a person, place, theory or experiment related to the history of the atom. We all, students and faculty alike, greatly benefitted from these well-prepared presentations.

Student participants in "Traveling with the Atom: London and Paris", Allegheny College ACCEL Study Tour, 2002: John Krempecki, Julie Langsdale, Colby Mangini, Andrea Price, Colleen Riley, Charlie Ruggiero, and Jennifer Sexton.

Student participants in "Traveling in the Liberal Arts Tradition: Berlin, Leipzig, Warsaw and Prague, Allegheny College ACCEL Study Tour, 2004: Maria Batarce, Martin Bobak, Alexis Book, William Eckenhoff, Matthew Giordanengo, Derek J. Golna, Matthew W. Gonzalev, David Iberkleid, John T. Krempecki, Caroline Lang, Kimberly M. Lorenz, Beverly Lytle, Donald M. Marsh III, Kristin S. Marstellar, Cole M. Maxwell, Thomas J. Miller III, Maura Perry, Roger E. Pogozelski, and Colleen Zink.

Writing a book is one thing, identifying a publisher is another. Judy Mullins, a friend and freelance editor, helped me get started in pinpointing a market and writing a more succinct "elevator pitch" summarizing the project. Special thanks go out to Dr Peter Morris for his encouragement in the final phases of writing and his invaluable advice on publishing houses that would be the most receptive to this work. Professor Mary Virginia Orna at the College of New Rochelle also provided encouragement at a crucial time in the writing and was especially helpful about how to construct a comprehensive book proposal. Drew Gwilliams, Commissioning Editor, Books, at The Royal Society of Chemistry has been a positive force in publishing this book from the moment it crossed his desk. His quick, reassuring, and consistently encouraging correspondence have made writing for the RSC a most pleasurable endeavor. Katie Morrey, Editorial Assistant, RSC, has made handling the details of the publishing process much easier.

Support from family always makes a difference to an author, particularly one sometimes too engrossed in his projects. Daughters Jennifer, Emily, and Rebecca, now establishing their own careers and families, still encourage their father's writing endeavors, and what a blessing that is. My wife, Kitty, as detailed in the dedication, has always had time for seemingly countless conversations about the uncertainties, frustrations, challenges, joys, and rewards of an academic life. Now we continue to share a love of scientific/historical traveling and all the myriad aspects of life "on the road". *Traveling with the Atom* would never have come to pass without her encouragement, praise, love, and support.

Glen E. Rodgers
From the shores of a wooded, New Hampshire pond

Dedication

This book is dedicated to Kitty, my diligent and dedicated wife who makes every effort to ensure that we share the joys of a trip well-planned, well-traveled, and well-recalled, but also shares dealing with the inevitable adversities that accompany any trip no matter how well-planned it might be. She endeavors to make both ordinary and difficult things happen with an ease and sureness that defies how difficult or ordinary they might be.

This book is geared toward the scientist and his or her students, friends, and companions. Its goal is to give some insight to the men and women who devised and carried out experiments, who dreamed up ideas that helped explain these experimental results, and who collectively unveiled the mystery of the atom. As scientific/historical travelers, we hope to appreciate where and with whom they lived and worked, what equipment they used, how they looked at the world, and the nature of their contributions so that when we arrive at a site, be it rated "one-atom" or "five-atoms", we understand, as much as time and space permits, its significance. The author's explanations of these experiments and ideas intended for his "students, friends, and companions" have had to pass muster with Kitty, the author's ultimate friend and companion. Her willingness to read every chapter, to engage in conversations about these atomic ideas, and enthusiastically and critically comment on the explanations, have been invaluable.

Traveling with the Atom: A Scientific Guide to Europe and Beyond
By Glen E. Rodgers
© Glen E. Rodgers 2020
Published by the Royal Society of Chemistry, www.rsc.org

Traveling is best shared between people who share a passion for the good things of life, people who share the value that anything worth doing is worth doing as well as possible, that is, as well as it can be done – with aplomb, enthusiasm, balance, and patience. In our case, it is between two people who travel to places far and wide where we look forward to sharing the triumphs and struggles of the men and women who uncovered one of the greatest ideas developed by humankind, the atomic concept. Here we are lifting a toast to Ludwig Boltzmann, the martyr of the atom, in the Hofbrauhaus in Munich.

Contents

Chapter 1	**Traveling with the History of the Atomic Concept**	**1**
	1.1 An Overview of "European Travels with the Atom"	1
	References	8
Chapter 2	**Bookending the Atom: Boyle and Schrödinger (Southern Ireland and Dublin)**	**10**
	2.1 A Quick Look at Places to Visit "Traveling with the Atom" in Ireland	10
	2.2 Miletus, Kutna Hora, or New York	11
	2.3 Robert Boyle (1627–1691)	13
	2.3.1 Youghal (Boyle)	18
	2.3.2 Travel Sites in Dublin Related to Robert Boyle	18
	2.4 Erwin Schrödinger (1887–1961)	23
	2.5 Summary	28
	Additional Reading	28
	References	29

Traveling with the Atom: A Scientific Guide to Europe and Beyond
By Glen E. Rodgers
© Glen E. Rodgers 2020
Published by the Royal Society of Chemistry, www.rsc.org

Chapter 3 Pneumatists Set the Atomic Stage: Boyle, Hooke, Newton, Black, Cavendish, Priestley, and Davy (Western England and Northumberland, Pennsylvania) — 30

3.1 A Quick Look at Places to Visit "Traveling with the Atom" in Western England and Northumberland, Pennsylvania — 30
3.2 Robert Boyle (1627–1691) the "Chymist" — 32
  3.2.1 Robert Boyle Arrives in England — 32
  3.2.2 The Gate Piers of Boyle's Stalbridge House — 34
  3.2.3 Robert Boyle Moves to Oxford — 34
  3.2.4 Robert Boyle Moves on to London — 37
  3.2.5 Travel Sites in London Related to Robert Boyle — 39
  3.2.6 Travel Sites Related to Robert Hooke — 40
3.3 Isaac Newton (1642–1727) — 42
  3.3.1 Travel Sites Related to Isaac Newton — 43
  3.3.2 Travel Sites Related to Isaac Newton in London — 46
3.4 Scottish and English Pneumatic Chemistry: "Fixed", "Phlogisticated", and "Inflammable" Airs (1756–1772) — 47
  3.4.1 Joseph Black ("Fixed Air") — 48
  3.4.2 Travel Sites in Edinburgh Related to Joseph Black — 48
  3.4.3 Daniel Rutherford ("Mephitic or Phlogisticated Air") — 51
  3.4.4 Henry Cavendish ("Inflammable Air") — 52
  3.4.5 Joseph Priestley (1733–1804): Quintessential Pneumatic Chemist — 53
  3.4.6 Travel Sites Related to Joseph Priestley — 56
  3.4.7 Priestley Travels to Paris — 61
  3.4.8 Priestley Discovers "Dephlogisticated Air" — 62
  3.4.9 Joseph Priestley in Birmingham — 62
  3.4.10 Travel Sites in Northumberland, Pennsylvania Related to Joseph Priestley — 64

Contents　　　　　　　　　　　　　　　　　　　　　　　　　　　　xvii

	3.5 Sir Humphry Davy (1778–1829)	65
	3.5.1 The Pneumatic Institution and Maturation of Humphry Davy	68
	3.6 Summary	72
	Additional Reading	73
	References	74
Chapter 4	**Hard Spheres and Pictograms, The First Concrete Atomic Theory: John Dalton (Northern England and Manchester)**	**76**
	4.1 A Quick Look at Places to Visit "Traveling with the Atom" in Northern England and Manchester	76
	4.2 John Dalton (1766–1844)	77
	4.2.1 Dalton in Eaglesfield, Pardshaw Hall, and Kendal, England	78
	4.2.2 Travel Sites in Northern England Related to John Dalton	79
	4.2.3 Dalton's Early Work in Manchester	83
	4.2.4 Dalton's Atomic Theory	84
	4.2.5 The Reaction to Dalton's Atomic Theory	89
	4.2.6 Travel Sites in Manchester Related to John Dalton	92
	4.2.7 Travel Site in London Related to John Dalton	96
	4.3 Summary	97
	Additional Reading	97
	References	98
Chapter 5	**Electricity and the Atom: Davy, Faraday, Clerk Maxwell, and Thomson (England and Scotland, 1801–1907)**	**99**
	5.1 A Quick Look at Places to Visit "Traveling with the Atom" in England and Scotland	99
	5.2 Sir Humphry Davy (1778–1829)	100
	5.2.1 Travel Sites in London Related to Humphry Davy	102
	5.3 Michael Faraday (1791–1867)	103
	5.3.1 Travel Sites in London Related to Michael Faraday	107

	5.4 James Clerk Maxwell (1831–1879)	111
	5.4.1 Travel Sites in Edinburgh, Scotland Related to James Clerk Maxwell	117
	5.4.2 Other Travel Sites in Scotland Related to James Clerk Maxwell	121
	5.4.3 Travel Sites in Cambridge Related to James Clerk Maxwell	125
	5.4.4 Memorial to James Clerk Maxwell in Westminster Abbey, London	127
	5.5 Sir John Joseph ("JJ") Thomson (1856–1940)	128
	5.5.1 Terling, Essex (Rayleigh)	129
	5.5.2 Travel Sites in Cambridge Related to J. J. Thomson	135
	5.5.3 Travel Sites in London Related to J. J. Thomson	135
	5.6 Summary	135
	Additional Reading	137
	References	139
Chapter 6	**The Brits, Led by the "Crocodile" and His Boys, Take the Atom Apart: Ernest Rutherford (England, Scotland, Ireland, New Zealand, and Montreal)**	**140**
	6.1 A Quick Look at Places to Visit "Traveling with the Atom" in England, Scotland, Ireland, New Zealand, and Montreal	140
	6.2 Ernest Rutherford (1871–1937)	142
	6.3 Rutherford in Brightwater, Foxhill, Havelock, Nelson, and Christchurch, New Zealand	142
	6.4 Travel Sites in New Zealand Related to Ernest Rutherford	143
	6.4.1 Lord Rutherford Memorial Reserve	143
	6.4.2 Lord Rutherford Memorial Hall	144
	6.4.3 Rutherford-Pickering Memorial	144
	6.4.4 Nelson College	146
	6.4.5 The Rutherford Gallery	146
	6.4.6 Rutherford's Den, Christchurch	147
	6.5 Rutherford's Early Years at Cambridge University and the Move to Montreal	147
	6.6 Travel Sites in Montreal Related to Ernest Rutherford and Frederick Soddy	150
	6.6.1 The Rutherford Museum, Ernest Rutherford Physics Building, McGill University	150

6.6.2 Rutherford-Soddy Plaque, Schulich Library of Science and Engineering	152
6.6.3 Rutherford's Residence	152
6.7 Rutherford Moves to Manchester, England	153
6.8 Travel Sites in Manchester Related to Ernest Rutherford	158
6.8.1 Rutherford Building, University of Manchester	158
6.8.2 Schuster Laboratory (or the Schuster Building)	159
6.9 Rutherford Moves Back to "the Cavendish"	160
6.10 Travel Sites in England Related to Ernest Rutherford and "His Boys"	164
6.10.1 Cambridge	164
6.10.2 London	167
6.11 Some Other Places to Visit Related to Rutherford's "Boys"	168
6.11.1 Frederick Soddy (1877–1956)	168
6.11.2 Henry Gwyn Jeffreys Moseley (1887–1915)	169
6.11.3 Travel Sites in Oxford Related to Harry Moseley	170
6.11.4 Ernest Walton (1903–1995)	172
6.12 Summary	173
Additional Reading	175
References	175

Chapter 7 Scientists at the Heart of Westminster Abbey   176

7.1 A Quick Look at Places to Visit "Traveling with the Atom" in Westminster Abbey and Greater London	176
7.2 Summary	194
Additional Reading	195
Reference	195

Chapter 8 The New French Chemistry and Atomism: Franklin, Lavoisier, Berthollet, Gay-Lussac, Ampère (Paris I)   196

8.1 A Quick Look at Places to Visit "Traveling with the Atom" in Paris	196
8.2 Benjamin Franklin (1706–1790)	197
8.2.1 Paris (Franklin)	197

8.3	Antoine Lavoisier (1743–1794)	198
	8.3.1 Travel Sites in Paris Related to Antoine Lavoisier	206
8.4	Lavoisier's Successors	210
	8.4.1 Claude Louis Berthollet (1748–1822)	210
	8.4.2 Travel Sites Related to Claude Louis Berthollet	212
	8.4.3 Joseph Louis Gay-Lussac (1778–1850)	212
	8.4.4 Travel Sites in Paris Related to Joseph Louis Gay-Lussac	215
	8.4.5 Amedeo Avogadro (1776–1856) and André-Marie Ampère (1775–1836)	216
	8.4.6 Travel Sites Related to André-Marie Ampère	217
	8.4.7 Louis Pasteur (1822–1895)	218
	8.4.8 Travel Sites Related to Louis Pasteur	218
8.5	Summary	219
	Additional Reading	220
	References	221

Chapter 9 Atoms Go South: The Italians Volta, Avogadro, and Cannizzaro (Italy) — 222

9.1	A Quick Look at Places to Visit "Traveling with the Atom" in Italy	222
9.2	Alessandro Volta (1745–1827)	227
9.3	Travel Site Related to Luigi Galvani	231
	9.3.1 Galvani Statue	231
9.4	Travel Sites Related to Alessandro Volta	232
	9.4.1 Pavia, Italy	232
	9.4.2 Como, Italy	233
9.5	Amedeo Avogadro (1776–1856)	236
9.6	Travel Sites Related to Amedeo Avogadro	237
	9.6.1 Vercelli, Italy	237
	9.6.2 Quarenga, Italy	238
9.7	Stanislao Cannizzaro (1826–1910)	238
9.8	Travel Sites Related to Stanislao Cannizzaro and Karlsruhe	241
	9.8.1 The Museum of Chemistry "Primo Levi"	241
	9.8.2 Cannizzaro Tomb, Palermo Pantheon in the Church of San Doménico	241

Contents xxi

        9.8.3   The Baden Ständehaus   241
  9.9 Summary   242
  Additional Reading   243
  References   243

Chapter 10 **Questioning the Reality of Atoms on the Ground: Loschmidt, Mach, Boltzmann, and Ostwald (Germany and Austria)**   **244**

  10.1 A Quick Look at Places to Visit "Traveling with the Atom" Principally in Vienna, Graz, and Grossbothen   244
  10.2 Physical *Versus* Chemical Atoms (Chemical Atomists, Physical Atomists, Anti-atomists)   246
  10.3 Josef Loschmidt (1821–1895)   248
  10.4 Travel Sites Related to Josef Loschmidt   248
      10.4.1 Vienna, Austria   248
  10.5 Ernst Mach (1838–1916) *Versus* Ludwig Boltzmann (1844–1906)   249
  10.6 Travel Sites Related to Ernst Mach and Ludwig Boltzmann   254
      10.6.1 Bust of Ernst Mach in Vienna Rathauspark   254
      10.6.2 Physics Institute, University of Graz   254
      10.6.3 Duino Castle   254
      10.6.4 Boltzmann's Grave   256
      10.6.5 Café Landtmann   256
      10.6.6 The Hofbräuhaus   257
  10.7 Wilhelm Friedrich Ostwald (1853–1932)   258
      10.7.1 Wilhelm Ostwald Park and Museum   260
      10.7.2 Wilhelm Ostwald Monument   261
  10.8 Summary   261
  Additional Reading   263
  References   263

Chapter 11 **Lighting the Dark Path to Atomism: Spectroscopy Shows the Way: Fraunhofer, Bunsen, and Kirchhoff (Germany I)**   **264**

  11.1 A Quick Look at Places to Visit "Traveling with the Atom" in Munich and Heidelberg   264
  11.2 Joseph von Fraunhofer (1787–1826)   265

	11.3 Travel Sites Related to Joseph von Fraunhofer	267
	11.3.1 Benediktbeuren and Munich	267
	11.4 Robert Bunsen (1811–1899) and Gustav Kirchhoff (1824–1887)	270
	11.4.1 Plaque at Bunsen Residence	273
	11.4.2 Salinen/Gradierbau, Bad Dürkheim	274
	11.5 Bunsen and Kirchhoff Travel Sites in Altstadt Heidelberg	273
	11.5.1 Bunsen Statue	273
	11.5.2 Bunsen and Kirchhoff Plaque	275
	11.5.3 Bunsen's Laboratory and Residence	275
	11.5.4 Bunsen Plaque	276
	11.5.5 Stadthalle (Town Hall)	276
	11.5.6 Robert Bunsen Grave	276
	11.5.7 Bunsen Exhibits at the Hörsaal of the Chemistry Department	276
	11.5.8 Kirchhoff Exhibits at the Kirchhoff-Institut für Physik	277
	11.6 Other Scientific/Historical Sites in Heidelberg	277
	11.6.1 Philosopher's Way (Philosophenweg)	277
	11.6.2 Mendeleev Residence	278
	11.6.3 August von Kekulé Plaque	278
	11.6.4 Pharmacy Museum (Apothekenmuseum)	278
	11.6.5 University Library (Universitätsbibliothek)	279
	11.6.6 Carl Bosch Museum Heidelberg	279
	11.7 Spectroscopy's Role in Discovering More Elements and Their Atoms	279
	11.8 Summary	281
	Additional Reading	282
	References	283
Chapter 12	**The Danes Jump in: Ørsted and Bohr (Denmark)**	**284**
	12.1 A Quick Look at Places to Visit "Traveling with the Atom" in Denmark	284
	12.2 Hans Christian Ørsted (1777–1851)	285
	12.3 Travel Sites Related to Hans Christian Ørsted	288
	12.3.1 Rudkøbing, Island of Langeland, Denmark	288
	12.3.2 Copenhagen, Denmark	289

	12.4 Niels Bohr (1885–1962)	291
	12.5 Travel Sites Related to Niels Bohr	304
	12.5.1 Copenhagen, Denmark	304
	12.6 Other Science-related Sites in the Copenhagen Area	307
	12.7 Summary	309
	Additional Reading	310
	References	310
Chapter 13	**Röntgen Rays Revolutionize Physics and Lead to the Inner Atom: (Germany II)**	**311**
	13.1 A Quick Look at Places to Visit "Traveling with the Atom" in Central Germany (Lennup, Gießen, and Würzburg)	311
	13.2 Wilhelm Röntgen (1845–1923)	312
	13.2.1 Some Background and Perspective	312
	13.3 Travel Sites Related to Wilhelm Röntgen	320
	13.3.1 Deutsches Röntgen Birthplace and Museum	320
	13.3.2 Röntgen-memorial	322
	13.3.3 Röntgen Grave, Alter Friedhoff Cemetery	322
	13.4 An Added Attraction in Giessen	323
	13.4.1 Justus-Liebig-Museum	323
	13.5 Summary	325
	Additional Reading	326
	References	326
Chapter 14	**The Discovery That Atoms "Fly to Bits": Becquerel and the Curies (Paris and Warsaw)**	**327**
	14.1 A Quick Look at Places to Visit "Traveling with the Atom" in Paris and Warsaw Connected to the Discovery of Radioactivity	327
	14.2 Antoine Henri Becquerel (1852–1908)	330
	14.3 Travel Sites Related to Henri Becquerel in Paris	333
	14.3.1 Muséum d'Histoire Naturelle	333
	14.3.2 Henri Becquerel Plaque	333
	14.4 Marie Sklodowska-Curie (1867–1934) and Pierre Curie (1859–1906)	333

14.5	Travel Sites Related to the Curies in Paris	348
	14.5.1 Curie Residences	348
	14.5.2 Curie Workplaces	349
	14.5.3 Curie Graves	350
	14.5.4 Curie Artifacts and Tours	351
14.6	Travel Sites Related to the Curies in Ploubazlanec, l'Arcouest, Brittany	352
	14.6.1 Irène and Frédéric Joliot-Curie Memorial	352
14.7	Travel Sites Related to the Curies in Warsaw, Poland	353
	14.7.1 Maria Skłodowska-Curie Museum	353
	14.7.2 Monument to Maria Skłodowska-Curie	353
	14.7.3 Mural of Maria Skłodowska-Curie	354
	14.7.4 Commemorative Plaque	355
	14.7.5 Curie Statue	355
	14.7.6 Memorial Plaque	355
	14.7.7 Maria Skłodowska-Curie Institute of Oncology	356
	14.7.8 Maria Skłodowska-Curie Park and Statue	356
	14.7.9 Powązki Cemetery	356
	14.7.10 Mural Commemorating the Universe of Maria Skłodowska-Curie	357
14.8	Summary	357
	Additional Reading	358
	References	358

Chapter 15	**Quantum Mechanics Reluctantly Proposed: Planck and Einstein (Germany and Switzerland)**	**359**
	15.1 A Quick Look at Places to Visit "Traveling with the Atom" in Germany and Switzerland	359
	15.2 Max Planck (1858–1947)	360
	15.3 Travel Sites Related to Max Planck in Munich	366
	15.3.1 Planck Family Home	366
	15.4 Travel Sites Related to Max Planck in Berlin	366
	15.4.1 Planck Statue and Plaque at Humboldt University	366
	15.5 Travel Sites Related to Max Planck in Berlin-Dahlem	367
	15.5.1 Max Planck Residence	367
	15.5.2 Archives of the Max Planck Society	368

15.6	Travel Sites Related to Max Planck in Göttingen	369
	15.6.1 Max Planck's Grave	369
	15.6.2 I. Physikalisches Institut	369
	15.6.3 Göttingen, the Science City	369
15.7	Albert Einstein (1879–1955)	369
15.8	Travel Sites Related to Albert Einstein in Ulm, Germany	377
	15.8.1 Einstein Birthplace Monument and Memorial Plaque	377
	15.8.2 Einstein Window in Ulm Münster	378
	15.8.3 Einstein Fountain Sculpture	379
	15.8.4 Einstein House	379
15.9	Travel Site Related to Albert Einstein in Munich, Germany	379
	15.9.1 Memorial Plaque	379
15.10	Travel Sites Related to Albert Einstein in Bern, Switzerland	379
	15.10.1 Einstein House and Plaque	379
	15.10.2 Einstein Plaque at the Former Patent Office	381
	15.10.3 Café Bollwerk	381
	15.10.4 Einstein Exhibit at Historical Museum of Bern	382
15.11	Travel Sites Related to Albert Einstein in Zurich, Switzerland	382
	15.11.1 Einstein Memorial Plaque	382
	15.11.2 Bust of Einstein	382
	15.11.3 Memorial Plaque to Mileva Marić	382
15.12	Travel Sites Related to Albert Einstein in Berlin, Germany	383
	15.12.1 Einstein Memorial Plaque	383
	15.12.2 Plaque at Einstein Residence	383
	15.12.3 Great Synagogue	383
15.13	Travel Site Related to Albert Einstein in Caputh, Germany	383
	15.13.1 Einstein's Summer House	383
15.14	Travel Site Related to Albert Einstein in Brandenberg, Germany	383
	15.14.1 Einstein Tower at Telegraphenberg	383
15.15	Summary	384
Additional Reading		385
References		385

Chapter 16	**Quantum Mechanics Brings Uncertainty to the Atom: de Broglie, Schrödinger, Heisenberg, Dirac and Born (France, Switzerland, England, Austria, and Germany)**	**386**

    16.1 A Quick Look at Places to Visit "Traveling with the Atom" in France, Switzerland, England, Austria, and Germany   386
    16.2 Prince Louis-Victor de Broglie (1892–1987)   388
    16.3 Travel Sites Related to Prince Louis-Victor de Broglie   390
        16.3.1 De Broglie Grave   390
        16.3.2 Rue Maurice et Louis de Broglie   390
    16.4 Erwin Schrödinger (1887–1961)   390
    16.5 Travel Sites Related to Erwin Schrödinger in Dublin   398
    16.6 Travel Site Related to Erwin Schrödinger in Zurich   398
        16.6.1 Schrödinger Residence   398
    16.7 Travel Sites Related to Erwin Schrödinger in Vienna   398
        16.7.1 Akademisches Gymnasium Plaque   398
        16.7.2 Institut fur Radiumforschung   398
        16.7.3 Schrödinger Zimmer (Office)   398
        16.7.4 Plaque at Schrödinger's Residence   398
        16.7.5 Schrödinger Bust with Equation   399
    16.8 Travel Sites Related to Erwin Schrödinger in Alpbach, Austria   399
        16.8.1 Grave of Erwin Schrödinger   399
        16.8.2 Schrödinger Hall   400
    16.9 Travel Site Related to Erwin Schrödinger in Warsaw   400
        16.9.1 Schrödinger Equation   400
    16.10 Werner Heisenberg (1901–1976) and Paul A. M. Dirac (1902–1984)   400
    16.11 Travel Sites Related to Max Born   405
        16.11.1 Max Born Grave   405
        16.11.2 I. Physikalisches Institut (I. Physics Institute)   406
        16.11.3 Göttingen, the Science City (Stadt die Wissenschaft)   406
    16.12 Travel Sites Related to Paul A. M. Dirac in Bristol, England   406
        16.12.1 Plaque at Childhood Home   406
        16.12.2 Dirac Second Home   406

Contents   xxvii

	16.12.3 Dirac Road	407
	16.12.4 "Small Worlds" Sculpture	407
16.13	Travel Site Related to Paul A. M. Dirac in Cambridge, England	408
	16.13.1 Residence	408
16.14	Travel Sites Related to Paul A. M. Dirac in Tallahassee, Florida	408
	16.14.1 The Paul A.M. Dirac Science Library	408
	16.14.2 Dirac Grave	408
	16.14.3 Paul Dirac Drive	408
16.15	Travel Site Related to Paul A. M. Dirac in London, England	409
	16.15.1 Memorial in Westminster Abbey	409
16.16	Travel Site Related to Werner Heisenberg in Helgoland	409
	16.16.1 Heisenberg Memorial Stone	409
16.17	Travel Sites Related to Werner Heisenberg in Munich	409
	16.17.1 Plaque at Heisenberg House	409
	16.17.2 Elementary School	409
	16.17.3 Maxgymnasium	409
	16.17.4 Werner-Heisenberg-Institute of the Max-Planck-Institute of Physics	409
16.18	Summary	410
	Additional Reading	411
	References	411

Chapter 17	**Nuclear Physics with "the Pope"; Fission and the Hahn/Meitner Controversy: Fermi, Hahn, Meitner, Heisenberg (Italy, Germany, Austria, Sweden, and Norway)**	**412**
	17.1 A Quick Look at Places to Visit "Traveling with the Atom" in Italy, Germany, Austria, Sweden, and Norway Related to Nuclear Fission	412
	17.2 A Little Review of What We Know About Atomic Nuclei	414
	17.3 Enrico Fermi (1901–1954)	415
	17.4 Travel Sites Related to Enrico Fermi in Rome	423
	17.4.1 Birthplace Plaque	423
	17.4.2 Campo dei Fiori	424

17.4.3 Centro Fermi	424
17.4.4 Fermi Collection, Physics Museum of the University of Rome, Enrico Fermi Building, Citta Universitaria	424
17.5 Travel Sites Related to Enrico Fermi in Florence	425
17.5.1 Enrico Fermi Plaque and Engraving, Basilica di Santa Croce	425
17.5.2 Museo Galileo	425
17.6 Travel Sites Related to Enrico Fermi in Pisa	425
17.6.1 Domus Galilaeana	425
17.6.2 Collegio Fermi	426
17.6.3 Palazzo della Carovana	426
17.7 Travel Sites Related to Enrico Fermi in Chicago, United States	426
17.7.1 "Nuclear Energy"	426
17.7.2 Enrico and Laura Fermi "Chicago Tribute" Marker	426
17.7.3 Fermi Grave	427
17.8 The Discovery of Nuclear Fission	427
17.8.1 Otto Hahn (1879–1968) and Lise Meitner (1878–1968)	427
17.9 Travel Sites Related to Lise Meitner in Vienna, Austria	434
17.9.1 Meitner Birthplace Plaque	434
17.9.2 Akademisches Gymnasium Plaque	434
17.10 Travel Sites Related to Lise Meitner, Otto Hahn, and the Discovery of Nuclear Fission in Mitte Berlin	434
17.10.1 Meitner-Hahn Plaque	434
17.10.2 Statue of Lise Meitner	434
17.11 Travel Sites Related to Lise Meitner, Otto Hahn and Fritz Strassman and the Discovery of Nuclear Fission in Dahlem (Berlin)	435
17.11.1 Fritz Haber Institute	435
17.11.2 Plaques in Honor of the Discovery of Fission	436
17.11.3 Hahn Monument	436
17.12 Travel Site Related to Lise Meitner and the Discovery of Nuclear Fission in Kungälv, Sweden	436
17.12.1 Plaque Honoring Meitner	436

Contents   xxix

17.13 Travel Site Related to Lise Meitner in Bramley, Basingstoke and Deane Borough, Hampshire, England	437
17.13.1 Lise Meitner Grave	437
17.14 Travel Sites Related to Lise Meitner, Otto Frisch, Otto Hahn and Fritz Strassman and the Discovery of Nuclear Fission in Munich and Göttingen, Germany	437
17.14.1 Deutsches Museum	437
17.14.2 Hahn and Meitner Laboratory Table	437
17.14.3 Meitner Bust	437
17.14.4 Otto Hahn Tombstone	438
17.15 Werner Heisenberg (1901–1976)	439
17.16 Scientific-Historical Sites Related to the German Atomic Bomb Project	442
17.16.1 The Lightning Tower in Berlin-Dahlem	442
17.16.2 Vemork Power Plant in Rjukan, Norway	443
17.16.3 The Atomkeller-Museum at Haigerloch	443
17.16.4 Heisenberg Grave in Munich	444
17.17 Summary	444
Additional Reading	445
References	445

Chapter 18 Mendeleev's and Our Path to the Periodic Table: Mendeleev, Meyer, and Winkler (Russia and Germany)   447

18.1 A Quick Look at Places to Visit "Traveling with the Atom" Relative to the Development of the Periodic Table	447
18.2 Introducing Dmitri Ivanovich Mendeleev	449
18.3 Early Proposals Leading to Mendeleev's Periodic Table	450
18.4 Travel Sites Related to Johann Wolfgang Döbereiner	453
18.4.1 Bug, Weißdorf, Germany	453
18.4.2 Jena, Germany	454
18.5 Travel Sites Related to John Newlands	454
18.5.1 Newlands Royal Society of Chemistry Plaque	454
18.5.2 Newlands Grave	455

18.6	Mendeleev's Periodic Table and His Famous Predictions	455
18.7	Eka-aluminum and Gallium (Paul Émile Lecoq de Boisbaudran)	457
18.8	Travel Sites Related to Paul Émile Lecoq de Boisbaudran	459
	18.8.1 Boisbaudran Home	459
	18.8.2 Rue Lecoq de Boisbaudran	459
18.9	Eka-boron and Scandium (Lars Fredrik Nilson and Per Teodor Clève)	459
18.10	Eka-Silicon and Germanium (Clemens Winkler)	459
18.11	Travel Sites Related to Clemens Winkler in Freiberg, the Silver City of Saxony	460
	18.11.1 The Clemens Winkler Memorial (Gedenkstätte)	460
	18.11.2 Clemens Winkler Collection	460
	18.11.3 Winkler Birthplace (Geburtshaus)	461
	18.11.4 Winkler Monument	462
	18.11.5 Terra Mineralia	462
	18.11.6 Hieronymus Theodor Richter and Ferdinand Reich Monument	462
18.12	Julius Lothar Meyer (1830–1895)	462
18.13	Travel Sites Related to Julius Lothar Meyer	463
	18.13.1 Meyer Birthplace Plaque	463
	18.13.2 Lothar-Meyer Bau	463
18.14	Mendeleev: the Later Years	464
18.15	Mendeleev and the Discovery of the Inert or Noble Gases – Where Do We Put Them in the Periodic Table?	465
18.16	Travel Sites Related to Dmitri Mendeleev in Tobolsk, Siberia	466
	18.16.1 Monument to Mendeleev	466
	18.16.2 Mendeleev Exhibit	467
	18.16.3 Mendeleev Mansion	467
18.17	Travel Sites Related to Dmitri Mendeleev in St. Petersburg	467
	18.17.1 Dmitry Mendeleev's Memorial Museum Apartment (D. I. Mendeleev Museum and Archives)	467
	18.17.2 Rosstandart Metrology Museum at the D. I. Mendeleyev Institute of Metrology	468

Contents   xxxi

    18.17.3 Mendeleev Statue and Three-
        Story-High Periodic Table   469
    18.17.4 Technologichesky Institut
        (Technology Institute) Metro Stop 469
  18.18 Mikhail Lomonosov (1711–1765)   469
  18.19 Travel Sites Related to Mikhail Lomonosov
    (St. Petersburg)   471
    18.19.1 Monument to Mikhail Lomonosov 471
    18.19.2 Lomonosov Collection   471
    18.19.3 Lomonosov Grave   472
    18.19.4 Lomonosovo (Formerly
        Oranienbaum)   472
    18.19.5 "Akademik Lomonosov"   472
  18.20 Summary   472
  Additional Reading   473
  References   473

Chapter 19 **Stockholm, the Atom, and the Nobel Prizes: Berzelius, Scheele, Arrhenius, and the Atomic Nobel Prizes (Sweden)**   474

  19.1 A Quick Look at Places to Visit "Traveling
    with the Atom" in Stockholm   474
  19.2 Jöns Jakob Berzelius (1779–1848)   476
  19.3 Travel Sites in Stockholm Related to Jöns
    Jakob Berzelius   480
    19.3.1 Berzelius Museum (No Rating)   480
    19.3.2 Berzelius Statue in Berzelius Park
        (Berzelii Park)   481
    19.3.3 Berzelius Grave   481
  19.4 Other Berzelius Travel Sites   481
    19.4.1 Berzelius Bust   481
    19.4.2 Berzelius Laboratory at Gripsholm 482
    19.4.3 Berzelius School with Bust in Front 482
  19.5 Carl Wilhelm Scheele (1742–1786)   482
  19.6 Travel Sites in Stockholm Related to Carl
    Wilhelm Scheele   483
    19.6.1 Scheele Statue   483
    19.6.2 Scheele Pharmacy   484
    19.6.3 The Crown Pharmacy in Skansen
        Park   484
  19.7 Svante Arrhenius (1859–1927)   484
  19.8 Travel Sites Related to Svante Arrhenius   487

	19.8.1 Arrhenius Bust	487
	19.8.2 Arrhenius Grave	487
19.9	The Nobel Prizes	487
19.10	The Early Days of the Nobel Prizes in Chemistry: Arrhenius Carries the Day	490
19.11	The Early Days of the Nobel Prizes in Physics: The Bias Against Theory	494
19.12	The Einstein Dilemma	497
19.13	More Nobel Prizes for Atomic Scientists	499
19.14	Quantum Mechanics Finally Honored	500
19.15	Nuclear Physics and the Hahn/Meitner Debacle	501
19.16	Travel Sites in or Near Stockholm Related to the Nobel Prizes	503
	19.16.1 Stockholm Concert Hall (Stockholms Konserthus)	503
	19.16.2 Stockholm City Hall (Stockholms Stadhus)	505
	19.16.3 The Nobel Museum (Nobelmuseet)	505
	19.16.4 Alfred Nobel Grave	506
	19.16.5 Alfred Nobels Björkborn	506
19.17	Summary and Concluding Remarks	507
Additional Reading		508
References		509

Appendix — 510

Place Index — 519

Subject Index — 533

CHAPTER 1

Traveling with the History of the Atomic Concept

1.1 AN OVERVIEW OF "EUROPEAN TRAVELS WITH THE ATOM"

This is a book about traveling in Europe to visit places related to one of the most carefully researched and enduring ideas in human history: the atomic concept. These places include: homesteads, birthplaces, graveyards, squares, statues, mines, universities, laboratories, museums, libraries, lecture halls, apartments, individual rooms, estates, cathedrals, abbeys, and even castles in some of the most picturesque rural areas, charming small towns and villages, ordinary working-class municipalities, and elegant and romantic cities in Europe.

The idea of atoms has been in the human consciousness for at least 25 centuries, and most likely longer than that. Starting in the fourth century BC, the Greek philosophers were among the first to conceive the atomic idea, but they were not of one mind on the nature of matter. That everything around us is a-tomic (made of in-divisible or un-cuttable particles called "atoms") was only one avenue of Grecian thought. Without experimental or observational evidence, their arguments could continue quite unfettered without hope of resolution.

Traveling with the Atom: A Scientific Guide to Europe and Beyond
By Glen E. Rodgers
© Glen E. Rodgers 2020
Published by the Royal Society of Chemistry, www.rsc.org

Alchemists, in the pursuit of the "philosopher's stone" (to produce gold and silver from a base metal such as lead) or the "elixir of life" (the fountain of youth) also tried to visualize the structure of matter in order to discern how they could best attain their goals. For several millennia they stirred many a pot, roasted every imaginable solid to red- and white-heats, and distilled solutions of every color, odor, and texture. They devised esoteric pieces of equipment (retorts, calcinators, alembics, mortars and pestles, for example) in pursuit of these goals and, in so doing, started to construct the experimental underpinnings of the discipline of chemistry. The atomic idea can be traced back through the often mystical shrouds of alchemical works, but the evidence is fleeting and tenuous, particularly to those of us who are not historians of science specializing in such times and practices.

The path from alchemy to today's ideas about atoms is anything but a straightforward march to an inevitable and final "truth". Indeed, atoms represent an idea that presently works to explain vast aspects of the nature of matter, but it is not inconceivable that it will be replaced by other ideas in the future. Perhaps people in 3020 AD will speak of atoms as an archaic idea that they regard somewhat pejoratively alongside geocentrism, animal magnetism, and phlogistonism (see Chapter 8 for a discussion of phlogistonism). The history of the atomic concept is not one of amassing one dry experimental fact after another, marching inexorably to the grand atomic idea. Rather, this pathway is characterized by ingenious experiments and clear-headed observations, as well as serendipitous accidents and irrelevant or ill-conceived manipulations; by great insights, as well as wrong-headed ideas; by persuasive arguments from humble men and women, as well as pig-headed opinions driven by big egos; by logical discussions, papers, and meetings, as well as personal attacks, stinging diatribes, and heated debates. This book, while principally about places to visit in Europe related to the history of atomic theory, will discuss these experiments and ideas, as well as the people responsible for them when such a discussion adds to the experience of traveling. We will explore these conceptual and experimental landmarks on the way to the atomic concept as they relate to the physical landmarks that have been preserved all over the European continent to commemorate this achievement, which is often compared to the most compelling of human ideas.

We travel for a wide variety of reasons, most of which are related to visiting places and people outside our usual experiences. We have always appreciated grand seaside or mountainous vistas, bucolic countrysides, ornate palaces, unspoiled villages and towns, and the great cities of the past and present. Many of us want to become thoroughly familiar with, even totally immersed in, the cultures, customs and languages of the European nations. By understanding people in other countries we hope to better understand our own country and, indeed, ourselves.

One of the most enjoyable ways to travel is to do so with a particular focus. Sometimes this is a geographical area – a country or a group of countries, or a river such as the Rhine, Thames, or Danube, or a mountain range such as the Pyrenees, Dolomites, or Alps. Increasingly, travelers delight in concentrating on an area of human endeavor – perhaps art, architecture, music or literature; perhaps wars, battlefronts, empires, or kingdoms; perhaps cuisines, wines, or (one of the author's favorites) beers. Others concentrate on physical activities such as biking, skiing, walking, or hiking. For many of us, traveling with one or several of these purposes in mind is the most attractive way for us to organize our trips. This book is about one of those purposes, which could be described under the general heading of "scientific/historical traveling – traveling with a focus on science and the history of science". *Traveling with the Atom* is scientific/historical traveling with an emphasis on the history of the atomic concept.

Traveling with a purpose starts with doing our homework, so that we know a great deal about the attractions available in the regions that we will be visiting. Of course, doing such research makes great economic sense. Traveling today is an expensive proposition, so it makes sense to maximize our experience by reading, web surfing, e-mailing, blogging, and talking with fellow travelers as we prepare for a trip. However, there's an even better reason for making such extensive preparations. For many people, doing this research is more than half the fun – it builds anticipation and enthusiasm for a trip so that it becomes one of the landmarks of our lives – something we will never forget. It doesn't really matter what the purpose is:

art, music, scenic geography, architecture, history, atomic history – whatever! The important thing is to choose a few topics and research them. This book about traveling with the atom will be a good resource in your personal travel research. Such research involves being sure that you know the significance of a discovery, experiment, or theory and the biographical background of the workers principally responsible for that work. This book will be a starting point but, to be truly prepared, it will be best to augment and sharpen the focus on your own. Use the internet, perhaps join a discussion group or two, and identify some like-minded folks in places you will be visiting and start a correspondence with some of them. The latter may sound a little risky but you will be sitting at your computer at home or in your favorite coffee shop writing a note to someone in a European country. You have nothing to lose. Be polite (remember *The Ugly American*) and perhaps just a little humble. Don't be afraid to share some of the results of your research with these new correspondents. Perhaps suggest that you meet at a particular site during your trip and/or that you share a cup of coffee or tea or even a meal together in a place of their choosing and perhaps as your treat. I have done this a number of times and the results always add to my enjoyment of the trip both for me and those who travel with me. The research that *you* do and the contacts that *you* make as a result will invariably enrich your traveling experience as well. My wife and I have been doing this for more than 20 years now. It is a great way to travel.

As you start your research in scientific/historical traveling, there are several books, pamphlets, articles, and series of articles that you should know about. Let's start with the books. Perhaps the best on general scientific/historical traveling is *The Scientific Traveler: A Guide to the People, Places and Institutions of Europe* by Charles Tanford and Jacqueline Reynolds.[1] It covers all the mainline sciences including archaeology, biology, chemistry, geology, medicine, and physics. Published in 1992, it is organized by countries grouped into regions of Europe. For example, the section on "Western Europe" has chapters on England, Scotland, Ireland, France, and the Netherlands. Each chapter starts with a chronological description and explanation of the science developed there and ends with a "Principal Places to Visit" section that includes street addresses and other guides useful to finding the site. In 1995, these two authors also published *A Travel Guide to Scientific Sites of the British Isles: A Guide to the People, Places and Landmarks of Science*.[2] It also is organized in two parts: the first on the scientific topics themselves and the second on "places to visit". These are fun books with much valuable information, although they are nearly 25 years old. Another book on the British Isles is *Our Scientific Heritage: An A–Z of Great Britain and Ireland (Science, Technology, Archaeology, Medicine, Engineering)*[3] by Trevor Williams, published in 1996. This is an alphabetical listing of places, with descriptions of birthplaces, museums, forts, mines, mills, stone circles, and so forth. It has longer sections on Cambridge, London, Manchester, and Oxford, as well as useful locator maps for various regions. Another small book that you might be able to obtain if you are lucky is *Guide to Museums with Collections on History of Chemistry*,[4] compiled by Jan W. Van Spronsen. Published in 1996, each page lists a museum, contact information, and a general description of the museum. More recently, there is *Science History: A Traveler's Guide*,[5] published in 2014 by the American Chemical Society and edited by Mary Virginia Orna. This is a collection of articles covering all of the sciences with chapters on the United Kingdom, Paris, Italy, Scandinavia, and Germany. Finally, there is the work of James and Virginia Marshall, working out of the University of North Texas. They have published a variety of CDs and books over the years. The best source of their comprehensive work is the "Rediscovery of the Elements".[6] (Although our topic is traveling in Europe, you might also enjoy *America's Scientific Treasures, A Travel Companion*[7] by Paul S. Cohen and Brenda H. Cohen published in 1998.)

Besides the above books on scientific/historical traveling, there is a seminal article published in *Chemtech*, entitled "Chemistry Museums in Europe" by John H. Wotiz,[8] that every traveler interested in the history of science should know about. Wotiz spent most of his career teaching at Southern Illinois University and led a series of European travel and study courses for the better part of 30 years. In the *Chemtech* article he describes the museums in Europe that he often included in those trips. One of the best elements of this article is a "Directory of European History of Chemistry Museums", which includes a short description and a few comments about each one and then, uniquely as far as I can tell, a rating system from one to five. A local science library may very well have the journal *Chemtech* available, or perhaps you can order it by interlibrary loan. Wotiz also published an excellent article in 1972 in the *Journal of Chemical Education* entitled "The Evolution of Modern Chemistry, an European Travel and Study Course".[9] If you would like to know a little more of what it was like to travel with the Wotiz group, see "A View from the Cockpit: A Mid-Summer's 'Flight' through Chemical Europe", by Leigh Wilson, in the ACS's *Science History: A Traveler's Guide*.[10] According to an obituary that appeared in 2001 after they died together in an automobile accident, John Wotiz and his wife Kay tried "to visit every place on the earth before they could no longer travel". A worthy ambition, that.

There is a continuing series of articles that has appeared fairly regularly in the journal *Physics in Perspective* that goes by the general title "The Physical Tourist". The articles cover history of science traveling sites in Ireland, Geneva, Copenhagen, Glasgow, New York City, Madrid, Berlin, Vienna, Cracow, "Lake Wobegon" (this is not a misprint), Paris, Munich, Los Alamos, and Cambridge, England. Some of these articles have been published as a book, *The Physical Tourist: A Science Guide for the Traveler*[11], edited by John S. Rigden and Roger H. Stuewer (Eds.), Birkhäuser Verlag AG, Basel, Boston, Berlin (2009).

The internet is also an invaluable source of travel information. As you are reading here you will no doubt want to "Google" some topics and start to amass the latest information and more details about the places you want to visit. Bookmarking these sites will soon yield a vast amount of information at your ready disposal. Many websites have been referred to in this book but these things change quickly so it will behoove you to use the key words and phrases you read about to do searches on your own.

The book is organized around countries, regions, and cities that have significant places – sometimes just a plaque or a statue or a grave, but other times truly wondrous places like a beautiful park, museum, apartment, institute, laboratory, or even a castle or cathedral – that honor, at least in part, some aspect of the development of the atom. Most often, these places are centered around a person or a small group of people. This person-centric approach to traveling with the atom is pretty much dictated by the way that the travel sites are set up on the ground. Chapters 2 through 7 are centered in the United Kingdom and Ireland and move from Southern Ireland (Chapter 2) to Western England (Chapter 3), Northern England (Chapter 4), and Scotland (Chapter 5). Specific cities are covered in various of these chapters. These include Dublin (Chapter 2), Leeds (Chapter 3), Manchester (Chapters 4 and 6), Glasgow and Edinburgh (Chapter 5), Cambridge (Chapters 3, 5, and 6), and London (Chapters 5, 6, and 7). Occasionally, we go to more remote places like the South Island of New Zealand and Montreal, Canada (Chapter 6). Chapter 7 is exclusively devoted to a scientific tour of Westminster Abbey.

Starting in Chapter 8 we move from the United Kingdom to the continent. Chapters 8 and 14 are centered in Paris and, occasionally, other parts of France. In Chapter 9 we move to Italy and Chapters 10 and 11 are centered in Germany and Austria. Chapter 12 is in Denmark; Chapters 13 to 16 take us to Germany, Poland, Switzerland, and France; Chapter 17 roams through Italy, Germany, Austria, and Sweden; and Chapter 18 is in Russia and Germany. At the end of the book, in Chapter 19, we will find ourselves in Stockholm, where the Nobel Prizes in the disciplines of Physics, Chemistry, Physiology, or Medicine (not the Peace Prize, which is given out in Oslo) have been awarded each year for more than a century. Our arrival there will allow us some opportunities to summarize the atomic story and our travels to see places that commemorate it in Europe and beyond. Also in this chapter, we will explore some of the complicated politics involved in awarding the Nobel Prizes relative to the atomic concept.

The two previous paragraphs give us a brief overview of the book's geographical organization. There are several other ways to get an overview of this organization. One of the best and quickest ways to get the lay of the land – or "the lay of the book", if you like – is by looking at beginning and end of each chapter. Each chapter starts with a "Quick Look at Places to Visit" covered in the chapter. Borrowing from Wotiz, these "Quick Looks" embody a rating system to indicate the relative importance and drawing power of each place. The rating system is set up from one ⚛ to five atoms ⚛⚛⚛⚛⚛. One-atom places are isolated markers, plaques, and some statues. Five-atom places represent the cream of the crop of "traveling with the atom" sites. If you have to choose, by all means hit the three-, four-, and five-atom spots first. This rating system for places is also incorporated in the narrative of each chapter. At the start of each chapter, the "Quick Looks" is immediately followed by a "gray box" which contains a summary of the relevant aspects of the history of the atom covered in that chapter. At the end of each chapter is a summary section where in one or two paragraphs the history and places covered in the chapter are discussed together.

Maps (called "charts" in this book) as well as longitude/latitude coordinates make a travel book more useful and therefore a better travel companion. Most chapters contain two, three, or even more charts of a given city or region. Coordinates are given when a given place to visit is discussed in the body of a chapter.

To make this book a good reference book, there are useful indices at the back. In addition to a general, comprehensive subject index, there is also a place index. The place index also indicates where places are shown on a chart. There is also a specialized appendix entitled, "A Traveler's Guide to the History of the Atom", providing a chronological listing of events and people crucial to the history of the atom.

REFERENCES

1. C. Tanford and J. Reynolds, *The Scientific Traveler: A Guide to the People, Places, and Institutions of Europe*, John Wiley & Sons, Inc., New York, 1992.
2. C. Tanford and J. Reynolds, *A Travel Guide to Scientific Sites of the British Isles*, John Wiley & Sons, Ltd., Chichester, 1995.
3. T. I. Williams, *Our Scientific Heritage: An A–Z of Great Britain and Ireland (Science, Technology, Archaeology, Medicine, Engineering)*, Sutton Publishing Limited, Phoenix Mill, 1996.
4. J. W. Van Spronsen, *Guide of European Museums with Collections on History of Chemistry*, Federation of European Chemical Societies, Antwerp, 1996.
5. M. V. Orna, *Science History: A Traveler's Guide*, American Chemical Society, Washington DC, 2014.

6. J. L. Marshall and V. R. Marshall, *Rediscovery of the Elements*, http://www.chem.unt.edu/~jimm/REDISCOVERY%207-09-2018/, accessed August 2019.
7. P. S. Cohen and B. H. Cohen, *American Scientific Treasures: A Travel Companion*, American Chemical Society, Washington D.C, 1998.
8. J. H. Wotiz, Chemistry Museums in Europe, *Chemtech*, April 1982, 221–228.
9. J. H. Wotiz, The Evolution of Modern Chemistry, a European Travel and Study Course, *J. Chem. Educ.*, 1972, **49**, 593–595.
10. L. Wilson, in *Science History: A Traveler's Guide*, ed. M. V. Orna, American Chemical Society, Washington DC, 2014.
11. *The Physical Tourist: A Science Guide for the Traveler*, ed. J. S. Rigden and R. H. Stuewer, Birkhäuser Verlag AG, Basel, Boston, Berlin, 2009.

CHAPTER 2

Bookending the Atom: Boyle and Schrödinger (Southern Ireland and Dublin)

2.1 A QUICK LOOK AT PLACES TO VISIT "TRAVELING WITH THE ATOM" IN IRELAND

The Atom in Southern Ireland – Miletus, Kutna Hora, and New York – The Robert Boyle Room ❋❋❋ in the Lismore Heritage Center – Lismore Castle ❋❋; the Birthplace of Robert Boyle – The Boorish Boyle Family Memorial ❋ in St. Mary's Collegiate Church, Youghal – The Outrageous Boyle Family Memorial ❋❋ in St. Patrick's Cathedral, Dublin – Boyle Bust in the Long Room of the Old Library, Trinity College Dublin ❋❋; Erwin Schrödinger's Kincora Road residence ❋; the plaque marking Schrödinger's Merrion Square workplace ❋ at the former site of the Dublin Institute for Advanced Studies; the "?What is Life?" sculpture ❋❋ in the National Botanic Gardens of Ireland.

Traveling with the Atom: A Scientific Guide to Europe and Beyond
By Glen E. Rodgers
© Glen E. Rodgers 2020
Published by the Royal Society of Chemistry, www.rsc.org

The Atom in Southern Ireland and Dublin

- Robert Boyle (1627–1691) is a figure who marks the transition from the shadows of alchemy to what we recognize today as modern chemistry.
- Not an alchemist, not a chemist, but a "sceptical chymist", Boyle believed that it was possible to transform base metals to gold but that the best chance for success was an open and well-documented experimental approach.
- He rejected the Greek view of the elements and advocated instead a more modern definition. He envisioned each element as composed of unique "corpuscles", or what we would call atoms.
- He interpreted his law ("Boyle's Law") relating the pressure and volume of a gas (ordinary air) in terms of corpuscles colliding with the sides of the container.
- Any account of the development of the modern atom should arguably start with the contributions of Robert Boyle, who was born in Southern Ireland.
- Erwin Schrödinger (1887–1961), an Austrian physicist who was born 260 years after Boyle, is a seminal atomist who developed a way of looking at the internal structure of the atom that we still use to this day.
- He lived and worked in Dublin from 1940 to 1956 at the then newly established Dublin Institute of Advanced Studies. He became a naturalized Irish citizen in 1948.

2.2 MILETUS, KUTNA HORA, OR NEW YORK

So, we promised in the Preface that we would gradually talk about how this great concept of the atom came into being. Where did this idea start and why do we begin with southern Ireland? After all, most of us know that many of the great ideas of human beings started in Greece. The atom is no exception to that general rule. A viable starting place for our "travels with the atom" journey would be a beach on the Aegean Sea, where we could imagine Leucippus and his student Democritus, perhaps around the year 430 BC, strolling along together and thinking about the concept of infinity. Leucippus is traditionally thought to have been born in the coastal town of Miletus (37.387733°, 27.256163°) at the mouth of the Maeander River in what is now the Aydin province of Western Turkey. We travelers identify with a word like "meander", which comes from the name of this river and refers to its winding pattern. Democritus was from the Greek coastal town of Abdera (41.003721°, 25.022089°) on the northern coast of the Aegean. We can imagine Leucippus and Democritus philosophizing that the concept of infinity was not appropriate when thinking about matter. As together they meandered along a smooth beach that they knew was composed of tiny particles of sand, they wondered how small those particles could ultimately be. Could they be divided indefinitely, that is, forever? Most likely not. They wondered if the sea water could be similarly divided into smaller and smaller units. They decided there must be a stopping place in the march toward infinity. In the case of matter, there must be parts that are partless and particles that are in-divisible or "a-tomic". They envisioned these "atoms" as homogeneous, solid, and indestructible hard spheres. The concept of the atom was born. Therefore, our travels with the atom adventure could very well start in seaside places like Miletus or Abdera.

Or perhaps we could start in a facsimile of an alchemists laboratory in Kutna Hora (49.952362°, 15.276914°), just east of Prague in the Czech Republic, where a precious metal like silver could be dissolved in "aqua fortis" (what we know today as nitric acid) producing a beautiful translucent blue solution (a result of the copper impurities in early samples of silver) and accompanied by a burst of red-brown gas (what we know today as nitrogen dioxide) that hovers over the solution. As students of science (including scientists, their students, friends, and companions), many of us are able to step back and see this as an almost magical process. The mystery of such a spectacular experimental beginning is enhanced when a series of procedures results in the regeneration of the silver metal. How are such wondrous processes to be explained? The idea that matter was composed of "corpuscles" was a good way to think about alchemical manipulations and can be traced back to at least the fourteenth century. Perhaps our travels with the atom should start by staring at the alembics and retorts of an alchemical laboratory.

Or, we could start in front of a Georges Seurat pointillistic painting such as "Entrance to the Harbor, Port-en-Bessin" at the Museum of Modern Art in New York (40.761480°, −73.977660°). This is shown in Figure 2.1. We are "scientific/historical travelers",

Figure 2.1 Port-en-Bessin, Entrance to the Harbor. 1888, Georges-Pierre Seurat. New York Digital Image © The Museum of Modern Art/Licensed by SCALA/Art Resource, NY.

yes, but many of us often like to travel with multiple purposes in mind. For some of us the focus of our travels also includes the pursuit of great art in the form of paintings, sculptures, and architecture, as well the performing arts of music and dance. Art sometimes expresses profound ideas in ways that science cannot. Pointillistic paintings have something in common with the human search for the fundamental particles of matter. These paintings *look* smooth and continuous if you are far enough away from them, but when examined closely one discovers that they are composed of tiny dots of paint that blend together to produce the overall effect. Matter too looks continuous to the casual observer, but up close (very close indeed!) we discover, somewhat non-intuitively at first, that it is composed of tiny dots of matter that we might call corpuscles or atoms.

2.3 ROBERT BOYLE (1627–1691)

> I now mean by elements ... certain primitive or simple, or perfectly unmingled bodies; which not being made of any

other bodies, or of one another, are the ingredients of which all those called perfectly mixed bodies are immediately compounded, and into which they are ultimately resolved. Robert Boyle[4]

So, we begin not in Greece, Turkey, the Czech Republic, or New York, but instead in Southern Ireland in Lismore, County Waterford. And what a beautiful place it is to start. Chart 2.1 shows the general location of Lismore and the other atom-related towns in this region. If you go to a more detailed chart of Ireland, you will see that Lismore is about halfway between Waterford and Cork or, farther afield, halfway between Rosslare Harbour on St. George's Channel to the east, and Killarney to the west. To the north, Lismore is bordered by the Knockmealdown Mountains and the River Blackwater that flows down to the harbor town of Youghal (pronounced "Yawl") on the Irish Sea to the south. Lismore is a microcosm of Ireland with the added benefit of being the birthplace of the first person who could, with some

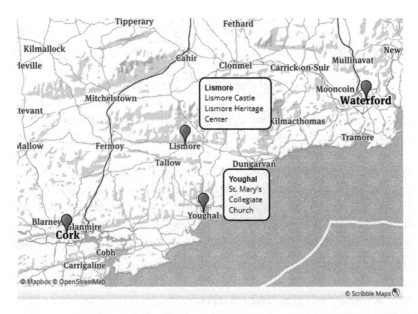

Chart 2.1 Travel sites in Southern Ireland where Robert Boyle was born and his family is memorialized. Boyle was born in a castle in Lismore. His family is memorialized in St. Mary's Collegiate Church in Youghal.

merit, be called the "father of modern chemistry". Lismore, the birthplace and childhood home of Robert Boyle, is an appropriate place to start our travels-with-the-atom adventures. Located in the heart of the country, it offers an accurate view of small-town Ireland (in contrast to the tourist traps of Killarney or, God forbid, the Blarney Stone). This clean and attractive town was designated the "Overall National Winner of TidyTowns 2004". Its Millennium Park is small and beautiful. The ice house, at the high point of the park, where salmon was stored during the winters of World War II, is particularly intriguing. Stand there and try to imagine living in this small town 75 years ago. The Lismore Heritage Centre (located in the former courthouse directly across the street from the park) (52.137146°, −7.932885°) uses colored artwork and multimedia presentations to describe the history of this little town, which was founded in 636 AD.

The center properly includes homage to its native son, Robert Boyle, universally regarded as one of the most important figures in the founding of modern chemistry and the history of the atomic concept. Opened in 2004, the Robert Boyle Room is beautifully decorated with several large colored and well-lit posters about Boyle and his work. A large portrait of Boyle (after that by Johann Kerseboom), shown in Figure 2.2, is enclosed within an elaborate wood and glass cabinet that folds out to reveal a timeline of science, with an appropriate emphasis on Irish scientists. All around the perimeter of the room above the cabinet and posters is an elegant border listing the important Boyle publications in chronological order. The cabinet also contains equipment and supplies for a variety of hands-on science activities (including one on Boyle's Law itself) for children aged 5 to 12 who are invited in from the surrounding schools. Given his determination to base science on experiments and observations, Boyle would definitely approve of this lovely and efficient room constructed and actively used in his honor. It is an excellent example of good science and history of science education.

In 1627, Robert Boyle was born in Lismore Castle (52.140502°, −7.932031°) (on Castle Avenue, just down the street from Millennium Park) as the 15th child (and seventh son) of Richard Boyle, the 1st Earl of Cork, 1st Viscount Dungarvan, 1st Baron Boyle of Youghal, Lord High Treasurer of the Kingdom of Ireland! Richard was a self-made man and his rise to great

Figure 2.2 The portrait of Robert Boyle that is displayed in the Robert Boyle Science Room in the Lismore Heritage Center. This is identical to The Shannon Portrait of the Hon. Robert Boyle F.R.S. by Johann Kerseboom, 1689. Used with permission of the National Portrait Gallery, London.

wealth (he was the richest man in the British Isles at the time of Robert's birth) is a bit beyond the scope of this book. Suffice it to say that he greatly benefitted from Britain's reconquest of Ireland at the Battle of Kinsale (southwest of Cork) in 1601 and the confusion and graft that ensued. He also married into wealth and in turn married off many of his children (but not Robert) to his great financial advantage. These were far from ordinary marriages: when 15-year-old Francis Boyle married, King Charles I gave the bride away, the Queen assisted the bride in preparing herself for bed, and both the King and Queen stayed to see the young couple into bed together! Lismore Castle was built in the late twelfth century by King John. The complex struggle to anglicize Ireland ultimately led to Sir Walter Raleigh, a favorite of Queen Elizabeth I, receiving a large land grant that included Lismore and Youghal. Raleigh, however, was preoccupied with being a queenly favorite and did little with his holdings. As a result, Richard Boyle purchased Lismsore Castle and other neglected properties from Raleigh in 1602. Until recently, the castle has been a traditional residence, but lately it has become available to private parties for some astoundingly high prices. Rentals come with the Duke's personal team of butlers "who shower guests with royal attention". The castle gardens are open to the public and there is a contemporary art gallery located in the west wing. Google "Lismore Castle Arts" and see what exhibits, or perhaps cinema might be playing when visiting the birthplace of Robert Boyle! According to Charles Tanford and Jacqueline Reynolds, in *The Scientific Traveler*, "one can get a fine view of the castle all year round from the road that enters Lismore from Clogheen to the north – a highly scenic road, crossing high moorland ..." and running through the Knockmealdown Mountains, which Tanford and Reynolds note is "... worth taking for its own sake".

Lismore Castle eventually passed into the hands of the Cavendish family, who were the Dukes of Devonshire. (The 4th Earl of Cork died in 1753 without a male heir. The ownership of the castle passed to his eldest daughter Charlotte Elizabeth Boyle (1731–1754) who married William Cavendish (1720–1764), the 4th Duke of Devonshire. As a consequence the Devonshire family still owns the castle.) The Dukes of Devonshire are of the same lineage as the scientist Henry Cavendish (1731–1810), who discovered what he called "inflammable air" but today is known as hydrogen gas (see Chapter 3, p. 52), and also William Cavendish, the Seventh Duke of Devonshire who founded the great Cavendish Laboratory of Physics at Cambridge in 1874 (see Chapter 5, p. 116 and Chapter 6, pp. 160–167). There is also an American connection to Lismore and its castle. It seems that Lord Charles Cavendish married Adele Astaire and they lived in Lismore Castle from 1932 to 1944. Their marriage ended the brother/sister dancing partnership of Fred and Adele Astaire, but Fred was a frequent visitor to Lismore. They are reputed to have enjoyed dancing together on the bridge on the east side of town near the Brothers School. Fred's next dancing partner was, of course, Ginger Rogers.

The Robert Boyle Summer School, established in 2011, is a celebration of the life, works, and legacy of Robert Boyle. This four-day event, meant for both scientists and non-scientists, includes an introduction to the life and works of Boyle, costumed recreations of Boyle's most famous experiments and a guided tour of the castle gardens. A Winter School was added in 2018. For current information, contact http://www.robertboyle.ie/

2.3.1 Youghal (Boyle)

The elder Boyle had a very high opinion of himself. To get an idea just how high, take the wooded scenic route that runs south from Lismore to Youghal. Part of the route runs alongside the River Blackwater that was at the heart of the Boyle "empire". The young Robert most likely traveled it quite often. Once in Youghal, with its colorful main street and gate, seek out St. Mary's Collegiate Church (51.955053°, −7.853548°), where there is a memorial to Richard Boyle in the south transept, also known as The Boyle Chapel. The 1st Earl of Cork lies recumbent with nine of his 15 children below him. Robert had not been born when this rather garish Italian-style memorial was erected (see Figure 2.3). Youghal was the headquarters for Sir Walter Raleigh, and the Elizabethan mansion where he lived is now a private residence on the grounds adjacent to the church.

2.3.2 Travel Sites in Dublin Related to Robert Boyle

2.3.2.1 St. Patrick's Cathedral. There is another memorial to the Boyle family in the National Cathedral and Collegiate Church of Saint Patrick, Dublin, otherwise known as St. Patrick's Cathedral (53.339571°, −6.271472°). See Chart 2.2. Depicted in Figure 2.4, the memorial is 40 feet tall and variously described as gaudy, tasteless, monstrous, garish, and utterly vulgar. Therefore, try to see it if you can! Perhaps you would like to add another adjective to the above list! Nevertheless, this memorial has small statues at its base representing 11 of the Boyle children, including the young Robert who is shown in the central arch of the lowest tier of the monument. Incredibly, it originally stood behind the high altar in a position that worshippers, sitting in the nave, could see it as they looked through the archway at the front of the cathedral. Richard Boyle erected this memorial to honor his wife, Catherine, her parents, and grandparents. Catherine was the mother of 14 of the 15 Boyle

Figure 2.3 The memorial to Sir Richard Boyle, 1st Earl of Cork, at St. Mary's Collegiate Church in Youghal, Ireland. It is located in the South Transept, also known as The Boyle Chapel. Reproduced from ref. 1 [https://www.geograph.ie/photo/1392224] under the terms of a CC BY-SA 2.0 license [https://creativecommons.org/licenses/by-sa/2.0/], Copyright Mike Searle.

children, of which Robert was the last. He was born early in 1627; she died early in 1629 in her late 40s. Catherine, her father, and grandfather are all buried in the altar area with many others (in the order of 600 others). (Richard is buried at St. Mary's Collegiate Church in Youghal.) The monument was completed in 1632 when Robert was only 5 years old.

Somewhat surprisingly, given its size and "grandeur", the memorial only occupied its original location for two years. In 1634, over the vehement objections of the 1st Earl of Cork himself, it was summarily removed and boxed. This would have been no mean task lightly ventured, but apparently the Archbishop of Canterbury, William Laud, was concerned that if it remained directly behind the altar the people would not know whether they were worshipping God or the 1st Earl of Cork. The monument remained in storage for 229 years, during which time it

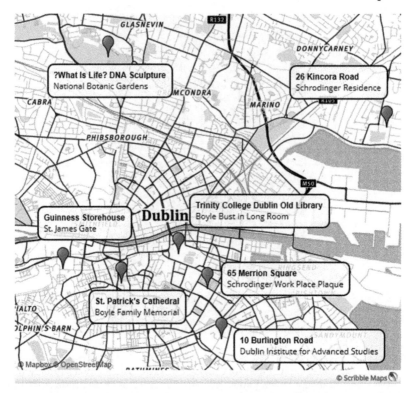

Chart 2.2 Travel sites related to Robert Boyle and Erwin Schrödinger in Dublin. These include the location of the elaborate Boyle family memorial in St. Patrick's Cathedral as well as Schrödinger's residence and workplace and the new statue inspired by Schrödinger's book *What is Life?*

suffered significant decay, resulting in the loss of the effigies of several family members. Jonathan Swift, author of the satirical "Gulliver's Travels" and dean of the cathedral from 1713 to 1745, was responsible for much of its restoration, but the Boyle memorial was not placed in its present position until 1863. The public entrance to the cathedral is on the southwest side of the cathedral. Once you enter, turn left and you will be face to face with the Boyle Memorial, right next to the gift shop area. Jonathan Swift himself is buried in the west end, very close to the Boyle memorial.

Incidentally, St. Patrick's Cathedral was fully renovated from 1860 to 1865 by the third-generation owner of the Guinness

Figure 2.4 The Boyle Memorial in St. Patrick's Cathedral, Dublin. The small figure in the lower tier is the 5-year-old Robert Boyle, surrounded by a number of his 14 siblings. His mother and her family are buried here. Reproduced from ref. 2 [https://commons.wikimedia.org/w/index.php?curid=25144654] courtesy of Andreas F. Borchert under the terms of a CC BY-SA 4.0 [https://creativecommons.org/licenses/by-sa/4.0/].

Brewery, who was at that time the richest man in Ireland. There is a statue of Sir Benjamin Lee Guinness on the grounds of the cathedral directly in view when you enter the church. On one side of the plinth it says, "St. Patrick's Cathedral by Him Restored A.D. 1865". He agreed to fully fund the renovation of the cathedral under the condition that he have complete control. In his honor, you might walk seven or eight blocks from the cathedral to the nearby famous Guinness Storehouse at St. James's Gate (53.341677°, −6.286883°), go up to the Gravity Bar, have them properly pour you a perfect stout and raise a toast to the Guinness family, Jonathan Swift, the 1st Earl of Cork Richard Boyle, and his little boy Robert, one of the first and most important figures in the modern history of the atomic concept!

2.3.2.2 The Long Room, Trinity College Library, Dublin.

The Trinity College Library (53.343939°, −6.256838°) houses the largest collection of manuscripts and printed books in Ireland. This runs to over three million volumes housed in eight buildings. Three areas of the Old Library are open to the public. These include a Treasury that displays medieval gospel manuscripts, most famously the *Book of Kells*. The Long Room is the main chamber of the Old Library and houses more than 2 00 000 of the library's oldest books. Both of these are impressive places and well worth the price of admission, but of particular interest to travelers-with-the-atom are the 38 marble busts of "men eminent for learning" that stand guard on either side of the library. These include the expected familiar names such as Aristotle, Cicero, Homer, Plato, Shakespeare, and Socrates, as well as John Milton, John Locke, Jonathan Swift, Francis Bacon, and, of course, Isaac Newton. But, perhaps a little unexpectedly, we find the 1st Earl of Cork's 14th child, the little boy whom we found standing a bit sheepishly at the center of the grand memorial in St. Patrick's Cathedral, Robert Boyle. Walk along the broad aisle of this 65-meter-long hall, find the oldest Irish harp in existence, look up into the stacks housing a quarter of a million books, and then stand and admire the bust of Robert Boyle.

As we might readily imagine, there was nothing easy about being the 7th son of the 1st Earl of Cork. However, there *were* certain advantages. For one, Robert didn't have to worry about being involved in political, military, and public affairs – his older brothers were expected to cover all these areas of endeavor! At the same time, he enjoyed the benefits of the great wealth amassed by his father. So, for example, in 1639 (at the age of 12) he and his brother Francis (who, as described above, had just been married – only 4 days before!) were sent with a tutor on an incredibly expensive trip abroad that included exotic places like Paris, Geneva, and Florence. In Geneva, where he and his brother studied for nearly two years, Robert found a rich man's sport that he liked – tennis! He also found himself in the middle of a fearsome thunderstorm that inspired a strong religious faith in which he would find comfort all his life. In Boyle's own words, he "was suddenly waked in a Fright with such loud Claps of Thunder ... & every clap ... both preceded & attended with Flashes of lightning so numerous ... & so dazling, that [he] began to imagine ... the

Day of Judgment's being come". He resolved during that storm to devote his life to promoting God's work on earth. Now, lots of us, particularly in our younger years, make such promises when we think we are in the valley of the shadow of death, but Robert Boyle, even with great wealth at easy command most of his life, kept his promise.

In 1641, during the height of Irish rebellions that left his father's lands devastated and two of his brothers dead from battle wounds, Robert was safely tucked away in Florence. He was still there in 1642 when Galileo died after being under house arrest in nearby Arcetri for the last nine years of his life. Galileo Galilei (1564–1642) is recognized today for his clear descriptions of his innovative, quantitative experimental methods and for his mathematical expressions of the results. Galileo stressed an inductive approach to science whereby carefully designed experiments lead to generalizations (variously dignified as axioms, hypotheses, theories, or laws) about nature. Such generalizations must be consistent with further experiments or be replaced with others that are. His quantitative approach and emphasis on inductive methods so revolutionized the approach to the study of natural phenomena that he is often called the "Father of Science". Robert carefully studied Galileo's work (involving projectiles, pendulums, inclined planes, thermometers, telescopes, and so forth) and methods, but there is no evidence that they actually met. Nevertheless, as we will see in the next chapter, the young Robert Boyle was definitely influenced by the Galilean ideas of how to conduct scientific studies.

2.4 ERWIN SCHRÖDINGER (1887–1961)

> Science is a game – but a game with reality, a game with sharpened knives .... If a man cuts a picture carefully into 1000 pieces, you solve the puzzle when you reassemble the pieces into a picture; in the success or failure, both your intelligences compete. In the presentation of a scientific problem, the other player is the good Lord. He has not only set the problem but also has devised the rules of the game – but they are not completely known, half of them are left for you to discover or to deduce. Erwin Schrödinger (quoted in ref. 3, p. 348)

Robert Boyle and Erwin Schrödinger are, to some extent, bookends in the story of traveling with the atom and both have roots in Dublin. Boyle was not only one of the first close-to-modern atomic scientists, he was a theologian and a great believer in the Christian God. Schrödinger, nearly at the other end of the atomic timeline, was a penultimate purveyor of the modern image of the atom, an image we still use and teach today, but, dramatically unlike Boyle, despite the above quote, he was a non-believer extraordinaire.

Schrödinger is usually described as an Austrian physicist. He was born and studied in Vienna and received his PhD in Physics at the university there in 1910. He was an artillery officer in the Austro-Hungarian Military Forces during World War I. After that he held various positions in Stuttgart, Jena, Breslau, Zurich, and Berlin. Much of his seminal contributions to the atomic concept were carried out at the University of Zurich between 1921 and 1927. (He treated electrons in atoms as if they were waves. Using "wave mechanics" modern physicists and chemists picture electrons in atoms as occupying "atomic orbitals" abbreviated as 1s, 2s, 2p and so forth. We will consider Schrödinger's contributions to the modern picture of the atom in detail in Chapter 15.) From 1927 to 1933 he held an important position in Berlin where he and Albert Einstein became colleagues and friends. His interest in wave mechanics was sparked by a footnote in a paper by Einstein that cited the PhD thesis of Prince Louis de Broglie. (See Chapter 16 on de Broglie.) Both Einstein and Schrödinger worked extensively on a "unified field theory" for the next 20 years and occasionally corresponded back and forth about their progress. Neither was able to complete such a theory.

When Hitler came to power in Germany in 1933, Schrödinger felt he had to leave and spent three years in Oxford, England, where during his first week there he learned that he had been awarded the Nobel Prize in Physics. In 1936, the pull of his native Austria was too strong and, even with the impending unease, he found a position in Graz. When Austria was annexed by Germany in 1938, however, he was curtly dismissed for his "political unreliability". Leaving most of his belongings, his money, and his Nobel Prize gold medal behind, he and his wife Anny escaped to Italy, where they were met by Enrico Fermi (see Chapter 17), who lent them some money. (The University of Graz was soon renamed

Adolf Hitler University and, under Nazi auspices, offered courses in "racial studies", chemical warfare, and the proper use of extermination agents and facilities.)

During that time Eamon de Valera, the Prime Minister of the newly established (in 1937) Eire, that is, the Irish Republic, was looking to establish a Dublin Institute for Advanced Studies to be modeled after the Institute of Advanced Study at Princeton that was founded in 1930. He invited Schrödinger, who was already world famous due to his much-exalted wave theory of the atom, to be the first professor of theoretical physics at the institute. After a circuitous and sometimes harrowing journey that included short stays in Oxford and Gent, the Schrödingers arrived in Dublin, a city they so enjoyed that they lived there for 17 years until Erwin retired and they moved back to Austria. They settled at 26 Kincora Road, Clontarf, Dublin (53.362833°, −6.203406°), near the medieval village close to Clontarf Castle.

It is an understatement to say that Schrödinger's lifestyle was radically different from that of his colleagues – and certainly a world away from that of Robert Boyle. Erwin and Anny Schrödinger (who married in 1920 and were still together when Erwin died in 1961) had dubious views of traditional marriage and both, but particularly Erwin, engaged in numerous and open extramarital affairs. When they arrived in Dublin, Erwin and Anny were accompanied by Erwin's mistress Hilde and their young daughter Ruth. Erwin gave instructions that Anny and Hilde were to be treated with exactly the same respect. Anny and Hilde evidently saw it this way as well at least most of the time, supervising household duties on a rotating basis, with Anny doing the honors one week and Hilde the next. One can imagine that this was not always a comfortable situation. However, Schrödinger was not content there. During his time in Dublin, he had several other lovers and fathered children by two of them. One of these new lovers was pregnant while he was carrying on an affair with another who also became pregnant. Occasionally, Erwin would bring one of these women to Clontarf for tea. Anny would entertain them, Ruth was friendly but, apparently, Hilde was not so welcoming.

Schrödinger became a naturalized Irish citizen in 1948. There is a plaque erected by the Irish-Austrian Society at the Kincora Road house that says Schrödinger lived here . You might stand looking at the house and think about the unorthodox – perhaps even by today's standards – "Clontarf Household", as it was sometimes called, that once existed here during World War II and afterwards.

While in Dublin, Schrödinger continued his work on a unified field theory that would encompass both gravitation and electromagnetic fields and other projects in theoretical physics, none of which was as important as his wave theory. However, he also contributed some thoughts on the biology of inheritance. In his little book *What is Life?*, published in 1944, he suggested that chromosomes contain a message written in a "code-script" – what today we call the genetic code – that is faithfully replicated time after time, in an organism. Occasionally, a sudden and distinct change or mutation occurs during these replications. (These mutations and natural selection are intimately involved in Darwin's theory of evolution.) Schrödinger was convinced that the mutations occurred on the molecular level, that is, they were a rearrangement of atoms in the chromosome. Both biologists and physicists who read *What is Life?* were persuaded that their disciplines had more of an overlap than they had realized. For example, the book encouraged and influenced both James Watson and Francis Crick, who went on to discern the helical structure of DNA. Crick had a degree in physics, but was inspired to apply this knowledge to biological problems. Watson, a biologist by training, said that after he read *What is Life?*, he was "polarized toward finding out the secret of the gene" by applying ideas drawn from physics. In 2013, an elaborate aluminum sculpture called "?What is Life?" was dedicated in the National Botanic Gardens of Ireland (53.372578°, −6.271918°). Shown in Figure 2.5, it celebrates the 60th anniversary (in 1953) of the discovery of the DNA double helix by James Watson and Francis Crick. This multi-faceted sculpture mounted on a green hill in these beautiful botanic gardens represents the transmission of genetic information among DNA, RNA, and proteins.

The Dublin Institute for Advanced Studies was located in two interconnected eighteenth-century Georgian houses on the south side of Merrion Square in Dublin. Initially it included the School of Celtic Studies and the School of Theoretical Physics. Schrödinger was the initial director and first professor of the latter. There is a plaque to this effect at 65 Merrion Square

Figure 2.5 The "?What is Life?" sculpture at the National Botanic Gardens of Ireland. Photograph by Glen Rodgers.

(53.338626°, −6.249437°); "Erwin Schrödinger, Creator of Wave Mechanics, worked here 1940–1956". There is not room on the plaque to detail his other activities. The French Embassy is right next door and the Georgian garden Merrion Square Park is across the street. The poet and novelist Oscar Wilde lived here and there is an amazing multi-colored statue of Wilde on the northwest corner of the square.

The Dublin Institute for Advanced Studies (now located at 10 Burlington Road) (53.331322°, −6.245717°) continues to flourish. It still has the Schools of Celtic Studies and Theoretical Physics but now has added a School of Cosmic Physics. The nearby Trinity College Dublin hosts an annual "Schrödinger Lecture Series" in October or November.

Schrödinger was an accomplished cyclist and almost always biked the 6 km to work from his Kincora Road residence over the River Liffey to Merrion Square even during the winter months. Schrödinger was also a devoted theater-goer. He liked the Abbey Theater (26 Abbey Street Lower) (53.341677°, −6.286883°) and the Gate Theater (Cavendish Row, Parnell Square) (53.353369°, −6.262022°) that offer numerous performances each year. The Abbey and Gate theaters offered controversial content during the 1920s and 1930s. For this reason, they were known as the Sodom and Begorrah (not Gomorrah – Google this for some history that you can use to entertain your friends). If you attend a performance at one of these theaters, think of Erwin and Anny, or Erwin and Hilde, or Erwin and whomever attending one together. They probably had good reasons for loving these theaters, begorrah.

There are numerous plaques to famous personages all over the city of Dublin. Included in this august list are Oscar Wilde, James Joyce, Bernard Shaw, William Butler Yeats, Ernest Shackleton, Bram Stoker, and, yes, Erwin Schrödinger. If Schrödinger's inclusion in this list delights you and your traveling friends, you might be interested in trying to find a science walking tour in Dublin. There have been some good examples of these over the years although, presently, they are difficult to find. On these walks, you might even learn more about Boyle's gaudy family memorial or Schrödinger's unorthodox love life!

2.5 SUMMARY

Where does the road to traveling with the atom begin? We might have started on an Aegean Sea beach in 460 BC. We could have started in a traditional alchemist's lab in the Czech Republic. We even could have started in front of a pointillistic painting representing how something continuous like a beach or the sea can be represented with a series of tiny paint dots. Instead we have started in the emerald isle of Ireland with the birth of a "sceptical chymist", who later stated his belief in atoms and elements roughly as we view them today. Certainly, the atomic idea started slowly. Sometimes it was advanced one step and then retreated two, but gradually this idea started to gain ground. The first viable landmark in our "travels with the atom" adventure is the birth place of Robert Boyle, a sceptical chymist, born in Lismore Castle in beautiful southern Ireland, the 7th son of the 1st Earl of Cork. Also in Southern Ireland, we can see impressive but gaudy memorials to the Boyle family in Youghal and Dublin. Dublin also was a place that another important atomist spent considerable time. Later in this book we will learn in more detail of Schrödinger's considerable contributions to the atomic concept. Here in Dublin we can see where Erwin Schrödinger, an Austrian physicist but a naturalized Irish citizen, lived, loved, and worked.

ADDITIONAL READING

- G. Murchie, in *The Mystery of Matter*, ed. L. B. Young, The Atom is Conceived, Oxford University Press, New York, 1965, pp. 18–19.

- Alchemy Museum, Kutna Hora, Czech Republic, https://www.atlasobscura.com/places/muzeum-alchymie, accessed March 2019.
- J. Gribbin, *The Scientists, A History of Science Told through the Lives of its Greatest Inventors*, Random House, New York, 2002, pp. 126–144.
- C. Tanford and J. Reynolds, *The Scientific Traveler, A Guide to the People, Places & Institutions of Europe*, John Wiley & Sons, Inc., New York, 1992.
- J. Gribbin, *Erwin Schrödinger and the Quantum Revolution*, John Wiley & Sons, Inc., Hoboken, New Jersey, 2013.

REFERENCES

1. St Mary's Collegiate Church, Youghal (Sir Richard Boyle Memorial), https://www.geograph.ie/reuse.php?id=1392224, accessed June 2019.
2. Dublin St. Patrick's Cathedral Nave Boyle Monument 2012, https://commons.wikimedia.org/w/index.php?curid=25144654, accessed June 2019.
3. W. Moore, *Schrödinger: Life and Thought*, Cambridge University Press, Cambridge, 1989.
4. R. Boyle, *The Sceptical Chymist*, J. Cadwell, 1661.

CHAPTER 3

Pneumatists Set the Atomic Stage: Boyle, Hooke, Newton, Black, Cavendish, Priestley, and Davy (Western England and Northumberland, Pennsylvania)

3.1 A QUICK LOOK AT PLACES TO VISIT "TRAVELING WITH THE ATOM" IN WESTERN ENGLAND AND NORTHUMBERLAND, PENNSYLVANIA

Robert Boyle, the lions of his Stalbridge House ▨*; University College, Oxford, plaque for Boyle and Hooke's Laboratory* ▨*; Oxford University Museum of Natural History* ▨▨*; Corpus Christi Roman Catholic Church, London, built on the site of Boyle and Ambrose Godfrey's phosphorus factory* ▨*; the Royal Society on Carlton House Terrace, London* ▨▨*; Robert Hooke's birthplace at Hooke Hill, the Isle of Wight* ▨▨*; grave in St. Helen Churchyard, Bishopsgate, London* ▨*; memorials to Hooke in Westminster Abbey* ▨ *and St. Paul's Cathedral* ▨*, at the Monument to the Great Fire of London* ▨▨*; Isaac Newton's Woolsthorpe Manor* ▨▨▨ *(where the apple fell), his King's School* ▨ *and the Grantham Museum* ▨▨*; Site of Newton's*

Traveling with the Atom: A Scientific Guide to Europe and Beyond
By Glen E. Rodgers
© Glen E. Rodgers 2020
Published by the Royal Society of Chemistry, www.rsc.org

alchemical laboratory ※※ at Cambridge University; Newton's statue in the Trinity College Chapel ※※※ Cambridge; several Newton Sites ※ near Leicester Square, London – Joseph Black Sites in Edinburgh; Portrait ※ in the Scottish National Portrait Gallery, Portrait in the Raeburn Building ※; Balance and glassware ※※ in the National Museum of Scotland, Royal Society Plaque near the Joseph Black Building ※※ in the King's Buildings section with its Chemistry Collection ※; Black's elaborate grave ※※ in the locked Covenantor's Prison section of Greyfriars Kirkyard, Black's residence at Sylvan Place ※; Joseph Priestley Mill Hill Chapel ※※※; Priestley ※※ and James Watt ※ statues, Leeds Library ※※※ (founded by Priestley and others); Joseph Priestley's birthplace in Fieldhead ※; statue in Birstall ※※; picture and plaque at public house ※; Priestley's house ※ on the Green in Calne and the nearby pond named after him ※ ; International Historic Chemical Landmark in the Bowood House where Priestley discovered oxygen ※※※; Priestley statue ※※ in Birmingham; Priestley House ※※※※ and grave ※ at Riverview Cemetery in Northumberland, PA; Penzance statue ※※ of Humphry Davy; Beddoes' Pneumatic Institution in Bristol ※※.

The Atom in England before Dalton

- In Oxford, from 1654 to 1668, Robert Boyle (1627–1691), one of the earliest practitioners of the modern experimental method, and his famous assistant, Robert Hooke (1635–1703), used their air pump to demonstrate that air pressure (what Boyle called "the spring of the air") is plausibly due to the action of "corpuscles" slamming against the sides of a container. He formulated what is still known as Boyle's Law that states that the volume occupied by a gas is inversely proportional to the pressure put upon it at constant temperature.
- In his *Sceptical Chymist*, published in 1661, Boyle presented his definition of an element and strongly expressed his belief in a corpuscular or atomic hypothesis.
- Isaac Newton (1642–1727), although universally known for his work in physics and mathematics, was a devoted alchemist fixated on finding the philosopher's stone and the elixir of life. He was, however, also an avowed atomist, believing that matter is composed of solid, massy, hard, and impenetrable particles.

(continued)

- Early pneumatic chemists Joseph Black (1728–1799), Daniel Rutherford (1749–1819), and Henry Cavendish (1731–1810) isolated what they called "fixed air" (carbon dioxide), "phlogisticated air" (nitrogen), and "inflammable air" (hydrogen), respectively.
- Joseph Priestley (1733–1804) was the quintessential pneumatic chemist isolating eight new airs including, most famously, dephlogisticated air (oxygen).
- Humphry Davy (1778–1829), in his days at the Beddoes' Pneumatic Institute, was most famous for his work with nitrous oxide, but also concentrated on gathering data on a variety of gaseous compounds including their relative mass ratios.
- These natural philosophers were more comfortable referring to their sets of experimental data rather than speculating about the nature of atoms. A concrete atomic theory had not yet been proposed but the vital data needed to formulate and verify such an idea was accumulated by these pneumatic chemists.

3.2 ROBERT BOYLE (1627–1691) THE "CHYMIST"

It seems not absurd to conceive, that at first production of mixt bodies, the universal matter, whereof they among other parts of the universe consisted, was actually divided, into little particles, of several sizes and shapes, variously moved. Robert Boyle (from ref. 13, Part I, Proposition I).

3.2.1 Robert Boyle Arrives in England

Recall from Chapter 2 that Robert Boyle was born in Lismore, Ireland, but embarked on a trip to Europe at an early age. In 1644, Boyle, still only 17, finally arrived in England (see Chart 3.1). He had been abroad for five years, his powerful father had died amidst the wars in Ireland, leaving his estate in disarray, and civil war had broken out between Charles I and Parliament. Living at first in London with his widowed sister, Katherine, Robert worked to put his share of his father's estate in order. Happily, Katherine had many intellectual friends and many of them were just then starting to gather together to discuss science (then still called natural philosophy) and its power to change the world. Indeed,

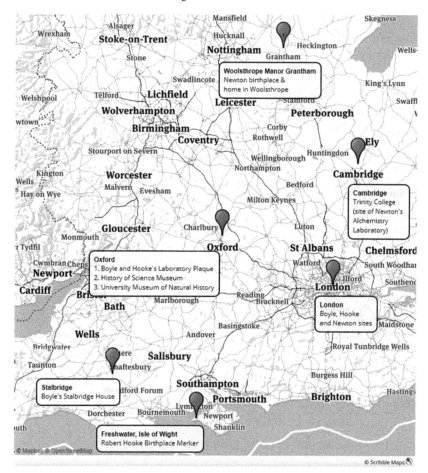

Chart 3.1   Travel sites in England related to Robert Boyle, Robert Hooke, and Isaac Newton. These include birthplaces, residences, laboratory sites, and memorials in their honor.

they were followers of Francis Bacon (1561–1626), who famously stated that "knowledge is power". As the youngest member of the "Invisible College", as Boyle later called the group because it did not have a regular meeting place, Robert began his life-long interest in science. In 1645, he moved into his father's manor at Stalbridge in Dorsetshire, England where, now firmly convinced (like Bacon and Galileo before him) that observations and experiments were the cornerstones on which scientific investigations had to be built, he set up his first laboratory.

Contrary to the traditional chemistry lore found in many textbooks, Boyle cannot be regarded as someone who magically transformed alchemy into chemistry. He is more properly viewed as a transitional figure – not a traditional mystical alchemist, certainly not a modern chemist, but rather someone in between. He was responsible for dropping the "al-" of alchemy and indeed he may be best referred to as a "chymist", a term marking his place between alchemist and chemist. Boyle was devoted to "chrysopoeia" all his life. Chrysopoeia means gold-making, specifically by transmutating lesser or base metals to produce it. We who live in the twenty-first century are quick to ask why anyone could possibly believe in the transmutation of metals. However, it was not uncommon in those days to believe that plants and animals were not the only things with life cycles – it was thought that minerals and metals also had them. Aristotle taught that all things in nature seek their perfection, so it followed that base metals would, in time, metamorphose into silver or gold. It was hoped that alchemy might provide a way to hurry along the natural processes, using a "catalyst" (given many names like "philosopher's stone" or the "elixir of life"). Boyle's commitment to chrysopoeia is a more accurate description of his activities than speaking of his devotion to alchemy in a generic sense. It was in Stalbridge that he first set up a chrysopoetic laboratory.

3.2.2 The Gate Piers of Boyle's Stalbridge House

(50.963229°, −2.381122°) A357, Stalbridge, Sturminster Newton. This old Elizabethan house where Boyle lived and started his experimental work was torn down in 1822. Only the gate piers remain, surmounted by heraldic beasts, in this case lions! Perhaps these figures, steeped in mythology and magical powers, their golden coats changed to stone, are fitting monuments to Boyle's chymistry.

3.2.3 Robert Boyle Moves to Oxford

In 1654, Boyle moved to Oxford – to High Street next to University College, in the house of the apothecary John Crosse. Once again, he lived with his sister Katherine. This house in Oxford no longer exists but there is a plaque (51.752725°, −1.252010°) on the outer wall of University College that shows the location of Robert Boyle and Robert Hooke's laboratory. (The plaque is on the outside of the Shelley Memorial, a marble statue in honor of the poet Percy Bysshe Shelley. It is on High Street opposite the entrance to All Souls' College.) It was during Boyle's 14 years in Oxford that he reached the height of his scientific productivity. By then he had solidified his inheritance enough that he could hire private assistants – the most famous being Robert Hooke – to help in his scientific work. In 1657, with Hooke's assistance and at great cost, Boyle devised his own air pump, which had originally been invented in 1650 by Otto von Guericke of Magdeburg, Germany. An air pump, or "pneumatic engine" as it was originally called is what we know today as a vacuum pump. Using his pump, Boyle was the first to observe the relationship between the volume of a gas (in this case, "ordinary air") and pressure. As shown schematically in Figure 3.1, a sealed lambskin pouch containing ordinary air expanded when the pump removed the air that surrounded the pouch. (On at least one occasion, the pouch

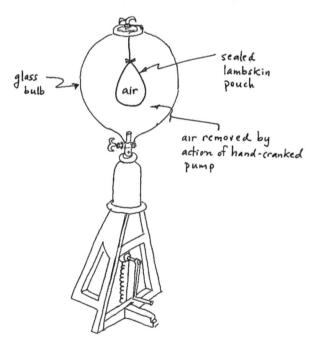

Figure 3.1 A schematic drawing of Boyle's air pump containing a sealed lambskin pouch filled with ordinary air. See also ref. 1.

exploded with a startling report.) In 1662, he formulated what is still known today as Boyle's Law (the volume occupied by a gas is inversely proportional to the pressure put upon it at constant temperature). Boyle's experimental results lent great support to the idea that gases were composed of discrete, rapidly moving "corpuscles" (or what we would call today atoms or molecules) separated by a void. They collided with the walls of the container and thereby exerted a pressure. The decrease of the pressure of the air surrounding the pouch (caused by using the air pump to remove some of the surrounding particles) allowed the corpuscles of the air in the pouch to expand its volume. Boyle's work with air was published under the title *New Experiments Physico-Mechanicall, Touching the Spring of the Air and Its Effects*, usually referred to simply as *The Spring of the Air*. The 1st edition (1660) did not mention what we know as Boyle's Law, but the 2nd (in 1662) did.

In 1661, in between the two editions of *The Spring of the Air*, Boyle published the enigmatic *Sceptical Chymist*. It was in this book that he published his definition of an element, and where he most strongly expressed his belief in a corpuscular or atomic hypothesis. Specifically, he stated that all matter seems to be "divided into little particles of several sizes and shapes variously moved". He also advocated atomism in his *Origins of Forms and Qualities According to the Corpuscular Philosophy* (1666). So, Boyle divided matter into elements and elements into unique atoms or corpuscles. All of this sounds quite enlightened and modern, right? The "al" of alchemy is gone and it is tempting to mark this change as the precise beginning of chemistry as we now know it. However, one must regard such a conclusion skeptically. Certainly, Boyle wanted to transform alchemy, to make it more scientific by framing good hypotheses based on sound experimental methods (that were reported with great accuracy and detail). But we must be careful not to attribute too much to Robert Boyle. For more about Boyle as a transitional figure between alchemy and modern chemistry, see Lawrence Principe's excellent book, *The Aspiring Adept: Robert Boyle and His Alchemical Quest*.[2] For complete translations of Boyle's many writings, consult "The Boyle Project", directed by Michael Hunter.[3]

Before leaving Oxford, the scientific/historical traveler is encouraged to visit the History of Science Museum ❋❋❋ in the Old Ashmolean Building (51.754531°, −1.255057°). Items to see here are described in Chapter 6. Also, one should take the opportunity to duck into the Oxford University Museum of Natural History on Parks Road (51.758700°, −1.255456°). Here, you can walk around the perimeter of the courtyard and view 19 large life-like statues of some of the greatest scientists of all time. These include many discussed in this book including Galileo Galilei, Roger Bacon, Isaac Newton, Hans Ørsted, Humphry Davy, and Joseph Priestley. If you continue out Parks Road you will come to the Clarendon Laboratory, where there is a plaque (51.75937°, −1.25653°) dedicated to Harry Moseley, whose X-ray work establishing atomic number is also described in Chapter 6, p. 169. Other Moseley sites are also described in that chapter.

3.2.4 Robert Boyle Moves on to London

Boyle moved back to London in 1668 and again lived with his sister, Katherine, Countess of Ranelagh, in the Pall Mall district. In 1663, the Invisible College had become "The Royal Society of London for Improving Natural Knowledge" and the charter of incorporation granted by Charles II of England named Boyle a member of the governing council. During this time, he was also intimately involved in the production of, and experiments with, phosphorus. Phosphorus was discovered in Hamburg, Germany, in 1669 by Hennig Brandt, a German alchemist and physician. Brandt was looking for a substance capable of converting silver into gold and chose, of all things, to investigate human urine, the "golden stream". A typical recipe called for letting 50 to 60 pails of urine stand in a tub until it putrefied. It was then boiled down to a paste and the vapors drawn into water, producing a white, waxy substance that glowed in the dark. If the material was removed from water, it burst into flame. Brandt called his product "cold fire." (Eventually, it was called *phosphorus* from the Greek word meaning "light-bearing".) Brandt kept the recipe for producing "cold fire" a secret for six years, but eventually he told others about his amazing discovery and his secret made its way to London where it came to the attention of Robert Boyle.

Robert Hooke (who had helped Boyle devise his air pump) was a frequent visitor in London. In 1676–1677, Hooke evidently "busied himself in a 'designe'" for altering Lady Ranelagh's house that included a laboratory at the rear of the property for Boyle and his sister Katherine. (Recently, evidence has come to light about the degree to which Katherine and Robert shared a love of scientific ideas, experimented together and edited each other's manuscripts. Given this accumulating evidence, it is appropriate that a plaque honoring Lady Ranelagh has recently been put in place at Lismore Castle, Lismore, birthplace of Robert Boyle.)

In 1679, Boyle, assisted by his apprentice Ambrose Godfrey Hanckwitz, who had been sent to Hamburg to confer with Brandt, discovered a method that produced high yields of elemental phosphorus using all the urine collected from the privies of the Ranelagh house! Boyle called phosphorus *aerial nocticula* – spirit of night light (no pun intended) – and, in 1680, put a description of his method into a sealed envelope that he instructed was not to be opened until his death. After leaving Boyle's service, Hanckwitz set up a laboratory that he called the "Golden Phoenix" and for many years was one of the few people in the world who could produce six or seven ounces of phosphorus on demand. (Phosphorus was not only a source of magical delights but also were used in early forms of matches. It has always been dangerous to be a matchmaker (pun intended) but these early forms were exceedingly dangerous, even fatal.) Hanckwitz produced both the pure "white" phosphorus and a mix with "oil of urine". Godfrey's fingers were often blistered and the sores healed only very slowly. On one occasion, a vial containing phosphorus (immersed in water) broke in Godfrey's pocket and the element ignited and burned holes in his breeches! This could not have been a pleasant experience.

Godfrey's laboratory was passed down through his family and one of the original furnaces was still being used to make charcoal as late as 1859. (In 1769, phosphorus was found in bones and its isolation from human urine was discontinued. However, its use in manufacturing matches made the demand for the element so great that the European battlefields were combed for human remains.) In 1872, Godfrey's "Golden Phoenix" where the Golden Stream was used to make the stunning element we call phosphorus was demolished to make room for the present-day Corpus Christi Roman Catholic Church.

3.2.5 Travel Sites in London Related to Robert Boyle

See Chart 3.2 for the location of these and other sites.

3.2.5.1 Corpus Christi Roman Catholic Church. ※ (51.510920°, −0.122523°), 1–5 Maiden Lane. A church built on the site of a phosphorus factory! The potential for enlightenment is stunning. See "The London of Robert Boyle" by Thorburn Burns[4] for an account of the placements of the various sections of the factory relative to the rooms in the present-day church. Burns says that if you "have a good imagination you can reverse the 'transmutation' and, in a way, visit Godfrey's laboratory".

3.2.5.2 The Royal Society. ※※※ (51.505865°, −0.132507°), 6–9 Carlton House Terrace. Robert Boyle FRS ("Fellow of the Royal Society") was, in 1660, a "Founder Fellow" of the Royal Society (RS). From 1662 to his death in 1703, Robert Hooke was "curator of experiments" at the RS. This is, indeed, a great place for

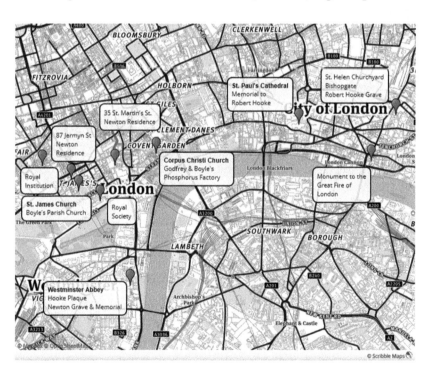

Chart 3.2 Travel sites in London related to Robert Boyle, Robert Hooke, Isaac Newton, and Humphry Davy. These include residences, laboratories, churches, memorials, and graves.

scientific/historical travelers to visit. For example, the RS holds a first edition of the *Sceptical Chymist* and, while it is very valuable, you can see a facsimile reprint of it in the Library. You might also search for "Robert Boyle's way of making phosphorus", "An Experimentall Account of the Compression of Aire Made by Mr Boyle", or "New Experiments Physico-Mechanicall, Touching the Spring of the Air and Its Effects" in the Archives. The Royal Society's website says that "The Library and Archives are open to researchers and members of the public. Access is free of charge. New readers are always welcome, and it is not necessary to make an appointment. On your first visit, you will need to complete a registration form and bring proof of address and photographic identification with you. You are welcome to contact us before your visit to ensure you get the most out of it." Their exhibits vary from year to year so be sure to be aware of what's on display when you plan to visit. However, just perusing the books in the general section of the library for various works by atomic scientists would be great fun!

3.2.5.3   St. James's Church (No Rating). (51.508874°, −0.136618°), 197 Piccadilly, St. James Square. The church is within easy walking distance of the Royal Society. St. James was Boyle's Parish church, but there is no marker there to indicate that. Katherine, Lady Ranelagh, died on December 23, 1691. Robert Boyle died only a week later, and they were buried together in the south chancel of St. Martin-in-the-Fields. However, when the old church was demolished in 1721, prior to the building of the present church, no systematic record was made of the dispersal of the remains of the bodies buried there. Therefore, most unfortunately, it is not possible to visit the grave of Robert Boyle.

3.2.6   Travel Sites Related to Robert Hooke

3.2.6.1   Robert Hooke Marker, Freshwater on the Isle of Wight. (50.682128°, −1.516286°). Hooke was born here. At the base of "Hooke Hill" at the junction with Afton Road (A3055) is a marker honoring Robert Hooke, "physicist, scientist, architect and inventor". The great British poet, Alfred Lord Tennyson (1809–1892), lived nearby at the Farringford House on the road between

Freshwater and Alum Bay. In their book, *A Travel Guide to Scientific Sites of the British Isles*, Charles Tanford and Jacqueline Reynolds[5] note that visitors should be sure to walk to the Needles Headland on Alum Bay.

3.2.6.2 Robert Hooke Grave. (51.514844°, −0.081646°), St. Helen's Church, Bishopsgate, London. St. Helen's is a medieval church located close to the bullet-shaped office tower that Londoners call "the Gherkin".

3.2.6.3 Robert Hooke Memorial. Westminster Abbey (51.499438, −0.127553). This simple memorial in the Abbey is located in the floor beneath the Lantern, near the pulpit. (See Chapter 7 for details.)

3.2.6.4 Robert Hooke Memorial. (51.510137°, −0.086133°) at the Monument to the Great Fire of London of 1666. Hooke, the Surveyor of the City of London after the great fire, and Sir Christopher Wren designed this Doric column, not only to honor the great fire but to also function as a scientific laboratory. It had a built-in "zenith telescope" and was a place designed to carry out gravity and pendulum experiments. The entrance to the under ground laboratory is directly below the ticket booth area but, unfortunately, is not available to the public. (It's fun to mention that you know about the laboratory when you buy your ticket.) It's worth hiking up the 311 steps (each six inches high) to view the city of London from that vantage point. Hooke used these steps to measure the effects of different heights on atmospheric pressure. Looking south, far below you will see St. Magnus the Martyr, a Christopher Wren church (mentioned in both Dickens's *Oliver Twist* and T.S. Eliot's *The Waste Land*) and, across the River Thames, the "Shard of Glass", currently the tallest building (95 stories) in the United Kingdom, which is designed to resemble a piece of glass roughly broken off.

3.2.6.5 Robert Hooke Memorial. (51.513852, −0.098340) at St. Paul's Cathedral. This is on the wall in the crypt of the cathedral near Wren's tomb.

See Chapter 7 for more information on the many scientists honored and buried at Westminster Abbey. Unfortunately, at the time of writing, this does not include Robert Boyle.

3.3 ISAAC NEWTON (1642–1727)

> It seems probable to me, that God in the Beginning form'd matter in solid, massy, hard, impenetrable, moveable Particles, of such Sizes and in such Proportion to Space, as most conduced to the End for which he form'd them....
> Isaac Newton, in ref. 14, Query 31, 375–376.

Isaac Newton was born on his mother's sheep farm, Woolsthorpe Manor, in 1642, the same year that Galileo died. Newton was not much interested in farming and, luckily for him, members of his family recognized this and sent him to Cambridge University in 1660. After he graduated in 1665, the plague hit London, the university was closed as a precaution, and Newton returned home. The year 1666–1667 is sometimes called his *annus mirabilis* or "year of miracles". (John Dryden was the first to use this expression in 1667. Many thought that the year 666 – the "year of the beast", see Revelations 13: 18 – plus 1000 was going to be a year of great calamity. This forecast seemed to have some truth to it as it was year of both the "Great Fire" and the "Great Plague" of London, but Dryden thought that God had miraculously intervened to prevent even greater disasters and wrote his poem "Annus Mirabilis" as a result.)

In 1666, back at Woolsthorpe, sitting alone in the garden next to the manor, Newton observed the falling of an apple and wondered if the force which made it fall was the same as that holding the moon in orbit around the earth, the planets around the sun, and all other heavenly bodies in their appointed places. He also purchased a prism at a local fair, made a hole in the shutter of the window of his room, admitted a small stream of sunlight, and found that white light was broken down into its component colors. As an added bonus, he fashioned a second prism that put the colors back together to produce white light once again. This was a most important initial step in the study of light. (It wasn't a universally acclaimed step, as some detractors said that Newton had "profaned nature" in this experiment.) We will see that light (not just the thin slice of visible light to which the human eye is sensitive) and its interaction with atoms will be an important factor in the identification of the atoms of a variety of elements and also in discerning atomic structure. Newton went on to invent calculus (as did Gottfried Leibniz independently), became the Lucasian Professor of Mathematics at Cambridge University (the same chair that Stephen Hawking held for 30 years, from 1979–2009), was a long-time President of the Royal Society, and the Master of the Royal Mint. But here let's concentrate on Newton's contributions to alchemy, chemistry, and the development of the atom.

Like his older friend Boyle, Newton was an avowed atomist. In his *Opticks* he stated, "It seems probable to me, that God in the Beginning form'd matter in solid, massy, hard, impenetrable, moveable Particles, of such Sizes and in such Proportion to Space, as most conduced to the End for which he form'd them ...." Sometimes Newton's view of the nature of matter is referred to by the rather fancy name, "dynamic corpuscularity". Are there forces that operate among these particles? Although Newton discovered the Universal Law of Gravitation, he did not believe that gravity played an important role in holding elementary particles together. Rather, in Book 2 of his *Principia*, he demonstrated that a gas composed of particles repelling each other by a force that decreases as they get farther apart, *i.e.*, a force inversely proportional to distance, would account for Boyle's observations, *i.e.*, Boyle's Law. These types of repelling forces have now been replaced by what is known as the "kinetic molecular theory of gases" (See Chapter 5, p. 112 and Chapter 10, pp. 250–251).

3.3.1 Travel Sites Related to Isaac Newton

3.3.1.1 Woolsthorpe Manor. (52.808805°, −0.630535°), Water Lane, Woolsthorpe-by-Colsterworth, near Grantham. This is a National Trust site. The house is furnished as it might have been in the mid-seventeenth century. Indeed, the famous apple tree was part of the orchard still located just outside the manor. From the ages of 12 through 17, Newton attended the King's School (52.915926°, −0.641024°) in nearby Grantham. As per the custom of the time, he carved his signature on a library window sill. No one knows for sure if the signature is authentic, but a replica of it is on display in the Grantham

Museum (52.910433°, −0.640146°). Grantham is also the home of Margaret Thatcher, so the museum has material about the first woman Prime Minister of the United Kingdom, as well as about one of the most famous physicists ever to live. Margaret Thatcher, by the way, started her professional life as a research chemist. She most likely would have appreciated our interest in traveling with the atom. There is also a wonderful statue of Newton at St. Peter's Hill in the town, near the Grantham Museum. Many, but not all, hope that a statue of Thatcher will be placed close by.

3.3.1.2 Trinity College, Cambridge (Newton). Despite his monumental achievements in mathematics and physics, Newton spent much of his life in pursuit of the philosopher's stone and the elixir of life. While his older friend, Robert Boyle, was a transitional figure between alchemy and modern chemistry, Newton, on the other hand, was a dyed-in-the-wool (an appropriate description – after all, he was born on a wool farm, Woolsthorpe) alchemist. Returning to Cambridge in 1667, it didn't take Newton long to establish himself. By 1669 he was fully established as the Lucasian Professor of Mathematics, but he could not avert his eye from the elixir and the stone. At Cambridge, very near to his rooms, and in a garden accessible only to him and his assistant, he established an elaborate laboratory where he carried out in excess of 400 alchemical experiments using a variety of equipment including furnaces, retorts, funnels, crucibles, and vials; various materials including the metals antimony, bismuth, iron, copper, and lead; and reagents including what we call today sulfuric (oil of vitriol) and nitric (*aqua fortis*) acids, and copper(II) sulfate (vitriol). He apparently recorded these experiments in a small leather-bound notebook now housed in the University Library, Cambridge. At one point in the 1680s, his assistant described a six-week period of time during which.

> "the fire scarcely going out either night or day, he sitting up one night, and I another, till he had finished his chemical experiments, in the performances of which he was the most accurate, strict, exact...." Sarah K. Bolton[6]

3.3.1.3 Site of Isaac Newton's Alchemical Laboratory. (52.207318°, 0.117749°). Chart 3.3 shows the location of the laboratory between the Great Gate of Trinity College and the Trinity College Chapel. (An apple tree propagated from the apple orchard at Woolsthorpe has been planted there.) Various tests, for example, a series of borings have established the probable location of the laboratory and a nearby rubbish pit. (For further information, see R. E. Spargo's article "Investigating the site of Newton's Laboratory in Trinity College, Cambridge".)[7] If you go there, find a spot out by Trinity Street, where you can see the Great Gate and the end of the magnificent chapel. (You might note that Henry VIII, who founded the college in 1546, is shown perched above the Great Gate and below the clock. If you look closely you will see that his sceptre has been replaced by a chair leg. Occasionally, this is replaced by other objects, for example, a bicycle pump, but eventually a chair leg is always reinserted. Tradition reigns

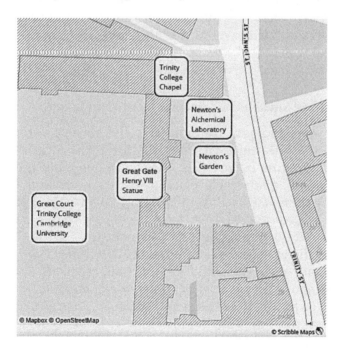

Chart 3.3 The probable location of Newton's alchemical laboratory between the Great Gate of Trinity College and the Trinity College Chapel.[8]

at Trinity College!) As you take in the scene, contemplate the great Isaac Newton, not only as one of greatest mathematicians and physicists of all ages, but as an alchemist trying to change lead atoms into gold atoms, or hoping to discover the secrets of everlasting life.

3.3.1.4 Statue of Isaac Newton in Trinity College Chapel.
Tickets to visit the Chapel can be purchased at the King's College Visitor Center on King's Parade, right opposite the Great Gate and the location of Newton's alchemical laboratory. If you are interested, Google William Wordsworth's description of this statue, which he describes as

> "The marble index of a mind for ever,
> Voyaging through strange seas of thought alone."[15]

There are a number of plaques in the chapel, including some of the other atomic scientists mentioned elsewhere in this book. These include Pyotr Kapitza, Ernest Rutherford, and J. J. Thomson, all of whom are discussed in Chapter 6. Part of Newton's personal library is located in the Trinity College library.

3.3.2 Travel Sites Related to Isaac Newton in London

Newton was appointed Warden (later, the Master) of the Royal Mint in 1696. During that time, the Mint was located in the Tower of London, between the inner and outer walls, but this proved a rather miserable place to live, and Newton did not like living there. He lived from 1696 to 1710 at 87 Jermyn Street (51.508176°, −0.137393°), St. James, where a plaque has been placed. He was a member of Parliament from 1701–1702 and served as the President of the Royal Society from 1703 until his death. Members of the public are welcome at the RS and, as noted above, can look up various books and articles about and by Newton. It is best to contact the RS first and let them know what you might be interested in. In 1704, *Opticks*, which he had written over 20 years previously and which presented his ideas on light and color, was published. (In 1687, he had published his *Principia*, a monumental work that explained his ideas on universal gravitation.) He was knighted in 1705. In 1710, he moved to 35 St. Martin's Street (51.90933°, −0.13069°), just a one-minute walk from Leicester Square. This house no longer stands, but a marker has been placed above eye level on a nearby wall . For a long time, there was a statue of Newton located in the southwest corner of Leicester Square near what is still known as "Newton Gate" (51.509939°, −0.130416°). At the time of writing, this statue is no longer in place. In 1725, he moved to Kensington where he lived until he died in 1727. He was buried in Westminster Abbey. His elaborate monument is described in detail in Chapter 7.

3.4 SCOTTISH AND ENGLISH PNEUMATIC CHEMISTRY: "FIXED", "PHLOGISTICATED", AND "INFLAMMABLE" AIRS (1756–1772)

Robert Boyle's air or vacuum pump became a primary tool for discovering and working with a variety of gases, known at that time as "airs". The term "gas" had been introduced by the Flemish physician and chemist Jan Baptista van Helmont around 1620 but was virtually ignored until it was reintroduced by Lavoisier at the end of the eighteenth century. Van Helmont regarded "airs" as forms of matter in complete chaos. Van Helmont is often regarded as the founder of what became known as "pneumatic chemistry". The term pneumatic comes from the Latin *pneumaticus* meaning "of the wind, belonging to the air". Pneumatic chemists were concerned with the production and isolation of gases but also their physical properties and their uses in medicine. This discipline was advanced in 1720 by Stephen Hales' invention of the "pneumatic trough", used for collecting gases in a bottle above a basin of water. Certainly, Robert Boyle must be regarded as one of the important early pneumatic chemists. Others include Joseph Black, Daniel Rutherford, Henry Cavendish, Joseph Priestley, Antoine Lavoisier, Humphry Davy, and John Dalton. Boyle had worked with what he called "common air" but he suspected "that there may be dispersed through the rest of the atmosphere some odd substance on whose account the air is so necessary to the subsistence of flame". Common air indeed turned out to be a mixture of odd substances.

3.4.1 Joseph Black ("Fixed Air")

In 1756, Joseph Black (1728–1799), a Scottish chemist and physician, while studying gout and kidney stones, discovered that heating limestone (or what we know today as calcium carbonate, one of the substances that compose kidney stones) produces lime and an "air". It appeared that this new air was immobilized or "fixed" in the limestone and freed only when heated. Without heat and in the presence of water, the "fixed air" would recombine with the lime to reform the limestone. Black followed these reactions by weighing the original limestone and resulting lime thereby calculating the mass of the escaped gas. The mass was regained when the limestone was reformed. In a telling experiment, he also found that "common air" combined with lime to produce limestone, meaning that common air contained some "fixed air". Black went on to note that a candle would not burn in the fixed air and that mice died in it. The modern name for "fixed air" is carbon dioxide.

3.4.2 Travel Sites in Edinburgh Related to Joseph Black

There are several Joseph Black sites located in Edinburgh (see Chart 3.4). There is a famous portrait of Black in the Scottish National Portrait Gallery 1 Queen Street (55.9533°, −03.1883°). He is shown lecturing on his fixed air. There is also a portrait of the young Joseph Black in the Raeburn Room (55.947526°, −3.186502°) of Old College, located on South Bridge. (Arrangements to visit to the Raeburn Room need to be made in advance.) In the Playfair Collection at the National Museum of Scotland on Chamber Street (55.9468°, −03.1904°), they have one of Black's balances that may very well have been used in his "fixed air" experiments. There are some beautiful examples of the black-green glassware that Black used in his chemical demonstrations. Using this glassware, he could pour "fixed air" over a candle to extinguish it. (They also have a single-cylinder air pump devised by Robert Boyle.) This museum has recently undergone extensive renovations and is a beautiful facility to visit for many different reasons.

At the Joseph Black Building, David Brewster Road (55.9240°, −03.1762°) located in the King's Buildings (the main campus of the College of Science and Engineering of Edinburgh

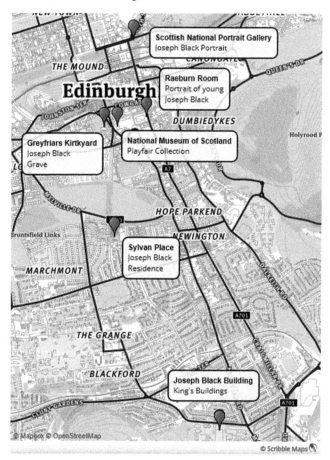

Chart 3.4 Travel sites in Edinburgh related to Joseph Black. These include portraits, balances, glassware, recently discovered artefacts, and his gravesite.

University), there is a Royal Society of Chemistry Plaque dedicated to Black. The Chemistry Collections inside the building is worth a visit but is not presently open to the public. Recently, an archaeological dig at the University uncovered some items that belonged to Joseph Black and were made by Josiah Wedgwood. Inquire about these items with the curator of the Collections, Dr Neil Robertson, or contact the school of chemistry office at 0131 650 7546. During our visit, we were escorted to a faculty office, where a chair belonging to Joseph Black was displayed. The King's Buildings Campus include buildings named after a variety of important scientists including Michael Faraday, Daniel Rutherford, and James Clerk Maxwell, among others. Black lived in what is now the Sylvan House, on an unnamed road near Sylvan Place, Edinburgh (55.938488°, −3.190769°), where a stone plaque notes him as "discoverer of carbon dioxide and latent heat". Latent heat is, for example, the amount of heat needed to convert a solid to a liquid at the melting point or to convert a liquid to a gas at the boiling point. Black introduced this term in 1762.

Black is buried in the Covenanters Prison section of Greyfriars Kirkyard (55.946667°, −03.192222°). (As you approach Greyfriars, be sure to look for the "Greyfriars Bobby", the bronze statue of the loyal Skye terrier who reportedly guarded his master's grave for 13 years.) As shown in Figure 3.2, the Covenanters Prison is a locked area of Greyfriars Kirkyard. The Scottish Covenanters were a group of Presbyterians who carried out an anti-government revolution in 1679 (about a century before Black's time), attempting to establish Presbyterianism as the sole form of religion in the country. The revolution failed, with the result that about 1200 Covenanters were locked away here. Due to severely brutal conditions, many died after only four

Figure 3.2 The locked entrance to the Covenanter's Prison in Greyfriars Kirkyard where Joseph Black's grave and elaborate tombstone are located. Courtesy of Katherine F. Montgomery.

months. The area is locked for a variety of reasons but the official one is because the gravestones are fragile and therefore dangerous to the public. With some effort, however, the scientific/historical traveler can gain admission to this area. It's probably best to go into the Kirk and ask when would be a good time (hopefully within an hour or two) to have someone take you into the area. The National Museum is nearby, and you could take in the Playfair Collection or just have some tea or coffee and a small treat.

The gravestone is certainly worth the effort to get in to see it. It is well preserved, given that it is now well over two centuries old. (Most likely, it has held up well compared to other graves due to its orientation relative to the elements of weather.) The epitaph is in Latin but is extraordinary in its detail and praise of Black. If you get in, stand and consider the epitaph, which reads as follows: "Joseph Black, Doctor of Medicine, born in France, but a British subject, his father being a native of Ireland, and his mother of Scotland. First a student in the University of Glasgow, and afterwards in that of Edinburgh, was a most distinguished Professor of Chemistry in both Universities; a felicitous interpreter of nature; acute, cautious, and skillful in research; eloquent in description; the first discoverer of carbonic acid and latent heat; died in the 71st year of his age, AD 1799. His friends, who were wont to esteem his worth and abilities, have sought to mark out the spot which contains his body by this marble, as long as it shall last." What an extraordinary tribute to one of Scotland's earliest chemists. Not many chemists are so honored.

You might want to note that James Hutton, whom many consider the founder of modern geology, is also memorialized close by to Black. This is fitting as they were good friends. Hutton died several years before Black and was buried in an unmarked grave.

3.4.3 Daniel Rutherford ("Mephitic or Phlogisticated Air")

When a candle stops burning in "common air" and then the "fixed air" (both that originally present as well that created by the burning process) is removed, a sizable volume remains. In 1772 this remaining air was characterized by Black's young student, Daniel Rutherford (1749–1819), as part of his doctoral thesis. He called this gas "mephitic air" or, later, as we will discuss, "phlogisticated air". (The term mephitic comes from the Latin, *mephitis*, a pestilential exhalation.) Like fixed air, a candle will not burn in mephitic air and a mouse cannot survive in it. The modern name for mephitic air is nitrogen.

3.4.4 Henry Cavendish ("Inflammable Air")

In 1766, in a rare report to the Royal Society of London, Henry Cavendish (1731–1810) reported that he had dropped zinc pellets into "muriatic acid" (hydrochloric acid) and watched another previously unknown "air" bubble out. Later, he repeated this experiment with iron and tin with the same results. A candle caused his new "air" to burn with a pale blue flame and so he called it "inflammable air". He mixed it with "common air" and found that it exploded upon ignition! As it turns out, Robert Boyle, in 1671, had prepared this gas by dissolving iron in hydrochloric or sulfuric acid. In 1700, Nicholas Lémery had noted that this air, when mixed with common air, will produce a "violent shrill fulmination". Lémery thought he had discovered the origin of thunder and lightning! Cavendish, however, was the first to systematically investigate its properties and therefore is credited with its discovery. The modern name for inflammable air is hydrogen.

Recall that we encountered the Cavendish family in Chapter 2 as it still owns Lismore Castle where Robert Boyle was born. The great Cavendish Laboratory of Physics at Cambridge University was founded in 1874 by William Cavendish, the Seventh Duke of Devonshire. Chatsworth House, the Cavendish Estate, the home of the present-day 12th Duke of Devonshire (born in 1944), is one of the most spectacular great houses of England. Henry Cavendish's surviving manuscripts and his extensive library can be seen by appointment at Chatsworth House ($53.264939°$, $-1.439422°$). Located near Chesterfield, Chatsworth House appeared in the 2005 film adaptation of Jane Austen's Pride and Prejudice. It represented Pemberley, Mr Darcy's home. Henry Cavendish did not live there, instead residing in London. Henry Cavendish is buried with many of his relatives in the graveyard

Table 3.1 The airs of the English and Scottish pneumatic chemists.

Discoverer	Place of discovery	Year	Original name	Modern name	Modern chemical formula
Joseph Black	Edinburgh, Scotland	1756	Fixed air	Carbon Dioxide	CO_2
Henry Cavendish	London, England	1766	Inflammable air	Hydrogen	H_2
Daniel Rutherford	Edinburgh, Scotland	1772	Mephitic air or phlogisticated air	Nitrogen	N_2
Joseph Priestley	Calne in Wiltshire	1774	Dephlogisticated air	Oxygen	O_2

of the Cathedral of All Saints, known as Derby Cathedral, 18–19 Iron Gate, Derby, England (52.924802°, −1.477389°).

So, as summarized in Table 3.1, the last half of the eighteenth century saw the discovery of some puzzling airs that were given some strange but descriptive names that have long since disappeared from modern chemical vocabulary. By the way, it's difficult to know which was the more puzzling and strange, Henry Cavendish or his inflammable air. Cavendish, painfully shy to the point that he rarely if ever spoke to a group of people, was incredibly wealthy; wealthy enough that he built a separate entrance to his London house so that he would not have to encounter anyone as he came in and out. He was terrified of women to the point that he could not bear to look at one and studiously avoided speaking to them. If he did encounter a woman, she was immediately dismissed from his employment! So, obviously, he didn't know very many women and only a few more men. One of his few acquaintances, met through his occasional attendance at Royal Society meetings, was his contemporary, Joseph Priestley. (Priestley was only one and half years older than Cavendish.)

3.4.5 Joseph Priestley (1733–1804): Quintessential Pneumatic Chemist

> The feeling of it to my lungs was not sensibly different from that of common air; but I fancied that my breast felt peculiarly light and easy for some time afterward .... Hitherto

> only my mice and myself have had the privilege of breathing it. Joseph Priestley writing about breathing his "dephlogisticated air", in ref. 16, p. 102.

Joseph Priestley was the greatest of the English pneumatic chemists. Although known as an educator, controversial theologian, and political writer, he also had an early interest in "natural philosophy". (The modern terms "science" and "scientist" were not used until William Whewell coined them around 1840. Whewell also invented the term "physicist".) In 1769, at the age of 29, Priestley met Ben Franklin, who arranged for him to join the Royal Society and encouraged him to pursue his interest in electricity. Franklin, though viewed primarily as an inventor and statesman by Americans, had sold his printing press, newspaper, and almanac in 1748 and used the resulting wealth to further his investigations of the new science of electricity. In 1767, with Franklin's encouragement, Priestley published his "History and Present State of Electricity". This publication repeated some of Franklin's experiments but added new ones that Priestley conducted. It established his reputation in the sciences. I know you are dying to ask if Priestley tried Franklin's kite experiment and what happened. He did, in 1766, but his wife Mary "would not suffer him to raise" the kite any higher than his head, so nothing happened. Incidentally, the Benjamin Franklin House in London, where Franklin lived from 1757 to 1775, is now a dynamic museum that includes a Student Science Centre. It is certainly worth visiting when in London. It is at 36 Craven Street (51.507645°, −0.124880°), not far from Trafalgar Square and St. Martin-in-the-Fields.

Priestley was an English "rational dissenter". In general, English dissenters included anyone who did not accept the official doctrine of the Church of England and included Congregationalists, Presbyterians, Methodists, Baptists, Unitarians, Quakers, and Roman Catholics. *Rational* dissenters believed that Christ was the head of the church and that scripture was its ultimate authority, but they were also of the strong opinion that the Christian faith should be demystified and informed by human reason. Priestley and other "radical thinkers" of his day did not believe in the divinity of Christ, the Trinity, or the concept of original sin.

Dissenters, though viewed as heretics by the Church of England, were a growing influence on the English intellectual landscape. Prior to 1774, Priestley had been the pastor of three dissenting churches in Needham Market (Suffolk), Nantwich (Cheshire), and Leeds, as well as a teacher at the Dissenting Academy at Warrington.

In the late 1760s, while serving as the minister of the Mill Hill Chapel in Leeds he found a dense "air" that hovered above the surface of the fermenting liquid (the "wort") of the neighboring Jakes & Nell public brewery. He soon confirmed that this was Black's "fixed air" and found a way to mix it with water to produce the tart, bubbling, and rather appealing sparkling or soda water. Sometimes his experiments went astray – like the time that they caused an entire batch of beer to be ruined and he was "forever banned" from further experimentation at the brewery. Those who mess with good English beers are not easily forgiven! He reported his discovery to the Royal Society, where he and other members incorrectly speculated that soda water might be a cure for scurvy, then a debilitating problem for the British navy. Why would they propose such a premise, you ask? A report by a physician had suggested that fixed air might control putrefaction. Since scurvy was thought to be a type of rot, Priestley went on to suggest that drinking fixed air in a water solution (sparkling water) might be a cure for scurvy. As a result, Priestley wrote "Directions for Impregnating Water with Fixed Air", so that the home brewing of what became known as "windy water" (drinking it, of course, quickly produced the wind) became very popular. British seamen loved their windy water, but it didn't solve their scurvy problems – that was left to lime juice (a good source of vitamin C) and thus British sailors became known as "limeys". A report on Priestley's work preparing soda water was sent to France where it was investigated by a young man named Antoine Lavoisier.

Thus began a series of investigations carried out by Priestley that over the next 25 years would yield at least eight new "airs" including "alkaline air" (ammonia, NH_3), "marine acid air" (hydrogen chloride, HCl), "diminished or dephlogisticated nitrous air" (nitrous oxide, N_2O), "nitrous air" (nitric oxide, NO), "phlogisticated nitrous air" (nitrogen dioxide, NO_2), "vitriolic acid air" (sulfur dioxide, SO_2), and "dephlogisticated air" (oxygen, O_2). Priestley successfully isolated these gases because he collected them in a pneumatic trough over liquid mercury rather than water, in which many of these are soluble. This was a technique that Cavendish had also used. Priestley also was the first to observe that "common air" that had been exhausted by the burning of a candle could be regenerated by plants. He placed mint (and other plants such as balm and spinach) in the exhausted air for ten days and found that a candle would burn in it once again and mice could live in it as well. Although he was unaware of the role of light in this process he had discovered photosynthesis. For those of us trying to limit our carbon footprint, take note that Priestley was one of the first to recognize the role of plants in that effort.

3.4.6 Travel Sites Related to Joseph Priestley

See Chart 3.5 for the location of these and other sites.

3.4.6.1 Leeds. Mill Hill Chapel, where Priestley was the minister from 1767 to 1772, faces onto the City Square of Leeds (53.796464°, −1.547802°). There is a statue of Priestley on the square that shows him holding a magnifying glass and a mortar and pestle. This equipment is relevant to his preparation of oxygen, which he did at the Bowood House (described below), but as "travelers with the atom" we can nod knowingly and argue that a Priestley statue in Leeds should show him drinking a sparkling water with a sprig of mint! There is also a statue of James Watt, inventor of the Watt steam engine, and not a bad chemist himself. At the back of the chapel itself, near "Priestley Hall", is a highly polished plaque showing that "Priestley L.L.D. F.R.S." was the minister 1767–1773. (Recall that F.R.S. stands for Fellow of the Royal Society.) Note that this is not the original chapel that Priestley worked in. That seventeenth-century chapel was replaced in 1848. (Note that it is now a Unitarian parish – and it is no longer heretical to be of such persuasion. Priestley would be impressed.) There is plaque outside that conveys this information and notes Priestley's tenure there. Nearby is the modern "Priestley House", which is on the site of the house that the Priestleys lived in while he was minister at the chapel.

Pneumatists Set the Atomic Stage 57

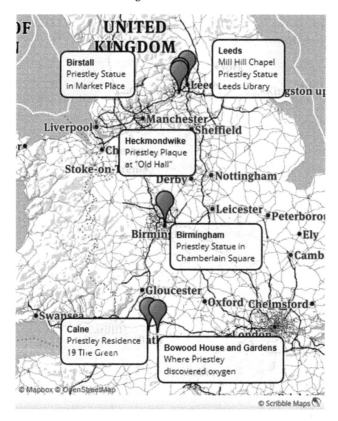

Chart 3.5 Travel sites related to Joseph Priestley.

While in City Square, the scientific traveler should venture into the nearby Leeds Library ◉◉◉ (founded in 1768 by Priestley and others). In this athenaeum you can view an early list of members in the Minute Book, 1768–1799, that includes Priestley's name. (It is no. 4. You might ask why there is a line through his name and the name W. Wood is penned over it. You can figure this out for yourself if you look carefully at the plaque at the back of the chapel.) One can also request to view Priestley's *A Short Account of a new Chart of History* and his *The History of Present State of Electricity with Original Experiments* (2nd edn, 1769). If you want to see these items, it might be a good idea to correspond with the library prior to your arrival. You might also find it useful to contact the Priestley Society at https://www.sciencehistory.org/

joseph-priestley-society. All of these items together make for a wonderful stop for the scientific traveler. ▓▓▓.

3.4.6.2 Birstall and Heckmondwike. While you are in the vicinity, you can visit the site of Priestley's birthplace in the difficult-to-find Fieldhead ▓ in the parish of Birstall, West Yorkshire. The original building at 5 Owler Lane (53.743246°, −1.660672°) where Priestley was born has been replaced by the Victorian era house that, at this writing, was in rather ramshackle shape. Look for a small plaque above this very old building. In nearby Birstall, at the Market Place (53.7319°, −1.6599°) is another statue of Priestley ▓▓. (It's not quite the same pose as the statue in Leeds but there's still no sparkling water with mint.) Also, see if you can find "Old Hall" in Heckmondwike (53.711257°, −1.676710°) which, at this writing is a Samuel Smith's public house where a plaque has been put in place. There is an unidentified picture of Priestley out front ▓. Priestley lived here from 1742 to 1752.

3.4.6.3 Calne, the Bowood House Years (1772–1780). We don't envision natural philosophers (later known as "scientists") as having patrons like artists or musicians (for example, Beethoven's Count Waldstein). Boyle and Cavendish were independently wealthy, but Priestley did not have this advantage. He was a man of ideas, however, and this drew the attention of the Second Earl of Shelburne (William Fitzmaurice Petty), a liberal statesman who shared many of those ideas. For example, Shelburne and Priestley both opposed George III's aggressive policies toward the American colonies. In 1772, Lord Shelburne took the unusual step of traveling up to Leeds to personally offer Priestley the post of science adviser, librarian, and general supervisor of the education of his two sons. The position came with a generous salary, a lifetime annuity, few official duties, two houses (one for the winter in London and another for the summer in Calne in Wiltshire close by Shelburne's elaborate Bowood Estate) and a stipend to support Priestley's pneumatic research. This research had recently been recognized by the Royal Society that awarded him the Copley Medal, which some have called the Nobel Prize of his day. With this recognition and with his wife Mary and their three children now comfortable in new surroundings, Priestley enthusiastically set up a laboratory near the orangery and library in a beautiful wing off the main house. According to Kenneth S. Davis in his book *The Cautionary Scientists*,[9] Joseph was slender, clear-eyed, energetic, cheerful, meticulously organized and well-motivated. We can imagine him rising early on a bright sunny, mid-summer morning, saddling his horse and riding over to his laboratory (less than 3 miles away) to carry out the experiments he had planned for the day.

We scientific travelers can see these places today. The house where Priestley lived sits at 19 The Green in Calne (51.435527°, −2.001897°) with a plaque that confirms that he lived here from 1772 to 1779. Priestley liked to walk some 300 yards down the hill to what is now known as "The Doctor's Pond", where there is plaque placed in his honor (51.4382°, 2.002°). While Priestley often rode his horse, we will have to drive from Calne over to Bowood. Park your car, buy a ticket for the Bowood House and Gardens (51.429027°, −2.037952°) and start walking briskly toward the main house. Depending on the season, you can visit a variety of beautiful gardens, 60 acres of rhododendron groves in May, and beautifully manicured pathways amongst swaths of trees. The house and grounds have changed since the late eighteenth century, but we can still imagine Priestley working here, satisfied and excited now that Lord Shelburne was his patron. The Bowood House is still owned by an Earl of Shelburne and was opened to the public in 1975. The striking main house is surrounded by elaborate plantings and statuary. Inside you will be tempted by the grand library (much altered since Priestley was managing it), the Orangery, and the Chapel. However, you will soon find, right next to the library, the small room with the brass plaques on the door. (The American Chemical Society joined the Royal Society of Chemistry in designating Bowood House, the site of Priestley's English laboratory, an International Historic Chemical Landmark in 2000.) This is the room where Priestley had his laboratory. Unfortunately, you will not see a burning glass, pneumatic trough or other Priestley equipment. All that was sold when Lord Shelburne came upon hard times toward the end of his life. However, his equipment is shown in Figure 3.3, taken from Priestley's *Experiments and Observations on Different Kinds of Air* (1774 and 1790). It bears little resemblance to the fancy equipment we find in chemistry laboratories today but in a way that increases its allure. Stand at

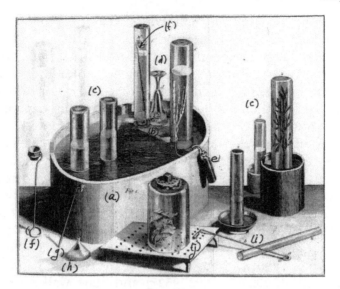

Figure 3.3 Pneumatic chemistry equipment of Joseph Priestley (a) Earthenware 8″ deep pneumatic trough, (b) a shelf for holding jars, (c) jars for collecting airs including one with a sprig of mint, (d) a beer glass containing enough air to sustain a mouse for 20–30 minutes, (e) a gas generator, (f) wire stand for supporting a measured amount of a substance, (g) a cork for closing a vial, (h) a funnel, (i) equipment used to introduce a candle with a flame into a jar of an air, and (j) a receiver for keeping mice alive. It is open at the top and bottom permitting air circulation. Sometimes it was kept in a warm place to avoid chilling the mice. A mouse was introduced to a jar containing an air by passing it quickly through the water. Reproduced from ref. 10 with permission from Elsevier, Copyright © 1939 Wiley-Liss, Inc., A Wiley Company.

the door, think (and perhaps read from this book) about what Priestley did before and after his trip to Paris to meet Lavoisier. This is where one of the greatest chemical discoveries of all time was made.

At Bowood, he acquired a large "burning glass" measuring a foot in diameter and supported by a wooden frame. He used it to focus the summer Bowood sunlight on various samples including wood, metals, and metal calxes, or what we would call metal oxides. Mercury calx, a beautiful red stone, was known to turn into the bright silvery mercury metal when heated. On August 1, 1774, Priestley aimed his burning glass on mercury calx and noticed that it also produced an air, and a most remarkable air at that. He collected it and quickly determined that it was not Black's "fixed air" because it would not readily dissolve in water. Perhaps it was Rutherford's "mephitic air". If it was it should snuff out a candle. He reached for a candle and put it in with the new air. To his amazement the flame of the candle flared up, producing an unanticipated amount of heat and light. What was this air? We can be confident that Priestley had a burning desire to know more!

3.4.7 Priestley Travels to Paris

There are distinct advantages to having a patron, but sometimes the benefactor makes a specific request of you. When this happens, you have to essentially drop everything and do his bidding. This is what happened to Priestley. He knew he was on to something big, but Lord Shelburne wanted him as a companion on a continental tour. Like the young Robert Boyle had before, Shelburne and Priestley set out for the continent in late August, 1774. They stopped in Belgium, Amsterdam, Germany, Strasbourg, and Paris. Joseph missed Mary and the children but Paris (like it is for many of us) was alluring. He had contacts there and had even been made an honorary member of the French Academy of Sciences in 1772. He also wanted to meet Antoine Lavoisier who had started to correspond with him. Priestley and Shelburne spent a month in Paris and Priestley met many times with members of the French Academy. Some of these meetings occurred in Lavoisier's home over sumptuous dinners. In this setting he found himself, in rather halting French, explaining the discovery of his new air to Lavoisier and others. They were intrigued and asked many questions. Lavoisier too had been thinking about airs, but specifically about what in the air caused substances to burn. (For more on Lavoisier, see Chapter 8.)

Priestley also found himself embroiled in some intense theological discussions while in Paris. In England, he was considered a heretic for his radical beliefs about Christianity. Still, he was a Christian natural philosopher and believed that natural philosophers were revealing the workings of a clockmaker God. The French men of science, however, believed that their job was to liberate humankind from the superstitions of all religion. They believed that no man of science could truly believe in God. The English heretic was a deluded believer in the eyes of his French colleagues.

3.4.8 Priestley Discovers "Dephlogisticated Air"

Returning to Bowood, Priestley went back to work. He had obtained a pure sample of mercury calx in Paris and now repeated his experiments. He produced more of his new air and reluctantly decided to a put a mouse in it. He had done this with other airs he had discovered but mice often did not fare well in them. Moreover, his wife Mary captured and cared for his mice and Joseph wanted to treat them as humanely as possible. Joe Jackson, in *The World on Fire*,[11] one of the finest scientific biographies written, has a wonderful section on how Priestley may have conducted these experiments with Mary's mice. Jackson writes:

> "He picked up a mouse by the scruff of the neck, talking to it softly in hopes of allaying its fears" and placed it in a jar of his new air. He may have sat back, played his flute, and waited. But Mary's mouse was fine. It stared back at Joseph and wiggled its whiskers. Eventually, of course, the air ran out but Joseph skillfully rescued the mouse before anything untoward could happen to it. The next day he repeated the experiment with the same mouse who once again survived nicely until finally rescued. Joseph could not resist. He had to know what it felt like to breathe his air. He inhaled a fresh batch and wrote: "I fancied that my breast felt peculiarly light and easy for some time afterward …. Hitherto only my mice and myself have had the privilege of breathing it."

Jackson suggests that perhaps Joseph gave "his small co-discoverer his freedom, possibly with a piece of cheese before release in the field".

3.4.9 Joseph Priestley in Birmingham

So, what happened to these men: Shelburne and Priestley? Shelburne went on to serve as Prime Minister of England from 1782 to 1783 and negotiated the end of the American War of Independence. For his services he was made Marquis of Landsdowne. The present occupant of the Bowood House is the 9th Marquis.

Priestley left the employ of Lord Shelburne in 1780. He continued to think of himself as an "aerial philosopher" and a practitioner of a "doctrine of airs". He and Mary and the children moved to the estate he called "Fair Hill" on the outskirts of Birmingham. There they spent some happy years and Joseph was involved in the Lunar Society, so named because its members met at full moon so they could safely travel home after the meetings. Members of the Lunar Society included Erasmus Darwin (grandfather of Charles), James Watt and Josiah Wedgwood. However, his heretical views on religion combined with his support for the American and French Revolutions made him a ready target and the subject of incendiary political cartoons such as the one shown in Figure 3.4. By then the French Revolution was in full swing. In 1791, on the second anniversary of the storming

Figure 3.4 "Doctor Phlogiston" cartoon of Joseph Priestley; Anti-Priestley cartoon, showing him trampling on the Bible and burning documents representing English freedom; in his pockets are "Essays on Matter and Spirit", "Gunpowder" and "Revolution Toasts"; the caption reads, "Doctor Phlogiston, The Priestley politician or the Political Priest!"

of the Bastille, his home, library, and laboratory were destroyed by a "Church and King" mob, and they moved temporarily to London.

In 1874, to honor the centenary of Priestley's discovery of oxygen, the citizens of Birmingham raised a beautiful statue of him in Chamberlain Square (52.480069°, −1.904604°) in the heart of the city center. It shows Priestley at the age of 41 (in the happier Bowood days) directing the rays of the sun through a lens onto the calx of mercury contained in a mortar. This pose is similar to that of the statue in Leeds.

3.4.10 Travel Sites in Northumberland, Pennsylvania Related to Joseph Priestley

In the spring of 1794, realizing that he and Mary were no longer safe in England, they sailed to a place where freedoms of speech and religion were protected, the fledging United States, whose founding he had so fervently advocated. He built a house in Northumberland in central Pennsylvania, furnished it with a laboratory and died and was buried there in 1804. Priestley's house stands on Priestley Avenue in Northumberland, PA (40.890503°, −76.789950°). It is one of the most significant scientific/historical sites in America. The Priestley House at 472 Priestley Avenue is both a National Historic Landmark and a National Historic Chemical Landmark. A two-and-a-half story Federal-style house located on a rise above the north branch of the Susquehanna River, it is furnished appropriate to the 1794–1804 time period. Priestley's laboratory, on the north side of the house, contains his glassware and equipment. It was here in 1799 that Priestley discovered another "air", what we now call carbon monoxide. In 1874, 100 years after Priestley's discovery of oxygen, a meeting of chemists was held at Priestley's grave in Northumberland; there they proposed to establish the American Chemical Society. It was officially founded two years later. Each year the society awards its highest honor, the Priestley Medal, given for "distinguished service in the field of chemistry". A plaque denoting the Priestley House as a National Historic Chemical Landmark was installed in 1994.

Joseph Priestley is buried in Riverview Cemetery located at the corner of 7th and Orange Streets (40.896620°, −76.802756°), in Northumberland next to his wife Mary and their youngest son

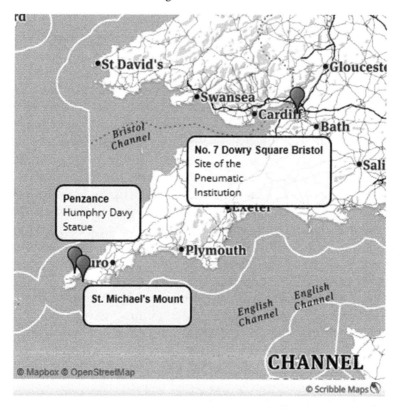

Chart 3.6 Travel sites in southwest England related to Humphry Davy. These include his place of birth and a prominent statue in Penzance, Cornwall and the site of the Pneumatic Institution in Bristol where he established his reputation as an experimental chemist, famously explored the uses and effects of nitrous oxide, and started his electrical experiments.

Henry, both of whom had died in Pennsylvania in the first two years after they moved from England. A map of the cemetery is available at the Priestley House.

3.5 SIR HUMPHRY DAVY (1778–1829)

> The moment after, I began to respire 20 quarts of unmingled nitrous oxide. A thrilling, extending from the chest to the extremities, was almost immediately produced. I felt a sense of tangible extension highly pleasurable in every limb .... Humphry Davy in ref. 17.

Penzance is in so-called "tinner" territory. As most scientific/historical travelers know, the Latin word for tin is *stannum* and the symbol for the element is Sn. Given these Latin origins, it comes as no surprise that Penzance is known as a stannary port with a Stannary Palace, a stannary prison, all governed by the stannary laws, stannary courts and stannary parliaments whose legislators were known as stannators. Cornwall, with its reliance on tin and other valuable metals, was the "wild west" of England. It was in Penzance, only about 10 miles from "Lands End" in this Celtic fringe of England that Humphry Davy, one of the most fascinating characters in the history of chemistry, was born in 1778. From a county dominated by ancient metals came a lively young man who was destined to isolate six lively new metals: sodium, potassium, magnesium, calcium, strontium, and barium and – ultimately, at the shiny new Royal Institution in London – become one of the best spokesmen for, and demonstrators of, the newly born science of chemistry.

Humphry Davy was born at Number 4, The Terrace in the center of Penzance. According to June Z. Fullmer in her book *Young Humphry Davy*,[12] the house stood "on Market Jew Street facing the left profile of the Victorian era statue (50.118673°, −5.537081°) erected in his honor". (See Chart 3.6 for the location of this statue and other sites related to Priestley.) The first-born of an oft unemployed stone carver and a loving and resourceful mother, the youthful Davy loved to fish for salmon and trout, hunt for woodcock and grouse, search for new minerals, and roam around the Lands End region exploring its intimate connection with the wild sea. The legendary sea wrecks of the area seem to have particularly fired his imagination. Penzance is on the west side of the south-facing Mount's Bay that stretches from the fishing village of Mousehole (pronounced "Mowzel") on the west to Lizard's Point on the east. A few miles east of Penzance the legendary St. Michael's Mount (reached by a causeway at low tide) (50.116830°, −5.477591°) rises from the sea and inspired Davy to write poetry in its honor. When in pursuit of early scientists who developed the new compounds that would soon be explained in terms of atoms, we scientific travelers explore beautiful places.

Davy's father speculated in the Cornish mines but died leaving the family in debt. The young Humphry benefitted from the attention of several important benefactors who, with his mother Grace, made sure he was well educated. Davy loved reading, storytelling, and writing poetry but was exposed to the wiles of chemistry fairly early on in the form of phosphorescent materials and fireworks (which came from one of his benefactor's apothecaries), that took a toll on his wardrobe. In 1795 he had been apprenticed to a surgeon and apothecary as medical men were always needed to treat both the seagoing population and the miners. In 1797, however, he started to study chemistry by reading (in French) the recently guillotined Antoine Lavoisier's *Traité élémentaire de chimie*. Although early on he considered himself a follower and devotee of "the immortal" Joseph Priestley, by then exiled to America, Davy accepted Lavoisier's chemical revolution and its new nomenclature. While still in Penzance (and still in his teens), he conducted a series of chemical exercises studying heat and light and even contested Lavoisier's inclusion of caloric (heat as a fluid or weightless gas) in the late Frenchman's list of 33 elements. (See Chapter 8 for more on Lavoisier, his life, his revolution, and his fate at the blade of the guillotine.)

Tuberculosis (or consumption as it was commonly known then) changed Davy's life and, arguably, the history of chemistry. Not because he was its victim but rather because the sons of prominent men were. James Watt, whose economic improvements to the steam engine played such a crucial role in pumping water from the Cornish tin mines that were going deeper and farther out under the sea, and Josiah Wedgwood, the master potter, both had sons who suffered from consumption. Both fathers were convinced that the mild Cornish winters would be good for their sons' health. Gregory Watt arrived in the winter of 1797/98 and lodged in the Davy boarding house in Penzance. Thomas Wedgwood also came that winter. Gregory and Humphry were the same age, and quickly realized their mutual love of chemistry.

The fathers were members of the Lunar Society of Birmingham that also included Joseph Priestley. The Davys were recommended to the Watts and Wedgwoods by Dr Thomas Beddoes, a former Oxford professor who was forced out of the university because of his far-left political views, akin to those of Priestley's, including the support of the French Revolution. He had been a student of Joseph Black's, had visited Lavoisier in Paris and was convinced that pneumatic chemistry could be used for medical purposes.

Beddoes had recently written an article about the treatment of consumption by "fractious airs". Members of the Lunar Society financially backed Beddoes idea of opening a "pneumatic institute" to study the effect of gases on improving human health. Watt had collaborated with Beddoes to design a machine to produce large quantities of various gases that could be used in the clinic. Beddoes was considering the best place to locate the institute and for assistants to staff it. In 1798 he visited Cornwall and, at the recommendation of Gregory Watt, met the young Davy who related the experiments he had been conducting. Beddoes was immediately impressed with the young man and started to recruit him to be the director of his institute. Although he was reluctant to give up his apprenticeship in order to join Beddoes, Davy, in October of 1798, just short of his 20th birthday, boarded a coach for the long trip to Bristol where Beddoes was building his institute. The Institute was to be short-lived, but Humphry Davy was on his way to chemical stardom!

3.5.1 The Pneumatic Institution and Maturation of Humphry Davy

Beddoes located his institution in Dowry Square, Clifton, overlooking Bristol. Clifton had a warm spring, Hotwell, that was located directly on the River Avon and attracted the type of clientele that Beddoes hoped would avail themselves of his "alternative medicine". Bristol was a busy port town specializing in tobacco, tin, wine, sugar, and the slave trade. It also had the advantage of being near the fashionable resort town of Bath. The buildings of the institute still exist today but are privately owned. No. 7 Dowry Square (51.451244°, −2.620221°) served as the ward for bed-patients while No. 6 served as the laboratory and Davy's quarters . Behind these buildings was a stable that Beddoes and Davy used for the bulk manufacture of their therapeutic gases. The most important of these gases, of course, was Priestley's dephlogisticated air that was becoming known as "oxygen". It was here, however, that Davy, in the tradition of the great pneumatic chemists who preceded him, replicated, and then improved upon, the production of Priestley's "dephlogisticated nitrous gas", then known, using the new Lavoisier-inspired nomenclature, as nitrous oxide.

Davy had the rather unsettling habit of systematically testing his materials on himself, so it is not surprising that early on he imbibed a bit of his nitrous oxide. Earlier workers had tentatively identified this gas as the cause of the plague, "capable of producing the most terrible effects, when respired by animals in the minutest quantities ...." Davy and Beddoes did not believe these reports and so, in March 1799, after preparing small quantities of the gas mixed with air, Davy demonstrated their fallacy by taking a small whiff for himself! Noting that he did not meet his demise, he gradually increased the amount he breathed to three or four, or even nine quarts at a time! This is a lot of gas for anyone to take in but for Davy, who was just a little man no more than 5' 5" in height with a 29" chest, it was indeed a prodigious amount. Its properties were astounding to say the least. He said that inhalation of this slightly sweet-smelling gas was "accompanied with the loss of distinct sensation and voluntary power, a feeling analogous to that produced in the first stage of intoxication." It also was "attended by a pleasureable thrilling, particularly in the chest and the extremities. The objects around me became dazzling and my hearing more acute". Predictably, he breathed greater quantities more and more often – until he was afraid he was becoming addicted and was able to back off.

Not unexpectedly, this discovery caused a sensation in Clifton and Bristol. Many associates famous and not so famous, rich and not so rich, sick and not so sick lined up for the new inhalation therapy. Davy had become friends with a broad cross-section of intellectuals from all walks of life. His good looks, personal charm, and story-telling abilities, combined with his startling discoveries, made him a rising star in Bristol. The young man fresh from Cornwall was surrounded by Lunar Society members who had funded the Institution, the grown children of Joseph Priestley, and a variety of politicians, industrialists, doctors, poets, authors, and journalists, some attracted by the Hotwell springs, some by association with Beddoes, and later some by word of Davy's enticing, intoxicating, and altogether dazzling experiments. Among these were William Wordsworth (Poet Laureate 1843–1850, author of "Upon Westminster Bridge"), Samuel Taylor Coleridge ("The Rime of the Ancient Mariner"), Robert Southey (Poet Laureate 1813–1843, author of "The Battle of Blenheim" and "The Three Bears", the famous childrens' story), and Dr Peter Mark Roget, a physician known mostly for his *Thesaurus*. Of these, Southey became one of his good friends during the Bristol days. Southey published five of Davy's poems (those conceived in 1795 and 1796 – when he was but a teenager) in the first volume of his *Annual Anthology*. Southey, however, had commented that most of Davy's poetry was "tedious and feeble" – in other words, don't quit your day job!

Perhaps he wasn't a poet up there with Coleridge, Southey, and Wordsworth, but all that reading and studying in Penzance paid off handsomely as these talented people made Davy's acquaintance. One described him as "a very extraordinary young man of twenty, profoundly acquainted with natural philosophy, and well-read in most kinds of learning & modest and unaffected in manners". Fullmer tells us that Southey regarded Davy as "by far the first in intellect". One of his acquaintances said, "What a pity that such a Man should degrade his vast talents to Chemistry." Ah, even broad-minded and perceptive men of letters can lapse into narrow-mindedness!

As many imbibers would chuckle, giggle and laugh out loud at the pleasurable effects of the gas, it didn't take Davy long to dub nitrous oxide "laughing gas". It certainly made an impression on Southey who wrote to his brother that Davy had "invented a new pleasure" and that "the air in heaven must be this wonder-working gas of delight". Davy was quick to notice that it had an anesthetic property. Once when experiencing pain from a wisdom tooth he breathed in three large doses and found that it reduced the pain significantly. Of its potential as an anesthetic he wrote:

> As nitrous oxide in its extensive operation appears capable of destroying physical pain, it may probably be used with advantage during surgical operations in which no great effusion of blood takes place.

Somewhat strangely, no one looked further into this potential for about 50 years, when it was first employed as anesthetic in Hartford, Connecticut. It took another century and a half before it was commonly used as anesthetic in dental practices.

Davy's inhalation of nitrous oxide did not do him any harm – although he may have been a little lucky because impure samples of N_2O can contain nitrogen dioxide (NO_2), which is a severe lung irritant. Unfortunately, he went on to imbibe other gases not so innocuous. For example, he self-tested nitric oxide, NO, which severely burned his tongue, palate, and teeth – to say the least a painful and scary experience. He tried various doses of hydrogen and carbon dioxide and also "hydrocarbonate" or "water gas", a mixture of carbon monoxide and hydrogen. This dangerous inhalation produced severe chest pains and nearly killed him. One of his associates wrote that "He seemed to act as if in case of sacrificing one life, he had two or three others in reserve on which he could fall back in case of necessity .... [O]ccasionally .... I half despaired of seeing him alive in the morning." In fact, Davy's habit of self-experimentation did have lasting effects. While it did not kill him overnight, it did shorten his life considerably. He was not well for much of the second half of his life and died shortly after his fiftieth birthday.

Despite these stories of what we would regard as foolhardy self-experimentation, it was in Bristol that Davy matured into an excellent chemist, or what he called a "chemical philosopher". He meticulously recorded the production, delivery and effects of various gases. As part of this work he carefully measured and recorded human and animal lung capacities. Although his nitrous oxide results were certainly scintillating and brought him great fame, it is important to note that he carefully isolated, purified and characterized the various nitrogen oxides, ammonia gas and both nitrous and nitric acid. All of this was published in 1800 in his book *Researches, Chemical and Philosophical; chiefly concerning nitrous oxide or dephlogisticated nitrous air, and its respiration*.[15] These data would prove most important in the hands of John Dalton as he formulated and defended his first concrete atomic theory in the next decade.

Humphry Davy's reputation soared as a result of his work at the Pneumatic Institution. However, he soon realized that Beddoes' clinic was not going to last long. Inhalations of nitrous oxide, while exhilarating, "cured" very few patients. Fortunately for Davy a new field came on the horizon just at the right time. Around 1796, Alessandro Volta in Pavia, Italy, prepared the first chemical battery (a "Voltaic pile") capable of producing a sustained and predictable current. (See Chapter 9 for details and many Volta sites to visit in the lake district of northern Italy.) In early 1800 Volta wrote a letter to Sir Joseph Banks of the Royal Society in London. News of this discovery spread fast and soon Davy was trying his own electrical experiments in his laboratory in Bristol. Not surprisingly, one of the first things he did was to build a large battery and shock himself to gage its effectiveness! Immediately, Davy realized that his future lay in electro-, and not pneumatic, chemistry.

In January of 1801 Davy was invited to be a lecturer at the newly formed Royal Institution (Ri) (51.509861°, −0.142778°) in London. He delivered his first public lecture there in April and went on to make his mark in the scientific world. In 1802, he was appointed Professor of Chemistry at the Ri and in 1807 was made a fellow of the Royal Society. For a dozen years he regularly produced a series of immensely popular lectures on the latest discoveries in and applications of chemistry. Active in research, he discovered six elements in two years (1807–1808), conclusively disproved Lavoisier's hypothesis that oxygen was present in all acids (1810), and "discovered" and launched Michael Faraday into a stellar career in chemistry and physics. In 1812, he was knighted and in 1815 invented the coal miner's safety lamp that saved many lives. We will cover his accomplishments at the Ri in Chapter 5.

3.6 SUMMARY

As we come to the end of the eighteenth century, where does our travel adventure related to the history of the atomic concept stand? Robert Boyle and Isaac Newton had earlier expressed their confidence that matter is atomic or "corpuscular". Joseph Priestley is in America and Humphry Davy is headed to London and chemical stardom. What about the atom? As the millennium turns, it certainly has no official standing, but many chemical philosophers are comfortable with the idea. Davy was using words such as "atom", "corpuscle" and "particle" to refer to the ultimate constituents of matter. According to June Fullmer, "Beddoes, Davy and many of their associates attuned their universe through their ardent belief in the existence of material corpuscles yoked in equilibria poised between attractions and repulsions." Davy certainly believed that matter was made of particles and that heat is the motion of these particles. The pneumatic chemists, most notably Joseph Priestley, had accumulated data on a variety of gases and the various elements that composed them. The natural philosophers of the day were wary of making unnecessary assumptions regarding the mechanical structure of the nature of matter. We are ready for someone to actually state a testable concrete atomic theory rooted in solid experimental data. That "someone" is John Dalton, and we will discuss his life, his theory, and the travel stops related to his career in the next chapter.

As for travels with the atom sites to visit, this chapter has detailed many. We have traced Robert Boyle from Stalbridge to Oxford and then to London, where the Royal Society on Carlton House Terrace is a must-stop. Robert Hooke's birthplace at Hooke Hill on the Isle of Wight is a delightful stop and then we can go on to his grave in London and various memorials there including in Westminster Abbey and St. Paul's Cathedral. Isaac Newton has a magnificent tomb in the Abbey as well, but his Woolsthorpe Manor in Grantham and the site of his alchemical laboratory and his statue in the Trinity College Chapel at Cambridge University are also unique sites for scientific/historical travelers. In Edinburgh, we can visit sites related to Joseph Black, one of the early pneumatists of note. We can follow Joseph Priestley starting in his modest birthplace in Fieldhead and then proceed to mark his career with the statues in Birstall, Leeds, and Birmingham. Of particular importance is the International Chemical Landmark in the Bowood house where he discovered oxygen and the modest house on the green in nearby Calne. Finally, we can follow Priestley to America, where he built a magnificent house in Northumberland, PA. For Humphry Davy, we can see the statue in his birthplace in Penzance and then follow him to the site of Beddoes' Pneumatic Institution in Bristol. In the next chapter we will catch up with Davy at the Royal Institution in London.

ADDITIONAL READING

On Boyle:

- J. Emsley, *The 13th Element, The Sordid Tale of Murder, Fire, and Phosphorus*, John Wiley & Sons, New York, 2000.

- R. Lomas, *The Invisible College, The Royal Society, Freemasonry and the Birth of Modern Science*, Headline Book Publishing, London, 2002.
- M. DiMeo, "Such a Sister Became Such a Brother": Lady Ranelagh's Influence on Robert Boyle, *Intellect. Hist. Rew.*, 2015, **25**, 21–36.

On Black:

- R. G. W. Anderson, in *Science History: A Traveler's Guide*, ed. M. V. Orna, American Chemical Society, Washington, DC, 2014, Scientific Scotland.
- J. L. Marshall and V. R. Marshall, Rediscovery of the Elements: Joseph Black, Magnesia and Fixed Air, https://digital.library.unt.edu/ark:/67531/metadc501453/, accessed June 2019.

On Davy:

- J. Z. Fullmer, *Young Humphry Davy: The Making of an Experimental Chemist*, American Philosophical Society, Philadelphia, 2000.
- D. Knight, *Humphry Davy: Science and Power*, Cambridge University Press, Cambridge, 1992.

On Priestley:

- M. E. Bowden and L. Rosner, *Joseph Priestley, Radical Thinker*, Chemical Education Foundation, Philadelphia, PA, 2005.
- K. Baker, *Joseph Priestley, Friends and Foes: Remarkable Lives in an Age of Revolution*, H. Charlesworth & Co Ltd 2009.

REFERENCES

1. J. B. West, Robert Boyle's landmark book of 1660 with the first experiments on rarified air, *J. Appl. Physiol.*, 2005, **98**, 31–39.
2. L. M. Principe, *The Aspiring Adept: Robert Boyle and His Alchemical Quest*, Princeton University Press, Princeton, 1998.

3. The Boyle Project directed by Michael Hunter, http://www.bbk.ac.uk/boyle, accessed June 2019.
4. The London of Robert Boyle, Thorburn Burns, http://www.bbk.ac.uk/boyle/media/pdf/TheLondonofRobertBoyle.pdf, accessed June 2019.
5. C. Tanford and J. Reynolds, *The Scientific Traveler, A Travel Guide to Scientific Sites of the British Isles*, John Wiley & Sons, Inc., New York, 1992.
6. S. K. Bolton, *Famous Men of Science*, CreateSpace Independent Publishing Platform, 2015.
7. R. E. Spargo, *S. Afr. J. Anim. Sci.*, 2005, **101**, 315–321.
8. Investigating the Site of Newton's Laboratory in Trinity College, Cambridge, http://www.dspace.cam.ac.uk/handle/1810/198285, accessed June 2019.
9. K. S. Davis, *The Cautionary Scientists, Priestley, Lavoisier, and the Founding of Modern Chemistry*, G. P. Putnam's Sons, New York, 1966.
10. M. E. Weeks, Discovery of the elements, *J. Am. Pharm. Assoc. (1912–1977)*, 1939, **28**, 712.
11. J. Jackson, *A World on Fire: A Heretic, an Aristocrat, and the Race to Discover Oxygen*, Viking, New York, 2005.
12. J. Z. Fullmer, *Young Humphry Davy: The Making of an Experimental Chemist*, American Philosophical Society, Philadelphia, 2000.
13. R. Boyle, *The Sceptical Chymist*, J. Cadwell, 1661.
14. I. Newton, *Opticks*, 2nd edn, Book 3, 1718.
15. W. Wordsworth, *The Prelude, or Growth of a Poet's Mind*, Edward Moxon, London, 1850.
16. J. Priestley, *Experiments and Observations on Different Kinds of Air*, London, 1775, vol. 2.
17. H. Davy, *Researches, Chemical and Philosophical*, London, 1800.

CHAPTER 4

Hard Spheres and Pictograms, The First Concrete Atomic Theory: John Dalton (Northern England and Manchester)

4.1 A QUICK LOOK AT PLACES TO VISIT "TRAVELING WITH THE ATOM" IN NORTHERN ENGLAND AND MANCHESTER

John Dalton Eaglesfield birthplace house ✦✦; *Pardshaw Hall Quaker Meeting House memorial stone* ✦✦; *Stramongate School* ✦ *and the Quaker Tapestry Museum* ✦✦ *in Kendal; In Manchester: Manchester Town Hall statue* ✦✦✦ *and Ford Madox Brown mural* ✦✦✦; *John Rylands Library collection of original writings* ✦✦✦; *John Dalton Building statue and original grave marker stones* ✦✦; *Manchester Literary and Philosophy Society* ✦; *Faulkner Street plaque* ✦; *Portico Library and Gallery* ✦✦; *Museum of Science and Industry collection of objects* ✦✦; *The Science Museum Dalton display in London* ✦✦.

Traveling with the Atom: A Scientific Guide to Europe and Beyond
By Glen E. Rodgers
© Glen E. Rodgers 2020
Published by the Royal Society of Chemistry, www.rsc.org

John Dalton and the First Concrete Atomic Theory

- John Dalton (1766–1844) started his scientific career as a meteorological observer and chronicler but soon expanded his interest to investigate its component gases.
- He formulated the Law of Partial Pressures and the Law of Multiple Proportions.
- His atomic theory incorporated Lavoisier's Law of Conservation of Mass and the Greek ideas of indivisible atoms. In 1808 he published *A New System of Chemical Philosophy* that stated that (1) all matter consists of tiny, indivisible atoms that cannot be created or destroyed, (2) atoms of one element cannot be converted into atoms of another element, (3) atoms of an element are identical in mass and other properties and are different from the atoms of any other element, (4) what we now call molecules result from the chemical combination of whole-number ratios of atoms, and (5) in chemical reactions, atoms are rearranged from one setting to another.
- Dalton devised his own symbols to represent atoms and molecules and constructed one of the first tables of atomic weights. He used his own data and that gathered by his predecessors Black, Davy, and Priestley to advance his ideas.
- Unfortunately, his adherence to a "rule of greatest simplicity" created controversy and significantly hindered the establishment of accurate tables of atomic weights for half a century. This adherence was compounded by his dogged lack of acceptance of Joseph Louis Gay-Lussac's Law of Combining Volumes and Amadeo Avogadro's Hypothesis that equal volumes of gases contain equal numbers of molecules.

4.2 JOHN DALTON (1766–1844)

"Matter, though divisible in an extreme degree, is nevertheless not infinitely divisible. That is, there must be some point beyond which we cannot go in the division of matter. The existence of these ultimate particles of matter can scarcely be doubted, though they are probably much too small ever to be exhibited by microscopic improvements. I have chosen the word atom to signify these ultimate particles ..." John Dalton[3]

4.2.1 Dalton in Eaglesfield, Pardshaw Hall, and Kendal, England

John Dalton was born into a poor, staunchly Quaker family in the small town of Eaglesfield in northern England. The youngest of three surviving children, his date of birth is somewhat uncertain as his parents did not register it. At the age of 10 he was sent to the nearby Pardshaw Hall Quaker School in the village of the same name. Here he was somewhat unevenly educated but benefitted from teachers such as schoolmaster John Fletcher who recognized Dalton's intellectual curiosity and tutored him in the evenings. Unfortunately, Fletcher resigned but, amazingly, Dalton, now 12, took it upon himself to assume the sole teaching duties there in a local Friends' Meeting House that still stands today. At the age of 15 he and his brother Jonathan started to teach and study at the Stramongate School in Kendal and remained there from 1781 to 1793. Dalton's first scientific love was meteorology. In Kendal he began to carry out and record meteorological observations, many with instruments such as thermometers and barometers that he built himself. The first entry commemorated the *Aurora Borealis*, 24 March 1787. He did this for 57 years, from the time he was 21 until the day he died, recording more than 2 00 000 observations. Figure 4.1 shows one his daily weather logs.

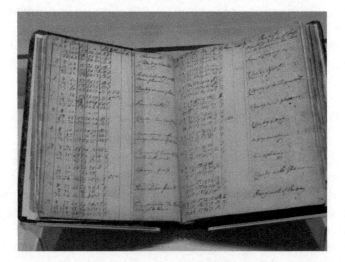

Figure 4.1 Daily weather log kept by John Dalton between 1803 and 1827. Source: Society of Friends. Displayed in Dalton exhibit, The Science Museum (Photograph: G. E. Rodgers).

Hard Spheres and Pictograms 79

Not surprisingly, some sources say, he always maintained that he never had time for marriage. In 1794, he was quoted as saying that "my head is too full of triangles, chymical processes, and electrical experiments, *etc.*, to think much of marriage". In 1793, at the age of 27, he was appointed a tutor in mathematics and natural philosophy at the New College in Manchester. (New College was the successor to the Warrington Academy, where Joseph Priestley had taught in the 1760s. See Chapter 3 for a review of Priestley as a "rational dissenter".) Also, in 1793, he published some of his early observations in his *Meterological Observations and Essays* which included a prominent section comprising the results of six years of auroral observations.

4.2.2 Travel Sites in Northern England Related to John Dalton

See Chart 4.1 to locate the following sites.

4.2.2.1 John Dalton Cottage. (54.641021°, −3.404884°), Dalton Lane, Eaglesfield Cumbria. The whitewashed bungalow where John Dalton was born is still standing. Traveling south from Cockermouth on the A5086 (also signposted Lamplugh), Eaglesfield is 2.6 miles (4 km) down the road. From the A5086, turn right (west) onto Hotchberry Brow. There are few marked streets in Eaglesfield but Dalton Lane is one of them. Googling "Dalton Lane Eaglesfield" will easily locate the cottage. A sign above the door reads "John Dalton DCL. LLD – The Discoverer of the Atomic Theory – was born here September 5, 1766, died at Manchester July 27 1844." The DCL refers to the honorary doctor of civil laws awarded him by Oxford University in 1832. The honorary LLD (doctor of laws) was awarded by the University of Edinburgh in 1834. In 2007, the Royal Society of Chemistry put in place a new blue plaque designating the Dalton Cottage as a National Chemical Landmark. It lists his positions and then specifically cites his work in discovering the laws of partial pressure and multiple proportions, recognizing color-blindness and revolutionizing chemistry through his atomic theory. After he became famous, John Dalton used to visit this little house and point out its features to his friends and visitors.

4.2.2.2 Dalton Memorial in Quaker Meeting House. (54.61245°, −03.394883°), Pardshaw Hall. The Quaker School here in Pardshaw Hall no longer exists but the Quaker Meeting

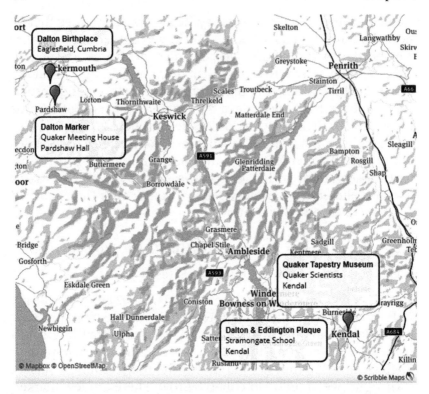

Chart 4.1 Travel sites in northern England related to the birth and early career of John Dalton. Centered on the Lake District National Park, this chart shows the location of Dalton's birthplace in Eaglesfield, the town of Pardshaw Hall, where he was a student and young teacher, the Quaker Meeting House in Pardshaw Hall that has a telling memorial plaque in his memory, the Quaker Tapestry Museum in Kendal and the Stramongate School also in Kendal where he taught for 12 years before moving to Manchester.

House does. Dalton attended the school but also started teaching his younger colleagues there at the age of 12. Inside the walled burial ground of the meeting house, there is a flat stone inscribed with a memorial to Dalton that reads: "He was not for an age, but for all time." To get to Pardshaw Hall (not to be confused with Pardshaw) from Cockermouth, continue south on the A5086 past the turn for Eaglesfield. Pardshaw Hall is the next turn on the opposite side (east, to the left). Once in Pardshaw Hall, the Meeting House is the first building on the left and the graveyard is hidden from view by a high wall. To get keys to the graveyard, telephone Colin Wornham at (0)1900

Hard Spheres and Pictograms 81

823531. (The number should be preceded by 011-44 if calling from the United States or Canada.) It is best to call ahead to make arrangements.

4.2.2.3 Plaque at the Stramongate School. (54.32910°, −2.74315°), 52 Stramongate, Kendal. The plaque notes that the Stramongate School was founded by the Society of Friends in 1698. The school has connections not only with John Dalton, founder of the atomic theory, who taught here from 1781 to 1793, but also with Sir Arthur Stanley Eddington, pioneer of stellar structure and author of *The Expanding Universe*. This primary school still operates, and its symbol includes a distillation apparatus (see Figure 4.2). At the age of 27, Dalton left the Stramongate School to accept a position at the New College in Manchester.

4.2.2.4 The Quaker Tapestry Museum and Exhibition. (54.328731°, −2.743187°). Across the street is the Stramongate House, the Kendal Quaker Meeting House, and the Quaker Tapestry Museum & Exhibition. The latter houses 77

Figure 4.2 The seal of the Stramongate School where John Dalton taught from 1781 to 1793. (Photograph by G. E. Rodgers).

embroidered tapestries (constructed *via* the contributions of 4000 men, women, and children from 15 countries between 1981 and 1996) honoring "the Quaker influence on the modern world". The Scientists panel, as shown below in Figure 4.3, honors three Quaker Scientists: the astronomer Sir Arthur Eddington (1882–1944), Dame Kathleen Lonsdale (1903–1971), and, of course, John Dalton (1766–1844). Dalton is cited as the "atomic chemist" who investigated "the minute structure of matter". During a total solar eclipse observed on an island off the west coast of Africa, Eddington performed the first experimental test of Albert Einstein's general theory of relativity when he recorded stellar positions that verified the bending of light around the sun. Lonsdale was a pioneer in X-ray diffraction

Figure 4.3 The Quaker Science Tapestry honoring Quaker scientists John Dalton, Sir Arthur Eddington, and Dame Kathleen Lonsdale. The 77 panels of the Quaker Tapestry is a modern community embroidery made by 4000 people from 15 countries. The exhibition of panels can be seen the Quaker Tapestry Museum in the Friends Meeting House in Kendal, Cumbia UK. Further information: www.quaker-tapestry.co.uk © Quaker Tapestry. Reproduced with permission.

methods, who was one of the first women elected as a Fellow of the Royal Society (FRS). The tapestry shows the structure of hexamethylbenzene in which the central benzene ring was confirmed to be planar. After visiting the Tapestry Museum, scientific travelers might enjoy the "Kendal and the Quakers: A Discovery Walk" booklet. Further information is available at http://www.quaker-tapestry.co.uk.

4.2.3 Dalton's Early Work in Manchester

In Manchester, Dalton soon joined the Manchester Literary and Philosophical Society. His first communication to the "Lit & Phil" society was on the topic of red-green "colourblindness", from which he himself suffered. This type of color-blindness is still referred to as "daltonism" particularly in French, Russian and Spanish ("Daltonismo"). The New School where he taught closed in 1799 but Dalton stayed in Manchester and became a private teacher. His love of meteorology logically led him to consider the composition of the air and the nature of its component gases (using the older nomenclature) including Joseph Priestley's "dephlogisticated air", Daniel Rutherford's "phlogisticated air", Joseph Black's "fixed air", and even, he stated, the gaseous form of water or "water vapor". He concluded that the atmosphere was a mixture of gases and that the total pressure of this mixture was the sum of the "partial pressures" that each individual gas would exert if it alone occupied a given container. This soon became known as Dalton's Law of Partial Pressures (1801). The pressure, he maintained, was due to the particles (or what he referred to generally as atoms – the term "molecule" was not formulated until a few years later by Amadeo Avogadro. What we would call a molecule Dalton called a "compound atom") of these gases slamming against the inside walls of the container in which the gases were held. As part of this argument he maintained, like Newton before him, that the atoms (or what Newton called "primitive particles") of a given component only repelled each other and not the atoms of the other component gases. This conclusion, now proven untrue and replaced by what is known as the "kinetic-molecular theory of gases", was widely discussed and it was in defending these ideas that he was led to his atomic theory. See Chapter 5, p. 112 and Chapter 10, pp. 250–251 for a discussion of the kinetic-molecular theory of gases, so strongly advocated by James Clerk Maxwell, Ludwig Boltzmann and others.

4.2.4 Dalton's Atomic Theory

So, let's look at the basic assumptions of Dalton's atomic theory, which he first presented in an 1803 paper delivered at the Lit & Phil and first published in his book, *A New System of Chemical Philosophy* in 1808.[3] First, he maintained that all matter (not just gases) consists of atoms: tiny, indivisible particles of an element that cannot be created or destroyed. (This assumption incorporates both the Greek idea of indivisible atoms and "The Law of Conservation of Mass" which had been definitively established by Lavoisier. See Chapter 9 for a description of Lavoisier's extensive experiments.) Second, the atoms of one element cannot be converted into atoms of another element. (So much for alchemical transmutations! Lead atoms cannot be transformed into gold atoms.) Third, atoms of an element are identical in mass and other properties and are different from the atoms of any other element. (That is, atoms are unique, particularly in their mass, to a given element.) Fourth, what we now call molecules result from the chemical combination of a specific whole-number ratio of atoms of different elements.

Recall from Chapter 3 (p. 64) that Joseph Priestley had discovered carbon monoxide gas in 1799 after he had moved to Northumberland, PA. Dalton rightly maintained that this gas was composed of molecules that contained one atom of carbon and one atom of oxygen. ("Oxygen" had become the name of the gas that Priestley had called "dephlogisticated air". The Frenchman Antoine Lavoisier had renamed it oxygen as part of his famous chemical revolution. See Chapter 8 for the details of Lavoisier's seminal work that revolutionized chemistry.) Given that the mass of carbon and oxygen atoms are unique and that they always combine in a one-to-one ratio, it follows that carbon monoxide always has the same proportion by mass. This is an example of the so-called "Law of Definite Proportion" or the "Law of Constant Composition". It had been first formalized by the French chemist Joseph Proust, also in the year 1799. (Proust had first demonstrated this law in the compound we call copper carbonate.) This law quickly became well accepted. Each of the

Hard Spheres and Pictograms 85

gaseous compounds discovered by the "pneumatic chemists" (discussed in Chapter 3) such as Black, Davy, and Priestley followed the Law of Definite Proportions.

Also recall from Chapter 3 (p. 48) that Joseph Black had discovered what he called "fixed air" in 1756 and Priestley had found it above the fermenting liquid (the "wort") at the Jakes & Nell brewery in Leeds. (He then found a way to mix it with water to produce "sparkling water".) Fixed air is what we know today to be carbon dioxide. This is a second carbon-oxygen compound, this one composed of molecules containing one atom of carbon and *two* atoms of oxygen. The existence of both carbon monoxide and carbon dioxide is an example of the Law of Multiple Proportions, which has been expressed in various convoluted ways but essentially says that two elements can form more than one compound, each of which has a definite proportion by weight. Dalton himself first stated this law in 1803 when he reported on his studies of the multiple combinations of oxygen that combined with Davy's "nitrous air" (what we now call nitric oxide, NO). Dalton, in one of his first purely chemical investigations, found that two different amounts of oxygen reacted with a set amount of the nitric oxide, establishing two different products. These results set him on a pathway toward establishing the Law of Multiple Proportions.

In the above paragraphs, we are using the modern symbols for atoms: C for carbon, O for oxygen, N for nitrogen and so forth. (These symbols were devised by the Swedish chemist Jöns Jakob Berzelius in 1813. See Chapter 19 for a description of Berzelius's many contributions to the atomic concept and some of the beautiful places to visit related to his and other Swedish chemists' atomism.) Even though he and Berzelius were contemporaries, Dalton never accepted these symbols. Instead, he preferred to visualize atoms as small hard balls of different sizes. He even commissioned the construction of small wooden spherical models so he could better demonstrate how they combined. Each ball had a number of holes drilled in it so it could be connected with varying numbers of other balls using small wooded rods. (The Science Museum in London often has several of these wooden spheres on display. See p. 96 for further details. The Manchester Museum of Science and Industry does as well. See below.) Dalton's models were a source of controversy in some quarters because they gave atoms the appearance of physical reality. Dalton and some of his colleagues believed that atoms were, in fact, physically real, whereas others preferred to believe they were merely a convenient way to think about the way various elements and compounds combine in chemical reactions. (Sometimes we say that Dalton believed in "physical atoms", whereas these others believed in "chemical atoms". Chapter 10 discusses these designations in further detail.)

Given Dalton's mindset about the physical reality of atoms as hard spheres, it is not surprising that he devised two-dimensional atomic symbols that represented atoms as circles. These symbols are sometimes called his "pictograms". To differentiate the atoms from each other, each circle was a little different. Oxygen was merely an open circle, carbon's circle was filled in (graphite, after all, is black), nitrogen's circle had a vertical line down through it, hydrogen's circle had a dot in the middle, and so forth. A few of Dalton's symbols are shown in Figure 4.4. Note his symbols for carbon and oxygen atoms and the carbon monoxide and carbon dioxide molecules. Also shown are the symbols for nitrogen, oxygen, and three nitrogen oxides: nitric oxide, nitrous oxide, and nitrogen dioxide. So far so good you say. All of this is making sense.

In the above examples, Dalton was using his "rule of greatest simplicity" that assumed that when atoms combine in only one ratio, they do so in the simplest way possible. So, if only one molecule of two different atoms (that is to say, one compound composed of only two elements) was known, Dalton assumed that it was a one-to-one combination. If there were two compounds they were 1:1 and 2:1, and so forth. So, for the carbon-oxygen compounds, one was CO and the second was CO_2. For the nitrogen-oxygen compounds, one was NO, one was NO_2, and a third was N_2O. The rule of simplicity makes some sense, of course, because there was no way of knowing what the exact atomic ratio was for a given compound. You had to start somewhere and starting with the simplest ratio was a reasonable assumption. Dalton had an additional rationale that the fewer mutually repellant atoms there are in a molecule, the greater the mechanical stability of the system. Occasionally, Dalton made an exception. For the carbon-hydrogen compounds that he knew about, as shown in the above figure, he assumed the carbon-hydrogen ratios were

Hard Spheres and Pictograms 87

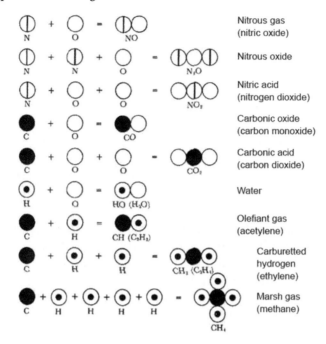

Figure 4.4 Dalton's symbols for atoms and simple compounds. Modern symbols appear beneath the Dalton symbols. The names in the right-hand column are Dalton's. When either the formula or the name of the product is different today, the modern formula or name is given in parentheses. Reproduced from ref. 1 with permission from the authors, R. E. Dickerson, H. B. Gray and G. P. Haight.

1:1, 2:1 and 4:1. (The data supported these assumptions well.) So, all of this seemed to be working out quite logically, but it was a bit fortuitous. The fly in the ointment turned out to be one of the simplest substances of all, water itself. Only one hydrogen–oxygen compound was known at the time and that was water. According to the simplest ratio rule, Dalton, logically enough, assumed that water was a one-to-one compound, that is, it was HO. Today, of course, we know it is H_2O.

Given all of the above and data involving other compounds, Dalton proceeded to construct a table of relative atomic weights. Again, an assumption had to be made. Hydrogen was the lightest element, so Dalton assumed that its relative atomic weight was exactly one unit. (By the way, the unit of atomic weight or mass is still known today, in some subdisciplines such as biochemistry, as a "dalton".) The oxygen:hydrogen weight ratio in water was known to be about 7 (the modern value is closer to 8). Therefore, if the weight of one hydrogen is assumed to be 1 and water is HO, the weight of one oxygen would be about 7. Using the same reasoning and the weight ratios known at the time for CO and NO, the weights of carbon and nitrogen could be calculated. Dalton's table of atomic weights is given in Figure 4.5. These weights differ fairly significantly from those known today because of Dalton's assumptions about the simplest atomic ratios (HO for water when it is actually H_2O and NH for ammonia when it is actually NH_3) and also because the weight ratios were not known accurately.

Be these difficulties as they may have been, Dalton had a table of atomic weights and his atomic theory made a lot of sense to many chemical practitioners. It organized a number of the assumptions and laws known at the time including the Greek ideas of indivisible atoms, Lavoisier's Law of Conservation of Mass, Proust's Law of Definition Proportions, and Dalton's own

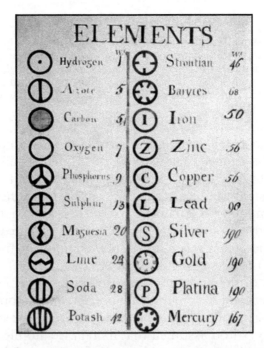

Figure 4.5 Table of John Dalton's atomic symbols (pictograms) and the atomic weights he assigned to each atom. Reproduced from ref. 2.

Law of Multiple Proportions. The mass ratios of a number of simple compounds, including the "airs" of the English pneumatic chemists like Black, Priestley, and Davy, were known and useful for supporting the atomic theory, even though these men were reluctant to speculate on the existence of atoms. (Humphry Davy, in particular, was most skeptical about the theory of his "learned friend", John Dalton, but even he, in the end, came to accept it.) Above all, this theory could be tested and its general assumptions held up well. Dalton had taken a concept that had been vaguely discussed for two thousand years and fashioned it into a working hypothesis that stood the test of time and became the atomic theory that revolutionized chemistry and science.

4.2.5 The Reaction to Dalton's Atomic Theory

The reaction in England and many other places to Dalton's atomic theory was favorable. Thomas Thomson (1773–1852), professor of chemistry at the University of Glasgow starting in 1817, was an early advocate of Dalton's atomic theory. He met with Dalton in 1804 and penned a concise summary of the theory in his journal. The first published description of Dalton's theory appeared in the third edition of Thomson's *System of Chemistry* (1807). He was confident that the theory would be quickly adopted. William Prout (1785–1850), an English physician, was interested in chemistry and was favorably impressed by Dalton's atomic theory. He is responsible in 1815 for what became known as Prout's hypothesis: all atomic weights are integral multiples of the weight of hydrogen and therefore hydrogen atoms are a fundamental building block of all atoms. Many people tried to prove this hypothesis, but ultimately it was abandoned only to be revived in a distinctly different form a century later with the discovery of the proton. William Wollaston (1766–1828), a wealthy and influential figure in the Royal Society, quickly became a strong advocate of Daltonian atomism as early as 1807. He talked with Berzelius when the Swede visited England in the summer of 1812 and shared with Berzelius a list of atomic weights at that time. Despite some reports to the contrary, Wollaston (although he was wary of overextending the claims about the nature of atoms) was essentially a strong supporter of Dalton's ideas.

Jöns Jacob Berzelius (1779–1848), the famous Swedish chemist, first heard of Dalton's Law of Multiple Proportions (by reading a paper by Wollaston in 1808) and provided additional examples. He devised more precise methods to measure and calculate relative atomic weights. He recognized that the atomic theory was a most important landmark in chemical history but was not particularly impressed by the experimental details in Dalton's *New System* when he received it from the author in 1812. He did not visit Dalton when he visited England in that same year. He introduced his now famous and useful symbolic notations in 1813. (Again, see Chapter 19 for more information on Berzelius.)

Humphry Davy (1778–1829), at the invitation of Count Rumford [Benjamin Thompson (1753–1814)], had joined the Royal Institution in 1801 and became a sensation there in a very short time. In 1807, when the still-young Davy (about 29) met with Wollaston and Thomson, he was skeptical of Dalton's atomic ideas. He was in the midst of his great discovery years (1807–1808), isolating alkali and alkaline earth metals from various salts by electrolysis using his voltaic pile as well as looking for hydrogen and oxygen in various sulfur-, phosphorus-, and nitrogen-containing materials. After some thought, Davy seemed to agree with Dalton's basic atomic principle but the details about the nature of atoms as hard, indivisible, spheres (*i.e.*, "physical atoms") seemed to him to be unsubstantiated and unnecessary speculation.

As a result of this high degree of support from a variety of quarters, John Dalton became one of the most famous men of his day. In 1800, he had been elected secretary of the "Lit & Phil" (Manchester Literary and Philosophical Society). In 1816, he was elected president, a position he held for the rest of his life. In 1804, he had been invited to present a set of lectures at the Royal Institution. He delivered another set of lectures on his atomic theory in 1809–1810 where he defended his ideas against, amongst others, Davy's skepticism. (He was not a very effective lecturer – certainly nowhere in the league with the handsome, well-spoken, popular, and dynamic Davy but his atomic ideas evidently carried the day.) Despite their different points of view about the physical reality of atoms, Davy nominated him for membership in the Royal Society in 1810, but Dalton refused the offer. In 1822, he was re-nominated and elected without his knowledge and subsequently received its first Royal Medal in 1826. Davy died at the early age of 50 in 1829 and a year later Dalton took Davy's place as one of the elite eight foreign associates in the French Academy of Sciences. Oxford University proposed to confer an honorary doctorate degree (the D.C.L. listed on the plaque at his birthplace, the John Dalton Cottage in Eaglesfield) on him in 1832, but the problem was that no Quaker could wear the scarlet robes associated with this degree. When asked about this problem, Dalton replied that the robes did not look anything like scarlet to him, but instead a dull shade of grayish green. After the doctoral presentation, he was presented to King William IV in the same robes. When Dalton died in July 1844, he was accorded a civic funeral with full honors. His body lay in state in the Manchester Town Hall for four days, while more than 40 000 people filed past his coffin. One hundred coaches followed the funeral cortège to Ardwick Cemetery. Suffice it to say that this type of honor has not been bestowed on many men or women of science.

Particularly as he got older, Dalton had difficulty accepting new procedures and ideas. He wanted to work out everything for himself and this inclination led him into errors of judgment that, using the 20/20 (presumably non-colorblind) vision of hindsight, we (sometimes too cavalierly) assume he could have avoided. As an example of this obtuse self-reliance, he once refused to take "a powder" prescribed by a doctor until such time as he could subject it to his own chemical analysis. Opinions vary as to Dalton's prowess in the lab. Davy called him "a very coarse experimenter". More modern evaluations are kinder to Dalton in that regard. Given that he could not readily accept the results of others, it is not surprising that when his not-particularly-accurate atomic weights were called into question, he adamantly refused to accept the more accurate values. In 1809, when Joseph Louis Gay-Lussac offered his Law of Combining Volumes that many others readily accepted, Dalton refused to recognize it. (See Chapter 9 for more on Gay-Lussac.) In 1811, when Amadeo Avogadro offered his hypothesis that equal volumes of gases contain equal numbers of particles, Dalton vigorously rejected it as well. (See Chapter 9 for more on Avogadro.) The combination of Gay-Lussac's Law and Avogadro's Hypothesis would have produced a strikingly more modern set of atomic weights in the second decade of the nineteenth century. As it was, it took nearly a half a century (until the First International Chemical Congress was held in Karlsruhl in 1860) for these ideas to be fully appreciated and utilized. (See Chapter 9 for more on the Chemical Congress and the role of Stanislao Cannizzaro.) Dalton became overly confident to a fault. In 1840, a paper he had written on phosphates and arsenates was refused by the Royal Society. His response? He published it himself and followed this up with four other papers. And, as noted elsewhere, he never could accept the chemical symbols devised by Berzelius even though most people found them much simpler and more convenient to use than his relatively awkward circular symbols.

4.2.6 Travel Sites in Manchester Related to John Dalton

See Chart 4.2 to locate the following sites.

4.2.6.1 Manchester Town Hall. (53.479149°, −2.244182°), Albert Square, Manchester. This grand gothic-style building with its magnificent clock tower was recently refurbished in 2014. In the foyer there are beautiful white statues of John Dalton and his student, James Prescott Joule, who is famous for his principle of the mechanical equivalent of heat and the unit of energy named after him. Dalton personally sat for the sculptor, Sir Francis Chantrey. From the main entrance on Albert Square two grand staircases lead up to the landing outside the Great Hall. In this large meeting room, there are twelve murals by Ford Madox Brown, one of which shows Dalton collecting fire-gas from a marsh, watched by a group of children and a cow peering curiously over a fence. "Fire-gas" or "marsh gas" is methane. Dalton is shown stirring up the mud of a stagnant pond, while an assistant (in this case a farmer's boy) catches the bubbles in a wide-mouthed bottle. In August 1804, Dalton discovered that his Law of Multiple Proportions applied to methane (CH_4) and acetylene or "olefiant gas" (CH according to Dalton, C_2H_2 by today's formula). These murals took 15 years to complete and give a panorama of the city's history. There are two science-related murals here, the other being William Crabtree observing the transit of Venus across the sun in 1639. The Dalton painting was done in 1887; Joule was still alive at that time so his contributions

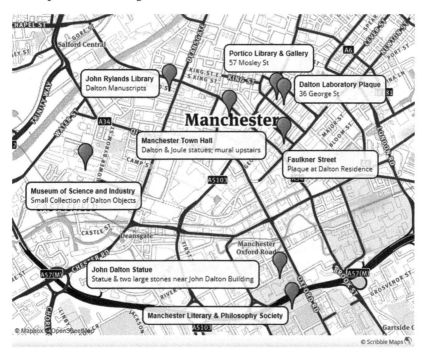

Chart 4.2 Travel sites in Manchester, England related to John Dalton. Includes statues at Manchester Town Hall and the John Dalton Building, the residence plaque at Faulkner Street, the John Rylands Library with its Dalton manuscripts, the Museum of Science and Industry with its small collection of Dalton objects, the Manchester Literary & Philosophy Society, and the Portico Library & Gallery.

to science could not be documented. Going west from Albert Square (toward Deansgate) in Lincoln Square is a striking statue of Abraham Lincoln (53.479685°, −2.247154°). The inscription notes Lincoln's gratitude for the support of the "working people" of Manchester during the embargo of southern cotton during the American civil war.

4.2.6.2 John Rylands Library. (53.480279°, −2.248763°) 150 Deansgate, Manchester. This neo-Gothic library now houses what is left of the John Dalton original manuscripts. (They had been housed in the Lit & Phil but this was destroyed in the 1940 "Christmas Blitz". A few manuscripts had evidently been stored in a metal box in the basement.) The grand Reading Room of the Library is touted as one of the finest of any library in the world. The Dalton documents have been treated and preserved and are generally available for viewing only if you can prove your professional status. (In 2003, an effort was made to make the building more accessible to visitors but it is not clear how this affects the availability of the Dalton materials.) In any case, the author was able to demonstrate professional status with a letter of introduction form the dean of the college where he worked for 30 years. The restored manuscript (encased in plastic) of Dalton's "Lectures on Chemistry, 1827" are his original hand-written notes for lectures in chemistry! The first two sentences were: "The science of chemistry is one of general utility and applicable to some degree to most of the concerns of life. The bread we eat, the wine, the beer, the tea & coffee we drink are all manufactured on the principles which it is the province of chemistry to explain." Reading such a document in a library as beautiful as the John Rylands, is pretty much unparalleled for a scientific-historic traveler. If you go, it would be an advantage to look at the John Rylands website beforehand.

4.2.6.3 *Manchester Metropolitan University*

4.2.6.3.1 John Dalton Building. (53.471758°, −2.239942°) corner of Oxford Road and Chester Street. Here we find another statue of Dalton as well as the two large stones that covered his grave at the former Ardwick Cemetery. This cemetery was converted to parkland in the 1960s and is now a school playing field known as Nicholls Field. A monument plaque stands at the entrance to Nicholls Field stating this was once Ardwick Cemetery. The statue is evidently a scaled-up version of the Chantrey sculpture (located in the Town Hall), but this one is in bronze. Be sure to see the John Dalton Assembly Hall in the John Dalton Building if possible.

4.2.6.3.2 Manchester Literary and Philosophical Society. (53.470714°, −2.239504°) Lower Chatham Street Loxford Tower, Manchester Area. The Manchester Literary and Philosophical Society, founded on 28 February 1781, is the second oldest learned society in Britain. Dalton joined the "Lit & Phil" in 1794, became its secretary in 1800, and was its president from 1817 until his death. The earliest meetings took place in a room in the original Cross St Chapel, but in 1799 the Society moved to 36

George Street, which remained its home until the Blitz of 1940. From 1960 to 1980, the Society enjoyed the facilities of its rebuilt home, but since 1981 has not had its own premises and today operates from an office and small library within Manchester Metropolitan University.

4.2.6.4 Dalton Laboratory Plaque. 36 George Street, Manchester (53.479555°, −2.239763°). The plaque notes that Dalton was the founder of the scientific atomic theory, president of the Manchester Literary and Philosophical Society, and that he had his laboratory here. This is only a block away from the Portico Library & Gallery.

4.2.6.5 Faulkner Street. (53.478615°, −2.239515°). Between Nicholas and Charlotte Streets and between two Chinese restaurants on the side of Faulkner Street facing the car park, a small black-and-white plaque commemorates Dalton's atomic theory. The theory was announced in 1803 while Dalton was lodging at 18 Faulkner Street. He lived at 35 Faulkner Street from 1804 to 1832 and after that, in the care of a housekeeper, he lived at 27 Faulkner Street.

4.2.6.6 The Portico Library & Gallery. 57 Mosley Street (53.479730°, −2.240675°). Many of the meetings of the Lit & Phil are still held in the Portico Library which goes back to the time of Dalton and Joule, both of whom were members. Its single large room, surmounted by a Georgian dome, has recently been restored and is open to the public. It opened in 1806. Dalton became the library's first honorary member: his official role was "Clockman". Evidently, he wanted to be a member of this independent library modeled after the Liverpool Library but could not afford it. He was asked to clean the grand clock *in lieu* of paying the membership fee. The clock was designed by the same person who made the clock for Big Ben. Dalton visited regularly at lunchtime to read the day's newspapers. The library is still based here, though much of the building is given over to The Bank pub. Dalton enjoyed a pint of ale in the evening quite often. In addition to the Bank, he probably would have enjoyed the Beer Garden at the Ape & Apple at 28–30 (wait for it) John Dalton Street.

4.2.6.7 Museum of Science and Industry. Liverpool Road, Castlefield Manchester (53.476796°, −2.255384°). Collections contain aircraft, computing, locomotives, and history of communications. This museum holds a small collection of objects belonging to John Dalton. These include his walking stick, a mountain barometer, thermometers, Leyden jars (for storing static electricity), flasks and beakers, a mercury barometer, a spark eudiometer (for measuring volume changes during reactions involving gases), and the commemorative gravestone from his grave in Ardwick Cemetery. Of special interest are five wooden balls made by Peter Ewart of Manchester, about 1810, used by Dalton for demonstrating his atomic theory. Amazingly enough, the museum also holds the remains of Dalton's eyes which are in very good condition and "have proven to contain useful amounts of DNA". In 1995 they were examined by a group of Cambridge physiologists and it was confirmed by DNA analysis that Dalton was a "deuteranope" (one who suffers from a form of color-blindness). The eyes, understandably, are available only to qualified researchers. James Prescott Joule established the mechanical equivalent of heat. The international unit of energy, the joule (J) is established in his honor. Most of Joule's scientific apparatus is also included in this museum's collections. It is advisable to contact the museum in advance if you would like to see these collections. The Museum of Science and Industry is a "sister site" of The Science Museum, London.

4.2.7 Travel Site in London Related to John Dalton

4.2.7.1 The Science Museum. Exhibition Road, South Kensington, London (51.497783°, −0.174566°). John Dalton's friend Mr Peter Ewart made him a number of equal-sized wooden balls, about an inch in diameter in 1811, which he used after that on many occasions. These spheres are owned by The Science Museum, but it is advisable to check in advance to see if they are on display. In 2016, the museum had an outstanding exhibit of Dalton materials to commemorate the 250th anniversary of his birth.

4.3 SUMMARY

John Dalton devised the first concrete atomic theory. It made a lot of sense to many chemical practitioners and received widespread support among his contemporaries. It organized a number of the assumptions and laws known at the time including the Greek ideas of indivisible atoms, Lavoisier's Law of Conservation of Mass, Proust's Law of Definition Proportions, and Dalton's own Law of Multiple Proportions. The mass ratios of a number of simple compounds, including the "airs" of the English pneumatic chemists like Black, Priestley, and Davy, were known and useful for supporting the atomic theory even though these earlier workers were reluctant to speculate on the existence of atoms. In conjunction with his theory, Dalton was able to construct one of the first-ever tables of atomic weights. Above all, this theory could be tested and its general assumptions held up well. Dalton had taken a concept that had been imprecisely discussed for two thousand years and fashioned it into a working hypothesis that stood the test of time and became the atomic theory that revolutionized chemistry and science.

There are number of striking places in England to visit that are related to John Dalton. These include the little white-washed house in Eaglesfield where he was born, the Quaker Meeting House in the little village of Pardshaw Hall, the town where he studied and taught as a very young man and which now honors him as a man for all time, the Stramongate School in Kendal where he and his brother taught, and the Quaker Tapestry Museum, also in Kendal. Finally, there are the many statues, libraries, plaques, paintings, museums, and collections of artefacts at various sites in the city of Manchester.

ADDITIONAL READING

- R. A. Smith, *Memoir of John Dalton and History of the Atomic Theory up to His Time*, Elibron Classics, 2005. This replica edition is an unabridged facsimile of the edition published in 1856 by H. Bailliere, London.
- F. Greenaway, *John Dalton and the Atom*, Cornell University Press, Ithaca, NY, 1966.

- A. J. Rocke, *Chemical Atomism in the Nineteenth Century: Dalton to Cannizzaro*, Ohio State University Press, Columbus, 1984.
- John Dalton's Manchester: A Short Walk around the City Centre, CHSTM Centre for the History of Science, Technology and Medicine, http://www.manchester.ac.uk/chstm, accessed March 14, 2019, Walk text by James Sumner, 2009.
- M. F. Lappert and J. N. Murrell, John Dalton, the man and his legacy: the bicentenary of his Atomic Theory, *Dalton Trans.*, 2003, 3811–3820.
- R. Dunn, Models and Molecules: Representation in the Work of John Dalton, *Kairos, Revista de Filosofia & Ciência*, 2015, **13**, 157–178.

REFERENCES

1. R. E. Dickerson, H. B. Gray and G. P. Haight, *Chemical Principles*, The Benjamin/Cummings Publishing Company, Inc., Menlo Park, CA, 3rd edn, 1979.
2. Dalton's Element List, https://commons.wikimedia.org/wiki/File:Dalton%27s_Element_List.jpg, accessed June 2019.
3. J. Dalton, *A New System of Chemical Philosophy*, 1808.

CHAPTER 5

Electricity and the Atom: Davy, Faraday, Clerk Maxwell, and Thomson (England and Scotland, 1801–1907)

5.1 A QUICK LOOK AT PLACES TO VISIT "TRAVELING WITH THE ATOM" IN ENGLAND AND SCOTLAND

The Royal Institution – one of the very best scientific/historical travel sites ✻✻✻✻✻: *lecture hall, exhibits, paintings, portraits, exquisite statue of Michael Faraday, and the Faraday Museum and Magnetic Laboratory – in London: the Faraday Memorial* ✻, *plaque* ✻, *statue* ✻, *the Grace and Favour House* ✻ *out by the Royal Palace at Hampton Court, the Cuming Museum* ✻, *Faraday's grave at Highgate West Cemetery* ✻✻ *– in Edinburgh: the James Clerk Maxwell Foundation at Maxwell's birthplace with its portraits, display cabinets and walking tour* ✻✻✻✻; *Edinburgh Academy* ✻, *the new (2008) statue in front of St. Andrew Square* ✻✻ *– in Corsock & Dumfries in southwest Scotland: the newly renovated Maxwell estate at Glenlair House* ✻✻✻, *Parton Kirk* ✻ *where Maxwell's family is buried, and the magnificent stained glass window in the Corsock and Kirkpatrick Church*

Traveling with the Atom: A Scientific Guide to Europe and Beyond
By Glen E. Rodgers
© Glen E. Rodgers 2020
Published by the Royal Society of Chemistry, www.rsc.org

*, and the Maxwell House * – in Cambridge: the Old Cavendish Laboratory *** grounds on Free School Lane with its original entrance, Rutherford crocodile on the Mond building exterior, plaque marking the discovery of the structure of DNA, and the Eagle Pub * – the Museum at the new Cavendish Laboratory ****, another of the best traveling with the atom sites.*

> **Electricity and the Atom**
> - In 1801, Humphry Davy (1778–1829) had moved from the Pneumatic Institution in Bristol to the Royal Institution (Ri) in London, where he was an instant celebrity. He drew large crowds to his flamboyant lectures including, in 1812, the young bookbinder's apprentice, Michael Faraday (1791–1867).
> - Although Davy had discovered six elements in less than a decade at the Ri, his most outstanding "discovery" was Faraday, who was the greatest experimentalist of his day.
> - Faraday's experimental work in electromagnetism combined with the theoretical work of James Clerk Maxwell (1831–1879) set the stage for the discovery of the electrical nature of the atom.
> - The work of William Crookes (1832–1919) soon led to the discovery of the electron by J. J. Thomson (1856–1940) in 1897. The electron was the first sub-atomic particle to be discovered.
> - Thomson also worked on the positive rays (so-called "canal rays") left over after a few electrons had been taken away from the gaseous atoms in his tubes.
> - Thomson proposed a model of the atom called the "plum pudding model".

5.2   SIR HUMPHRY DAVY (1778–1829)

When he saw the minute globules of potassium burst through the crust of potash, and take fire as they entered the atmosphere, he could not contain his joy – he actually bounded about the room in ecstatic delight. Edmund Davy (Davy's cousin) quoted in ref. 2, p. 109.

In Chapter 3 (pp. 65–72), we followed Humphry Davy from his birthplace in Penzance, in the Lands End area of southern England, to his time at Dr Thomas Beddoes's Pneumatic Institution in Bristol from 1798 to 1801. From there he was called to the newly minted Royal Institution (Ri), where he switched from being a flamboyant, almost foolhardy pneumatic chemist to a daredevil electrochemist on his way to chemical stardom. The change in his research interest was the result of Alessandro Volta's reports in 1800 of his invention of a chemical battery that, for the first time, provided for a continuous flow of electricity. Volta, an Italian physicist working in Como, Italy, had written a letter communicating his findings to Joseph Banks, president of the Royal Society and demonstrated his "voltaic piles" for Napoleon in Paris in 1801. (See Chapter 9 for more on the numerous scenic sites in Como related to the memory of Volta). Davy had a giant voltaic pile built at the Ri and proceeded to use it to discover six elements (potassium, sodium, magnesium, strontium, barium, and calcium) in three years (1807–1809). In 1810, he conclusively disproved Lavoisier's hypothesis that oxygen was present in all acids. In 1812 he was knighted and in 1815 invented the coal miner's safety lamp that saved many lives. This is an impressive list of accomplishments and discoveries, but perhaps the one that had the largest effect on chemistry and physics was his "discovery" of Michael Faraday.

Davy has been described as "full of mischief, with a penchant for explosions ... a born chemist" (see ref. 1). At the Royal Institution, he was an instant hit. His excellent lectures were well pitched toward the common citizen and augmented by striking demonstrations. Add to this that he was an extremely handsome young man with a gift for working audiences, and we can appreciate why he attracted so many people to his lectures, including many of the young ladies of that day. Many of the latter actually fainted in the presence of the dashing, young Davy. As Alan Hirshfeld notes in his book, *The Electric Life of Michael Faraday*,[2] one remarked that "it was impossible to have seen him and have supposed him an ordinary person". Another described his eyes as "radiant of genius, and the most bright and beaming expression of intellect that can possibly be conceived." Finally, coming perhaps more to the point, another ventured that "those eyes were made for something besides poring over crucibles". Figure 5.1 shows a portrait of the young Humphry Davy.

Figure 5.1 The young Sir Humphry Davy. Reproduced from ref. 3 [https://wellcomecollection.org/works/f48jat3q] under the terms of a CC BY 4.0 License [https://creativecommons.org/licenses/by/4.0/deed.en].

Although Davy was a devotee of "the immortal" Joseph Priestley, by then exiled to America, he also accepted Antoine Lavoisier's chemical revolution (See Chapter 8, pp. 198–206) and its new nomenclature. Accordingly, he referred to Priestley's two gases as oxygen and nitrous oxide. He remained skeptical about John Dalton's atomism although, according to Fullmer, he had an "ardent belief in the existence of material corpuscles yoked in equilibria poised between attractions and repulsions".

5.2.1 Travel Sites in London Related to Humphry Davy

5.2.1.1 The Royal Institution. 21 Albemarle Street (51.509590°, −0.142312°) see Charts 3.2 and 5.1 for location. The famous lecture hall upstairs in the Ri is where Davy, Faraday, and many, many others have spoken over the years. The lecture

Electricity and the Atom

hall is completely open so go inside and take a look. The Davy exhibit, including his original samples of sodium, calcium, magnesium, and chlorine is shown on the ground floor. The Ri holds two paintings of Humphry Davy, one of him as a young man and another after he was knighted. Take a close look at these paintings and see what you think of those eyes! There is also one of his Miners' Safety Lamps. "Heritage tours" can be arranged for groups from 10 to 40. The periodic table where you can try to identify the 10 elements discovered at the Ri (including the total of seven by Davy) is a fun treat.

5.3 MICHAEL FARADAY (1791–1867)

> "Let him wash bottles. If he is any good, he will accept the work; if he refuses, he is not good for anything". The advice offered to Humphry Davy when he extended an invitation to Michael Faraday to be his assistant at the Royal Institution.

> "When we consider the magnitude and extent of his discoveries and their influence on the progress of science and of industry, there is no honour too great to pay to the memory of Faraday, one of the greatest scientific discoverers of all time." Ernest Rutherford

Michael Faraday was one of ten children of a London blacksmith. In 1805, as a young teenager, he was apprenticed to a bookbinder, who allowed him to read the books in the shop on his off hours. Faraday had little access to a formal education, but was able to read about electricity in the *Encyclopedia Britannica*, chemistry in Lavoisier's revolutionary *Elementary Treatise on Chemistry* published in 1789, and Jane Marcet's *Conversations on Chemistry* which went through 18 British, 4 French, and 23 American printings. He also read *Improvement of the Mind* by preacher, writer, and hymn composer Isaac Watts, and credited it with teaching him how to think. (Watts most famously wrote the Christmas carol, "Joy to the World".) In 1812, a customer of the bindery (and a Ri member) gave Faraday a ticket to attend the last four of Humphry Davy's lectures at the Royal Institution. After these "farewell lectures", Davy intended to continue his research and then travel. The Ri was only a short distance from the shop where Faraday was an apprentice, Faraday was already a devotee of chemistry and Watts encouraged the importance of such direct observations, so young Faraday attended the lectures and, like so many others, was totally drawn into the master showman's presentations. He took careful notes, illustrated them with colored diagrams, and with his newly acquired book-binding skills, bound his 386 pages (!) and sent it first to the president of the Royal Society and, getting no response, to Davy himself! Davy was flattered and enthralled and, when a vacancy developed, invited Faraday to be his assistant at the Royal Institution. He took the job and, much to his surprise, soon found himself accompanying Davy and his new wife, Jane Apreece, on an 18-month-long European tour to France, Italy, and Switzerland. Sometimes things happen at a daunting pace when you are young, enthusiastic, and in the right place at the right time.

We scientific/historical travelers can identify with the grand European tour that Davy, his new rich and rather acerbic wife, her maid, and Faraday (acting as Davy's chemical assistant and valet) embarked upon in October, 1813. The trip was complicated by the fact that England and France had been at war almost constantly since 1793, *i.e.*, almost all of Faraday's life. (Davy's regular valet had backed out of the trip because he feared for his life.) However, this was not the problem that it could have been because Napoleon, being a great supporter of the sciences (see Chapter 8, p. 211, for more on Napoleon's high regard for the sciences), provided special passports for Davy and his fellow travelers. The trip was quite a shock for Faraday as he had never been more than 12 miles from home and knew not a word of French. Throughout the trip, Faraday kept a detailed journal, so we know many details. In Paris, Davy (and therefore Faraday) met Joseph Gay-Lussac, Claude Berthollet, and André-Marie Ampère, among others. (See Chapter 8 for more details on the contributions of these scientists to the development of the atomic concept.) The French-English conflict extended beyond politics and the battleground to the scientific arena. For example, the French thought oxygen was a component of all acids and the English did not. The French had beaten Davy to the discovery of boron by nine days and had used Davy's newly discovered potassium to do it.

Given these disagreements and others, it was not surprising that Davy and Ampère were soon disputing who should be credited with establishing that the newly discovered iodine was, in fact, an element. (Bernard Courtois had isolated this beautiful purple gas from seaweed in 1811.) After three months in Paris, the party set off across the Italian Alps (Faraday carried a barometer) and settled in Florence. Here Davy, using his portable laboratory, demonstrated that charcoal and diamond were both just carbon. He did this by burning diamonds with a giant burning glass and comparing the result with the burning of charcoal. (You can see this burning glass at the Galileo Museum in Florence. See Chapter 9 for details.) They went on to visit Genoa, Rome, Naples, and Mount Vesuvius, and then traveled north to Milan where they met Alessandro Volta, whose invention of the chemical battery had so influenced and enhanced Davy's career. From there they headed for Switzerland, where Faraday was astonished to meet Jane Marcet, the author of *Conversations on Chemistry*, the book that he had so admired back in his days as a bookbinder's apprentice.

Upon their return to England in April of 1815, Faraday assisted Davy in his invention of the miner's safety lamp. They continued to work together for four years but eventually had an acrimonious falling out, from which the relationship never fully recovered. Michael married Sarah Barnard in 1821 and they moved into a second-floor apartment at the Ri where they lived for 41 years. Given his humble beginnings, it is truly amazing that he went on to be the greatest experimental scientist of the early nineteenth century. Faraday's initial research at the Ri was in chemistry. He was an accomplished analytical chemist, determining the purity of a large variety of materials including clays, native lime, water, gunpowder, alloy steels, rust, and a variety of gases, liquids, and solids. He liquefied chlorine in 1823 (and later many other gases including ammonia, carbon dioxide, and nitrous oxide) and isolated benzene in 1825.

In 1820 came the news from Denmark that Hans Christian Ørsted (see Chapter 12) had discovered a relationship between a flowing electrical current and its effect on the magnetized needle of a compass. This was the beginning of the field of electromagnetism. Faraday eagerly took up this field and wondered if would be possible for a magnetic force (specifically a moving magnet as it turned out) to produce electricity. He was able to do this in 1831, and then went on to develop a number of electromagnetic principals including the concepts of "lines of force" and magnetic and electric fields. His work led directly to electric motors, transformers, and generators. All of this put him in competition with André-Marie Ampère once again, but they developed a strong friendship, and both made major contributions to the field of electromagnetism. Faraday's most famous book summarizing his work was *Experimental Researches in Electricity*.

Faraday also did work in the field of electrochemistry or, more specifically, electrolysis, a study of the amount of electrical energy needed to promote chemical changes. He constructed a set of laws that govern electrolysis and, as part of that work he helped to coin many familiar electrically related words including electrode, anode, cathode, anion, and cation. Like Ampère, his name is immortalized by having a unit named after him, in this case the "farad", a unit of electrical capacitance, and the Faraday constant used in electrochemistry.

He established two traditions at the Ri that still live on even today: the Christmas Lectures (1825) for children and the Friday Evening Discourses (1826) for adults. The latter grew out of the need for the Ri to get on an even financial footing. In 1821, Faraday had been appointed Superintendent of the House, a duty he took on in exchange for the Ri allowing him and his new wife to live there. As such, he was responsible for organizing lecture series, as well as managing the building. In 1826 he was appointed Director of the Laboratory. The Friday Evening Discourses aimed to bring science to the ordinary citizen in an attractive, entertaining, and instructive manner. Like his mentor Davy, Faraday was a terrifically entertaining scientific lecturer and his Friday Evening Discourses became the place to be among London's rich and famous. These discourses gradually raised enough money to put the Ri in a good financial position. Jane Marcet, whom Faraday called his "first instructress" because of her book *Conversations in Chemistry*, was a regular attendee.

In addition to the articulate, enthusiastic, and eloquent Faraday himself, there have been many distinguished speakers at the Ri over the years, including such atomic scientists as John Dalton (1834), James Clerk Maxwell (1861), William Crookes (1879), D. I. Mendeleef (1889), J. J. Thomson (1894), A.

Electricity and the Atom 107

H. Becquerel (1902), Pierre Curie (1903), E. Rutherford (1904), and F. W. Aston (1921).

Faraday was unconvinced about the physical existence of atoms. At best, we could call him an atomic agnostic. To him it was more likely that it was his lines of force that actually existed. After all, these had been observed. Atoms had not. Given his results, however, he *was* convinced that electrical forces were at work in Nature. He also was well aware that his contemporaries thought in terms of atoms so, that given, he wrote that "if we adopt the atomic theory or phraseology, then the atoms of bodies ... have equal quantities of electricity associated with them." As Forbes and Mahon note in their book *Faraday, Maxwell, and the Electromagnetic Field*,[4] Faraday went on to write that " ... for though it is very easy to talk of atoms it is very difficult to form a clear idea of their nature." He concluded (again, assuming the "atomic phraseology") that atoms can form ions that have a set charge associated with each of them. For example, sodium atoms form sodium cations (Na^+) with a +1 charge, magnesium atoms form magnesium cations (Mg^{2+}) with a +2 charge, and chlorine atoms form chloride anions (Cl^-) with a −1 charge. Furthermore, he found, all charges on ions are integral multiples of a single fundamental unit of charge. In addition to these conclusions derived from his investigation of electrolysis, his experimental work in electromagnetism was the basis for the theoretical work of the great Scottish physicist, James Clerk Maxwell. Maxwell developed the mathematical basis of Faraday's ideas on the relationship between light, electricity, and magnetism. By the turn of the next century it would be become apparent that the atom was composed of charged particles (the electron was discovered in 1897 and the nucleus was discovered in 1911), and that light interacted with atoms in a variety of ways. The work of Faraday and Clerk Maxwell cannot be underestimated in its contribution to the story of the discovery of the atom.

5.3.1 Travel Sites in London Related to Michael Faraday

See Chart 5.1 for the location of these sites.

5.3.1.1 Faraday Memorial. Newington Butts (51.495028°, −0.100681°), the Elephant and Castle roundabout. Here a drab, box-shaped "stainless steel sculpture" marks where

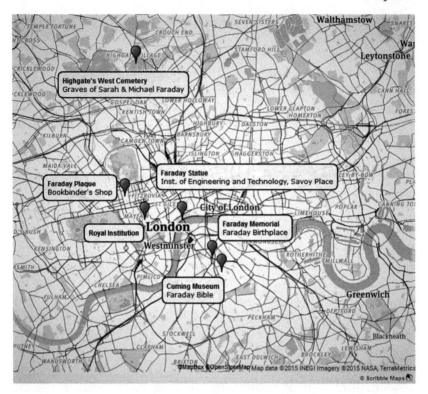

Chart 5.1 Travel sites in London related to Humphry Davy and Michael Faraday including The Royal Institution, Faraday Memorial (Newington Butts, the Elephant and Castle roundabout), Cuming Museum, Faraday Plaque (48 Blandford St), Faraday Statue outside the Institution of Engineering and Technology, and the Faradays' graves at Highgate's West Cemetery. The Faraday House at 37 Hampton Court Rd., located west of the city center, is not shown on this Chart.

Faraday was born. Changes in its appearance have made its connection to Faraday unclear, but there is a stone inscription in the sidewalk on the north side of the box that clarifies that it is the Faraday Memorial. There is an underground stop there of the same name.

5.3.1.2 Cuming Museum. Walworth Road (51.490968°, −0.097056°). Near the Faraday Memorial in the Elephant and Castle Roundabout, this museum has had a small display of Faraday material including the family Bible in which his birth is

Electricity and the Atom 109

recorded. Sadly, the museum was closed due to a fire in March of 2013 but reportedly their collections were preserved. The reader is encouraged to consult the museum's website to see what is presently available there related to Michael Faraday.

5.3.1.3 Faraday Plaque. 48 Blandford Street (51.518494°, −0.154158°), marking the site of the bookbinder's shop where Faraday was an apprentice. A brown plaque located between two second-story windows marks the spot.

5.3.1.4 Faraday Statue. Outside the Institution of Engineering and Technology, Savoy Place, Embankment. (51.50999°, −0.11888°). It was carved by Anglo-Irish sculptor John Foley who also carved the statue of Prince Albert in the Albert Memorial. From here there is a beautiful view of the London Eye. It is only a short walk down river to Victoria Embankment Gardens with its pleasant Embankment Café. If you have a little extra time, find the statue of the famous Arthur Sullivan (he of Gilbert and Sullivan fame). It's the one with the weeping woman, a muse, leaning forlornly against the plinth supporting the bust of Sullivan. The inscription on the monument asks, "Is life a boon?" and answers "If so, it must befall that Death, whene'er he call, must call too soon". Life was certainly a boon to Michael Faraday. It's worth sitting on the bench here and thinking about his remarkable life and contributions to the world as we know it.

5.3.1.5 Royal Institution. 21 Albemarle Street (51.509590°, −0.142312°). Entry is free. As noted earlier, the famous lecture hall upstairs in the Ri is where Davy, Faraday, and many, many others have spoken over the years. Visit the Faraday Museum here in the lower ground floor. Faraday's magnetic laboratory, shown in Figure 5.2, is displayed as it was in the 1850s. You might enjoy viewing the interactive tour at http://www.rigb.org/our-history/michael-faraday/magnetic-laboratory before or as part of your visit. Occasionally the exhibition is closed for private events, so it is a good idea to call ahead at Tel +44 (0)20 7409 2992. There is a good description of the founding of the Ri in Hirshfeld[2]. The Ri holds several paintings of Michael Faraday and there is a famous and striking statue of him located at the bottom of the grand staircase. Stand here as the author, his wife,

Figure 5.2 The magnetic laboratory of Michael Faraday in the basement of the Royal Institution. Photograph by G. E. Rodgers.

and one of their student groups have and get somebody to take your photograph with the great Michael Faraday. The Ri has a coffee bar downstairs and that might be a good place to relax and think about Davy, Faraday, and the other atomic scientists who have been associated with the Royal Institution (including James Clerk Maxwell, J. J. Thomson, and Ernest Rutherford) over the years.

5.3.1.6 Faraday House. (51.40472°, −0.34054°) Hampton Court Road, East Molesey, London. In 1848, due to the efforts of Prince Albert, Michael Faraday was awarded a Grace and Favour house near the Royal Palace at Hampton Court by Queen Victoria, free of all expenses except for upkeep. (The Royal Palace is where Henry VIII, Elizabeth I, James I and Charles I lived.) Faraday and Sarah moved here when he retired in 1858. After Faraday's death in 1867 in this house, it was renamed the Faraday House. It was formerly the Master Mason's house. There used to be a modern plaque at this location but, at this writing, there is only a stone engraving above the door that simply says, "**MICHAEL FARADAY** lived here, 1858–1867". Next door is a residence formerly occupied by architect Sir Christopher Wren. There is modern blue plaque at this location.

Electricity and the Atom

5.3.1.7 Highgate's West Cemetery. (51.568093°, −0.147609°). Michael and Sarah Faraday were buried in the "Sandemanian plot" – part of a dissenters' section – of this huge cemetery after turning down burial in Westminster Abbey. The Sandemanians were, as Faraday himself once said, "a very small and despised sect of Christians". Faraday was an Elder in the London congregation of the church. Attendance at services was strictly required of all members. In 1844, Faraday was absent from services one Sunday because he had been invited to tea at Windsor by Queen Victoria. He was immediately excluded from the fellowship and, although he was reinstated within weeks, it was 14 years before he was restored as an Elder. One did not cross the Sandemanians. (Despite his dissenter status, there is memorial plaque to Faraday in Westminster Abbey near Newton's tomb. See Chapter 7 for more on the many scientists buried and/or memorialized at the Abbey.)

Also buried in the West Cemetery is Jacob Bronowski, scientist and creator of the television series *The Ascent of Man* and the poet Christina Rossetti. In the East Cemetery are the graves of Douglas Adams, author of *The Hitchhiker's Guide to the Galaxy*, Henry Gray, anatomist and surgeon, and author of *Gray's Anatomy*, and Karl Marx, the philosopher, historian, sociologist, and economist. The West Cemetery is accessible by guided tour only, but the East Cemetery can be viewed on a self-guided tour.

5.3.1.8 Westminster Abbey. See Chapter 7 on the graves and memorials of scientists in Westminster Abbey. A memorial to Michael Faraday is here.

5.4 JAMES CLERK MAXWELL (1831–1879)

"From a long view of the history of mankind – seen from, say, ten thousand years from now – there can be little doubt that the most significant event of the 19th century will be judged as Maxwell's discovery of the laws of electrodynamics." Richard P. Feynman

"... the opinion seems to have got abroad, that in a few years all the great physical constants will have been approximately estimated, and that the only occupation which will then be left to men of science will be to carry on these measurements to another place of decimals .... But we have no right to think thus of the unsearchable riches of creation, or of the untried fertility of those fresh minds into which these riches will be poured." James Clerk Maxwell[7]

James Clerk Maxwell made two significant contributions to the development of the atom. First, he put Faraday's theories on electromagnetism on a firm mathematical foundation in the form of what became known as Maxwell's equations. Rather astoundingly, light turned out to be electromagnetic waves or radiation travelling through a medium. The interaction between light and matter, as we will explore later in this and other chapters, turns out to reveal a great deal about the inner structure of atoms. Second, in 1860 Clerk Maxwell worked on the mathematics of the kinetic molecular theory of gases. This theory pictures the atoms or molecules of a gas moving at great speeds, slamming into each other and the walls of its container billions of times per second, thereby exerting a pressure. Maxwell showed that, for any given temperature there is a distribution of molecular velocities. Some moved very rapidly and some very slowly. A most probable or so-called "mean" velocity could be calculated and this value increased with the temperature of the gas. Since velocity is related to kinetic energy (the energy of motion) by the formula $KE = \frac{1}{2} mv^2$, where KE = kinetic energy, m = mass, and v = velocity, a mean kinetic energy could also be calculated. This work was extended by the Austrian Ludwig Boltzmann and the distribution of velocities or kinetic energies is now referred to as the Maxwell–Boltzmann distribution. (See Chapter 10 for more on Boltzmann's work in support of the physical existence of atoms, his intense debate with Ernest Mach about atomism and places we can visit relative to this debate.)

James Clerk Maxwell, although his family was from rural south-western Scotland, was actually born in Edinburgh where his mother (almost 40 at the time) had been brought for the best possible medical care. James's father, John Clerk Maxwell, had the house at 14 India Street built so they could be closer to his sister Isabella, who lived nearby at 31 Heriot Row. When James was about two, his family moved back to "the gently rolling hills of Galloway" (ref. 4, p. 27) to live in the estate that his father had inherited. (As part of the somewhat bizarre negotiations surrounding this inheritance, the Clerk family, starting with John Clerk, had agreed to add Maxwell to their surname, which is properly Clerk Maxwell.) Their new house, designed by John, was called "Glenlair" and soon that name was applied to the entire estate.

From his earliest days, young James had an intense interest in how things work. "Anything that moved, shone, or made a noise drew the question 'What's the go o'that?' and if the answer didn't satisfy him he'd follow with 'but what's the *particular* go of it?' (ref. 4, p. 129). So, James spent his early boyhood here but these idyllic days came to an end with the death of his mother when he was just eight. After an unsuccessful year with a young tutor, James started to attend Edinburgh Academy and live with his Aunt Isabella back in Edinburgh. He had a rough start at the academy, soon acquiring the nickname "Dafty", but his reputation changed when he quickly mastered geometry and gained the respect of his teachers and classmates. He made some lifelong friends at Edinburgh Academy, including Peter Guthrie Tait, who, like Clerk Maxwell, became one of Scotland's great mathematical physicists. Among a myriad of other topics, Tait wrote extensively about the foundations of the kinetic theory of gases and, just incidentally, being an avid golfer, a paper on the influence of spin on the flight of a golf ball.

At the age of 16, Clerk Maxwell enrolled at Edinburgh University with initial thoughts of studying law but, as is often the case, came under the influence of a great teacher whose enthusiasm soon set him on the path toward a career in science. It wasn't long before he was setting up a laboratory in Glenlair, where he conducted all manner of experiments during his school holidays. Edinburgh University had been a great start but Cambridge University, where P. G. Tait was studying, seemed to be the place for aspiring scientists. He moved to Cambridge at the age of 19, graduated with honors in 1854, and decided to take a fellowship at Trinity College. This decision took some serious thought, as fellows were supposed to be ordained into the Church of England within seven years and remain unmarried. Clerk Maxwell knew that such life choices were unlikely for him but, in the meantime, he could continue his scientific studies. His initial primary interest was how humans perceive colors. His fascination with this topic went back to earliest days. When he was three years old someone had said to him "look at that lovely blue stone" and he responded, "but how d'ye *know* it's blue?" Given this life-long interest, he set his mind to the task.

Clerk Maxwell figured out that the primary colors were red, green, and blue. (These are the so-called "additive primaries" as opposed to the "subtractive primaries" of the artist which are red, *yellow*, and blue.) He devised what he called a "color top" using which he could add together various amounts of red, green, and blue light to produce any color he desired. If he used equal proportions of the three he generated white light. He tested color-blind people and found that many of them lacked the ability to perceive red light and therefore had difficulty telling red from green. With his color top he became an expert in the science of color vision. He and the color-blind John Dalton could have had quite a conversation. (See Figure 5.3 for a photograph of the young John Clerk Maxwell holding his color top.)

Figure 5.3 James Clerk Maxwell at age 24 holds his "color top", with which he demonstrated that the primary additive colors can be mixed together to produce white light.

Clerk Maxwell served as chair of natural philosophy at Marischal College in Aberdeen (which soon after merged with King's College to produce the University of Aberdeen) from 1856 to 1860. During this time, he investigated colors, the distribution of molecular velocities in a gas, the relationship between electricity and magnetism, and a variety of other topics, perhaps most famously at the time, an exceptionally well-regarded paper on the composition of the rings of Saturn. From there, still only 29, he took a position at King's College in London. There is a strong possibility that during this time (1860–1863) he visited with Michael Faraday who had retired from the Ri and was living at his "Grace and Favour House" in Hampton Court, still calling regularly at the Institution. Perhaps due to these visits, Clerk Maxwell was invited to give a talk at the Friday Evening Discourses and chose to discuss his ideas on color vision. As part of the talk, he and a colleague produced the first color projected image.

At various stages in his all-too-short life, Clerk Maxwell grappled with the topic for which he is most well-known: electricity and magnetism. He first read the works of Faraday and Ampère and, unlike many theoreticians, found Faraday's lines of force to be a useful concept. (To that point, the only other researcher who valued lines of force was William Thomson – eventually to be known as Lord Kelvin – of Glasgow University. Others dismissed them as "vague and varying" and of little use in further work.) Gradually, over a period of years using a variety of models and analogies, Maxwell converted the idea of electric or magnetic lines of force into what he called the electromagnetic field, which he defined as "that part of space which contains and surrounds bodies in electric or magnetic conditions." From there he expressed Faraday's original ideas in mathematical form and published his first paper on this in 1855. The title was "On Faraday's Lines of Force" and, in 1857, he steeled up his nerve and sent a copy off to Faraday himself who responded most favorably. In 1861 and 1862 he published a set of four papers *On Physical Lines of Force* and in 1864 his most important paper, entitled *A Dynamical Theory of the Electromagnetic Field* was published. In this revolutionary paper he summed up his proposals about electricity and magnetism in a series of 20 equations. (These were cumbersome and exceedingly difficult to understand. In 1884, a few years after Clerk Maxwell died, a man named Oliver Heaviside re-expressed them into four differential equations, which are now known as "Maxwell's Equations".) Electricity and magnetism were intimately connected in the idea of the electromagnetic field. Furthermore, he showed conclusively that light is electromagnetic radiation and his calculations predicted a value for the speed of light that was remarkably close to the known experimental value. Furthermore, light could come in forms beyond what humans could perceive. For example, he predicted the existence of what we now call radio waves. About twenty years later (after Maxwell had died), Heinrich Hertz did experiments that demonstrated the existence of these rays. Maxwell's calculations leave us with the model of light as a wave traveling through space with a rippling (or "oscillating") electric field always in conjunction (or "in-phase") with a perpendicular magnetic field that moves in step with it. This idea revolutionized physics and arguably is every bit as important as Newton's gravitational theory.

By the late 1860s, Maxwell was back at Glenlair working on enlarging the house that his father had designed and built and writing his *Treatise on Electricity and Magnetism*, which ultimately was published in 1873. (Remarkably, it is still in print.) Maxwell and his wife were content there, but he soon was called upon to do, as it turned out, one more great task. Cambridge University wanted to build a grand new laboratory of physics (perhaps to compete with the newly built Clarendon Laboratory at Oxford) and establish its first professorship of experimental physics. Maxwell was persuaded that he should be that first professor and therefore assume responsibility for designing and building the laboratory. The university's chancellor was William Cavendish, the 7th Duke of Devonshire whose great uncle had been Henry Cavendish, whom we discussed in Chapter 3, p. 52. Recall that Henry Cavendish was one of the important English pneumatic chemists and had discovered hydrogen gas, which he had called "inflammable air". Also recall that Lismore Castle where Robert Boyle had been born is still owned by the Cavendish family (See Chapter 2). It seemed appropriate, given this history and the chancellor of the university at the time that the facility should be called the Cavendish Laboratory of Physics or just "The Cavendish". It opened in 1874. When Maxwell was appointed, the Duke gave him an extensive collection of his great uncle's unpublished papers describing his

Electricity and the Atom 117

electrical experiments carried out from 1781 to 1791. When Clerk Maxwell went through these documents, he discovered that the painfully shy and reclusive Henry Cavendish had anticipated many of the great electrical works of Charles Augustin de Coulomb, Georg Ohm, and Alessandra Volta. He just had never communicated his results to others. It fell to Maxwell to summarize these works for all prosperity.

The Cavendish Laboratory operated from 1874 to 1974. The buildings that housed "The Old Cavendish" still stand and are now occupied by various other university departments. There are still some historic sites to see there, including the alligator carved in honor of Ernest Rutherford (See Chapter 6) and the Maxwell lecture theatre. In 1974, the Cavendish was moved to a new facility in West Canterbury. Maxwell's immediate successors as the Cavendish Professor of Physics included Lord Rayleigh, J. J. Thomson, Ernest Rutherford, and Lawrence Bragg, each one a winner of the Nobel Prize in Physics. The new Cavendish Laboratory is about a mile and a half west of the city center on Madingley Road. The Museum of the Cavendish Laboratory is located on the second floor (first floor if British) of the Bragg building.

5.4.1 Travel Sites in Edinburgh, Scotland Related to James Clerk Maxwell

See Chart 5.2 for the location of these sites.

5.4.1.1 James Clerk Maxwell Foundation. 14 India Street (55.955283°, −3.205654°). His birthplace is marked by a plaque on the building. ("James Clerk Maxwell Natural Philosopher Born here 13 June 1831") The house, purchased by the foundation in 1993, has been fully restored. The exhibition room on the main floor has many portraits of Clerk Maxwell, his family and his friend and colleague Peter Guthrie Tait. There are various exhibits of equipment and drawings, as well as a display explaining and depicting the first color projected image demonstrated by Clerk Maxwell at the Ri in 1861. In the Library are several wall panels describing Clerk Maxwell's other scientific achievements. Of particular interest are the panels describing the two friezes on either side of the plinth holding the new 2008 statue of Clerk Maxwell located in front of St. Andrew Square a short distance from here. Clerk Maxwell is recognized as one of the three

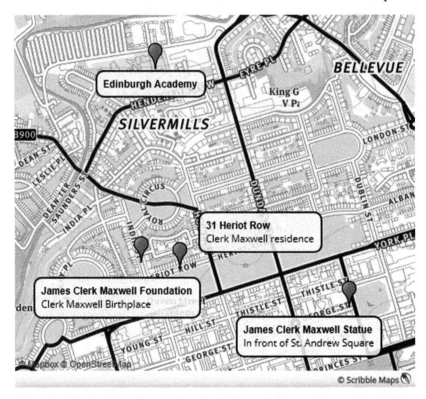

Chart 5.2 Travel sites in Edinburgh related to James Clerk Maxwell. The James Clerk Maxwell Foundation (14 India St), his residence at 31 Heriot Row, the Edinburgh Academy (Henderson Row), and the James Clerk Maxwell statue at the intersection of George and David Streets in front of St. Andrew Square.

most important physicists in the 300 years beginning with Isaac Newton and ending with Albert Einstein (roughly 1642 to 1955). To emphasize the Newton-Maxwell-Einstein connection, one frieze depicts Newton's separation of white light into its component colors while the other depicts Einstein's theory of relativity in which light beams are bent under the influence of large objects such as the sun. (Clerk Maxwell, as we know, fully explained the nature of light as electromagnetic radiation.) Photographs of these friezes are shown below in Figure 5.4. Also of interest in the Library is the cartoon by Ohto Koichi representing Einstein's response when asked if he "stood on the shoulders of Newton". He replied that "I stood on the shoulders of Maxwell". This cartoon is shown in Figure 5.5.

Electricity and the Atom 119

Figure 5.4 The two friezes on either side of the 2008 statue of James Clerk Maxwell located in front of St. Andrews Square. The top frieze represents Isaac Newton's splitting of white light into its component colors. The bottom frieze represents the bending of beams of light by large objects such as the sun. Photographs by G. E. Rodgers.

As one ascends the circular staircase to visit the room where Clerk Maxwell was born, one is treated to a gallery of portraits of "illustrious mathematicians and physicists arranged in chronological order". At the very top is a portrait of Oliver Heaviside, who reformulated Maxwell's equations to the form now used today. In the unadorned and simple birthroom there is a set of

Figure 5.5 The Ohto Koichi cartoon depicting Einstein standing on the shoulders of Maxwell is displayed in the Library of the James Clerk Maxwell Foundation Library at 14 India Street, Edinburgh. Photograph by G. E. Rodgers.

wall panels depicting James's childhood, early life, and career. On the fireplace wall are displayed a selection of water colors by Clerk Maxwell's cousin Jemima Blackburn. Her paintings show the early years of Clerk Maxwell's life including various family events like plays, picnics, and parties, James with his father and mother, and family arrivals and departures from Edinburgh and Glenlair.

Next door is a room with a beautifully displayed saying in Latin that translates to "From this house of his birth, his name is now widespread – across the entire terrestrial globe and even to the stars." To make an appointment to see Clerk Maxwell's birthplace, consult the Clerk Maxwell Foundation webpage. Also see "History Topic: A visit to James Clerk Maxwell's house".[5] Starting in this area is a most informative 3-mile walking tour designed by John Arthur that includes some of the following and many other sites. Google "Maxwell walking tour".

Electricity and the Atom 121

5.4.1.2   31 Heriot Row.   (55.95519°, −3.203499°). When he was about 10 years old Clerk Maxwell lived here with his Aunt Isabella Wedderburn during his days at Edinburgh Academy and the University of Edinburgh. James' cousin Jemina was a frequent companion, particularly during holidays at Glenlair. As noted above, Jemina became a fine artist specializing in water colors.

5.4.1.3   Edinburgh Academy.   (55.96016°, −3.204545°), Henderson Row. Both Clerk Maxwell and his friend P. G. Tait attended this school for boys and took various academic prizes. The original building is quite the same as it was in their day, but the academy has added new buildings including the James Clerk Maxwell Science Center.

5.4.1.4   James Clerk Maxwell Statue.   (55.954038°, −3.194294°). Located at the intersection of George and David Streets in front of St. Andrew Square. Erected in 2008, this is the only public monument to Clerk Maxwell. James is seated and holds his color top. His terrier Toby, to whom he reportedly often explained his work, is at his feet. The famous Maxwell Equations are on the back of the plinth and, as described earlier, on either side are the friezes depicting Newton and Einstein's paradigm-changing contributions to our ideas about light. These friezes frame Clerk Maxwell as one of the three most important physicists in three hundred years.

5.4.2   Other Travel Sites in Scotland Related to James Clerk Maxwell

See Chart 5.3 for the location of these sites.

5.4.2.1   Dumfries and Galloway, Scotland.   The following places are about 10 miles west of Dumfries, on roads that turn off the A75; Corsock on the A712, Parton on the A713; Glenlair on a lane off the B794, signposted to Nether Corsock.

5.4.2.2   Glenlair House.   (55.028745°, −3.943229°). This is beautiful country, well worth time spent here independent of the goals of a scientific traveler. The Maxwell estate, Glenlair House, is on a marked side road off route B794. John Clerk Maxwell designed the original house, which was built in about 1830. Several years after James was born in Edinburgh in 1831, the

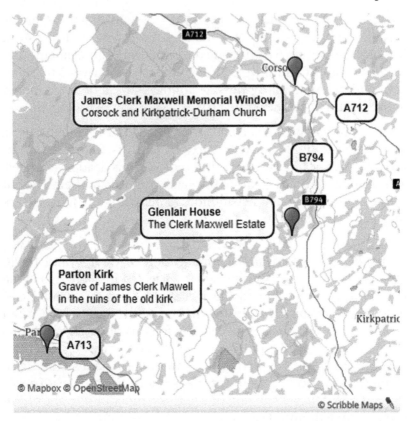

Chart 5.3 Travel sites in Scotland related to James Clerk Maxwell. This chart shows the Glenlair House (the Clerk Maxwell estate now being preserved by the Maxwell at Glenlair Trust), the Parton Kirk (where James Clerk Maxwell, his wife and his parents are buried in the ruins of the old kirk) and the Corsock and Kirkpatrick-Durham Church that houses the James Clerk Maxwell Memorial Window.

family moved here. In the late 1860s, after holding positions in Aberdeen (Marischal College) and London (King's College where he published his four seminal papers on electricity, magnetism, and electromagnetic fields), he returned to Glenlair and made plans for an addition to the estate. These included a vestibule and drawing room on the first floor and a bedroom and dressing room upstairs. James wanted to write his definitive book on electricity and magnetism (*Treatise on Electricity and Magnetism*) at Glenlair but the new addition was not quite ready in time. (One

Electricity and the Atom 123

can see the window of the room in the original building where he wrote it.) Unfortunately, the house was gutted by a kitchen fire in 1929 that forced the last of the Maxwell family out of the house. It lay in disrepair for many years. The Maxwell at Glenlair Trust, established in 2001, has undertaken to restore the estate. Their goals are to retain

> "for posterity the house where Maxwell returned every summer for holiday fun and adventures as a child, for 'anchoring' as a teenager, for refreshment as a professor, and to recollect and centre his private universe while he was deciphering the invisible universe with his formidable mathematical thought."

The improvements are stunning but ongoing. The refurbished addition is shown in Figure 5.6a. There is now a

Figure 5.6 On the left in Figure 5.6a is the partially refurbished James Clerk Maxwell addition to the original Glenlair estate. The door to the visitors' center is clearly shown. Maxwell wrote his treatise on electricity and magnetism in the second-floor room of the original building just to the right of the addition. Figure 5.6b shows the refurbished primary color tiles on the floor of the visitors' center. Photographs by G.E. Rodgers.

visitors' center on the original porch of the addition with many pictures and documents worth spending time looking over in detail. One of the highlights is the floor of the visitors' center where the original Minton Tiles were found under mountains of debris. Shown in Figure 5.6b, note that the tiles display the Clerk Maxwell primary colors, red, green, and blue. All but one of the original tiles were found to be intact. Visitors are requested to make an appointment by writing an email to fergie@glenlair.org.uk. It is an extraordinary stop for travelers with the atom.

5.4.2.3 *Parton (Near Dumfries).* Parton Kirk (55.006629°, −4.039241°). The Maxwell's regularly attended church here and, after Maxwell died in Cambridge just short of 48 years old, he was buried in the churchyard of the kirk. His father and mother and his wife were also buried here in the ruins of the "Old Kirk". (When approaching the present church, bear to the left to find the "Old Kirk"). There is a brass plate at the entrance to the churchyard that lists Maxwell's accomplishments but, perhaps more importantly, calls him "a good man, full of humour and wisdom".

5.4.2.4 *Corsock.* The Corsock Parish Church was built in 1839 largely due to the efforts of John Clerk Maxwell. Both the father and his son were elders of this church which was the original site of the James Clerk Maxwell Memorial Window. In 1947 the church was closed and the window transferred to the nearby Corsock and Kirkpatrick-Durham Church (55.062911°, −3.940028°). The church is normally locked but the keys can be obtained, at this writing, by inquiring of Betty Watson at her Kylelea Cattery (Tel 01644 440279) in Corsock or by asking at the village shop. It is certainly worth the effort to obtain the key and get into the church to see the amazing stained-glass window installed in the James's honor. (Whoa, come on now, by the by, how many other scientists do you know that have a stained-glass church window installed in his or her honor!) Betty will tell you that moving the window was not without controversy. Some of the members didn't want anything to do with the window and even expressed the sentiment that they would rather put a hammer through it then install it in their church.

Electricity and the Atom 125

As Tanford and Reynolds say "The motif of the window is the Magi following a brilliant Star of Bethlehem and a Greek inscription is embedded in the glass .... Loosely translated, it means 'All good giving and every perfect gift' – undoubtedly a reference to the Epistle of James from the New Testament 'All good gifts and every perfect gift comes from above, *from the Father of the lights of heaven.*' A plaque placed next to the window testifies to 'a genius that discovered the kinship between electricity and light and was led through the mystery of Nature to the fuller knowledge of God.'" Occasionally, science reveals aspects of the cosmos that are counterintuitive. The heliocentric nature of the solar system, the evolution of humankind from lesser species, and the idea that apparently solid matter is composed of infinitely small, indivisible particles called atoms, are just three examples. Add to this list that light, perhaps even the "lights of heaven" but certainly the ordinary light perceived by humankind, is electromagnetic radiation, a wave composed of coordinated (or so-called "in-phase") electric and magnetic fields, propagating through space without any supporting medium. James Clerk Maxwell took the experimental work of Michael Faraday and mathematically proved this counterintuitive idea. If you have a chance, sit in this quiet church as the author and his wife have, fall on your knees if you like, and contemplate the place, the man and his achievement, and the God who made the lights of heaven.

The Corsock Parish Church, where the Memorial Window was originally installed, is now a private residence called, appropriately, the Maxwell House (55.05903°, −3.93083°). Not far from the Corsock and Kirkpatrick-Durham Church, it still has a cross atop and directly below it one can readily see where the Memorial Window was originally placed.

5.4.3 Travel Sites in Cambridge Related to James Clerk Maxwell

See Chart 5.4 for the location of these sites.

5.4.3.1 The Old Cavendish Laboratory. Free School Lane (52.203391°, 0.119955°). (Free School Lane was so-named because there once was a boys' school there that did not charge fees.) Here, you can find the original entrance, above which is the

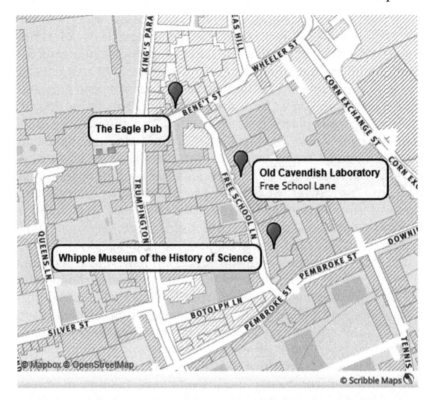

Chart 5.4 Travel sites in Cambridge related to James Clerk Maxwell, J. J. Thomson, and Ernest Rutherford. This chart shows the location of the Old Cavendish Laboratory on Free School Lane, the Whipple Museum of the History of Science, and the Eagle Pub. The Museum at the Cavendish Laboratory, located on JJ Thomson Ave off Madingley Road in West Cambridge is not shown.

badge of the house of Devonshire on the left and the University of Cambridge Seal on the right. Other things to see here are the Mond Building with its Rutherford crocodile, a plaque about the discovery of DNA, and the Maxwell Lecture Theatre, which is still used by the physics department for second-year lectures. Just down the lane (where it intersects with Bene't Street) from the Old Cavendish Laboratory is the Eagle Public House (52.203970°, 0.118042°). Given its proximity to the laboratory, this pub was frequented by many of its famous, and not so famous staff. Here, for example, Francis Crick and James Watson announced that

Electricity and the Atom

they had discovered the structure of DNA. There is a blue plaque next to the entrance to that effect. At this writing, the pub was offering Eagles's DNA Ale. If it's available, give it a try. In the back of the Eagles Pub, check out the RAF bar and its ceiling signed by World War II airmen. The pub is a treat for scientific travelers. Sit there with a Guinness and discuss with your guests and/or your family and friends all the amazing discoveries made near, and perhaps actually in, this pub. A good beer can get the creative juices going.

5.4.3.2 Whipple Museum of the History of Science. (52.202778°, 0.119199°). Just down Free School Lane from the old Cavendish, this museum houses many important collections. Most pertinent to our interests is the Cavendish Laboratory Collection, which includes many of the instruments used in the laboratory including optical and electrical instruments, X-ray tubes, vacuum tubes, and thermometers.

5.4.3.3 Museum at the Cavendish Laboratory. (52.209240°, 0.091759°). The New Cavendish Laboratory, JJ Thomson Avenue, off Madingley Road in West Cambridge. The museum is located on the first floor (second for Americans) of the Bragg Building. It contains many exhibits related to the history of the atom. There is a portrait of the founder William Cavendish on the stairs, a bust of James Clerk Maxwell and a list in Maxwell's hand-writing of the instruments in the laboratory when it opened in 1874. There's a cabinet that contains apparatus that belonged to Maxwell, much of which he used before becoming professor. There are guides available from the Cavendish reception. It's advisable to make prior arrangements to see this laboratory. Contact reception@phy.cam.ac.uk for further details. Also see the history of the Cavendish website.[6]

5.4.4 Memorial to James Clerk Maxwell in Westminster Abbey, London

5.4.4.1 Westminster Abbey. See Chapter 7 on the graves and memorials of scientists in Westminster Abbey. A memorial to James Clerk Maxwell is located here.

5.5 SIR JOHN JOSEPH ("JJ") THOMSON (1856–1940)

"We have in the cathode rays matter in a new state, a state in which the subdivision of matter is carried very much further than in the ordinary gaseous state: a state in which all matter ... is of one and the same kind; this matter being the substance from which all the chemical elements are built up." J. J. Thomson

Most people, even those who had intensively studied science and mathematics, were initially awed and dumbfounded by the theories of James Clerk Maxwell. It took several decades before they were understood with any degree of sophistication. Nevertheless, light as electromagnetic radiation, the role of electrical and magnetic fields in general, and the role of electricity in the structure of matter gradually came to be accepted.

It didn't take long for "The Cavendish" to become one of the preeminent laboratories of physics in the world. In 1879, after his sudden death, Clerk Maxwell was followed by Lord Rayleigh (John William Strutt) as the director. Rayleigh used the apparatus designed by Clerk Maxwell to test his theory of the electromagnetic field. There is a diagram in the Cavendish Museum showing Clerk Maxwell's proposal. Lord Rayleigh, one of the few members of the hereditary nobility to attain fame as an outstanding scientist, went on to study other phenomena that were related to the discovery of the atom, including the spectrum of so-called "black body radiation" that played a prominent role in the discovery of quantum mechanics (see Chapter 15). He also established what today is called Rayleigh radiation, which explains why the sky is blue. In addition, he played a prominent role in the discovery of the inert gases, particularly argon. In 1904, he received one of the first Nobel Prizes in Physics for these discoveries. His colleague in this research, Sir William Ramsay, won the Nobel Prize in Chemistry in the same year! (For more on the history of the Nobel Prizes, the first of which was awarded in 1901, see Chapter 19.) Although he was the second Cavendish Professor of Experimental Physics, he only held that post for five years. Most of his work was done at his own private laboratory at Terling Place, Essex, Witham. He died there in 1919.

5.5.1 Terling, Essex (Rayleigh)

3rd Baron Lord Rayleigh is buried in the large family plot in All Saints Churchyard, Terling Place (51.8021°, 0.57097°). Some reports indicate that special arrangements can be made to see his private laboratory. (Dr Gordon L. Squires, physics historian, Cavendish Museum).

In 1884, Rayleigh was succeeded by John Joseph ("JJ" as he was affectionately known by everyone) Thomson as the third Cavendish Professor of Physics. He was only 28 at the time of his appointment, leading one senior person to mutter "Matters have come to a pretty pass when they elect mere boys Professors." He held this post for 35 years, and played a large part in the discoveries that established the electrical nature of the atom. A word or two about Thomson and his leadership style at the Cavendish is in order. Francis Aston wrote of Thomson that "His boundless, indeed childlike, enthusiasm was contagious and occasionally embarrassing". He maintained a warm personal interest in his students and took great pride in their achievements, both scientific and otherwise. Lord Rayleigh said that "his relations with junior members of the college were very easy. He took the keenest delight in their athletic performances and nothing delighted him more than the opportunity of going to a good football match or watching a performance of Trinity men on the river." In 1893, he established the Cavendish Physical Society where informal discussions of recent work were held twice a month. In 1895, students (like Ernest Rutherford, who was the first of these) from other universities were accepted, the international flavor of the Cavendish blossomed. At about this time, at Mrs Thomson's suggestion, the practice of serving tea and refreshments promptly at 4 PM each afternoon was initiated. The research students themselves instituted the "Cavendish Dinners" that often involved various shenanigans. Light verse was composed for these occasions. One of these conveys an image of Thomson and the atmosphere that he established at the Cavendish.

> When the professor has solved
>   a new riddle,
> Or found a fresh fact, he's
>   fit as a fiddle.

He goes to the tea-room and
   sits in the middle
And jokes about everything under the sun.[8]

It must have been a lively and engaging intellectual atmosphere but also just plain fun at times. Another section of light verse expresses what it was like to work there.

The people are delighted with the wondrous
   things we do.
But few have any notion that we're such a
   jolly crew.
If some of them were here tonight I think
   we'd make it plain
We're not all just as dry as dust at the lab
   in Free School Lane.[8]

In the Cavendish Museum, there are some notes that describe this atmosphere. It is worth thinking about all this as you wander around the Old Cavendish grounds and again as you look at the exhibits in the Cavendish Museum and think about the contributions that this "jolly crew" made to the story of the discovery of the atom.

In 1895, Wilhelm Röntgen announced the discovery of X-rays in Würzberg, Germany. [See Chapter 13 for the amazing places to visit in Lennep (his birthplace), Giessen (his grave), and especially Würzberg (his laboratory) related to this man and his discovery.] Also in 1895, Cambridge University started to admit students from other universities to do research at "The Cavendish" and other colleges. Ernest Rutherford came all the way from New Zealand to be one of the first of these students. (See the next chapter for more on Rutherford.) In 1896, Henri Becquerel discovered radioactivity in Paris. (See Chapter 14 for the stories behind Becquerel and Pierre and Marie Curie's seminal work on radioactivity and the places to visit related to their work.) These discoveries changed the direction of physics, and Thomson and the young Rutherford were swept up in the challenges of the day.

Thomson was particularly interested in investigating so-called "cathode rays". The history of cathode ray tubes starts with

Electricity and the Atom

the work of Heinrich Geissler, a German inventor, and William Crookes. Geissler was an expert glass-blower who had devised a new, more efficient vacuum pump with which he could remove an impressive amount of the air from his expertly fashioned and air-tight tubes. (Evangelista Torricelli, a contemporary of Galileo, had done some of the earliest work in vacuum technology. Torricelli visited Galileo while the latter was under house arrest and near death in Arcetri near Florence. For a long while, the unit of pressure was commonly known as a "torr" in honor of Torricelli.) Tubes that were evacuated in this way were called "Geissler tubes". Physicists wondered if electrical charges could be sent through them. (Faraday had noted that fluorescence resulted when this was done but his vacuum technology was not good enough to allow more work to be done in that area.)

The English physicist Sir William Crookes did some of the best early work investigating cathode rays. Crookes perfected techniques for producing much better vacuums in his tubes, which are still called "Crookes Tubes" today. The air pressure in Crookes' tubes was 75 000 times less than that in Geissler's tubes. A diagram of such a Crookes tube is shown in Figure 5.7.

When an electrical power source is connected to the two electrodes (the negative cathode and the positive anode), an eerie green glow is seen around the cathode. Further experiments demonstrated that rays of some type were coming from the cathode and going toward the anode. Opinions varied as to whether these "rays" were a wave phenomenon or a stream of particles. Jean Perrin working in Paris in 1895 found that the particles carried a negative charge and were

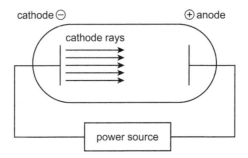

Figure 5.7 A schematic diagram of a Crookes tube.

deflected by a magnetic field. Crookes found that they travelled in straight lines like light and they even cast a shadow if an object (often a Maltese cross) was placed within the tube. However, unlike light, he confirmed that they were deflected by a magnetic field. (If light were deflected by all the magnetic fields around us, the world would be appear grossly distorted indeed.) Crookes also thought the rays might be a stream of negatively charged particles but when he tried to change their path with an electric field he was not successful. He was left puzzled and, in the end, referred to cathode rays as a fourth state of matter or an "ultra-gas" or "radiant matter", *i.e.*, he had no idea what these mysterious rays were. For many years, Crookes lived at 7 Kensington Park Gardens, London (51.511402°, −0.202538°). There is a blue plaque located there to honor him.

Enter Joseph John (J. J.) Thomson. Working at the Cavendish (which, as we have just seen, had been shepherded into existence by the first Cavendish Professor of Experimental Physics, James Clerk Maxwell, father of electromagnetic radiation), Thomson, the third Cavendish Professor, was drawn to study Crookes tubes because of this seemingly novel and mysterious form of radiation that was, in fact, not electromagnetic radiation. Again, the vacuum technology was improved and Thomson, in 1897, showed that the path of the cathode rays was, in fact (contrary to Crookes' results), deflected by an electric field. It was evident at that point that cathode rays were, in fact, streams of tiny little negative particles, that is, fundamental units of electrical current. Six years earlier, when such units had been merely hypothetical, they had been called "electrons" by George Stoney. Over Thomson's objections, this name was applied to the particles that composed cathode rays. He preferred to call them "corpuscles". Thomson had discovered what we all know today as the electron. Furthermore, using an elaborate tube in which he could deflect the rays by both fields, he counterbalanced the effects of the magnetic and electric fields and determined the ratio of the electric charge to the mass of the particle, the so-called e/m ratio. Using an estimate of the minimum charge on ions developed by Faraday as e, the mass of these particles was determined to be something like 1/1837 the mass of hydrogen atoms. (Some dozen years later, Robert Millikan in Chicago verified this charge using his famous "oil-drop experiment".) This

Electricity and the Atom

was a very tiny particle indeed. Crucially, Thomson used different initial gases and electrodes made of different metals (platinum, aluminum, copper, for example) in his tubes but the e/m ratio of the electrons produced was always the same. Thomson announced his discovery in a Friday Evening Discourse at the Royal Institution. For the first time, a fundamental component of atoms, a "sub-atomic particle", the electron, had been uncovered. The atom was no longer the unshatterable hard sphere envisioned by John Dalton.

So what role did electrons play in atoms? Thomson, for lack of additional evidence, supposed that electrons were the only sub-atomic particles. In 1904, he briefly advocated what has become known as the "plum-pudding model", or alternatively, the "currant-bun" or "watermelon model" of the atom. He envisioned the atom as being a sphere of massless positive electricity embedded with thousands of tiny electrons carrying the negative charge. In a neutral atom, the total positive charge was equal to the total negative charge. In one analogy, the massless positive charge is analogous to the bread of an English plum pudding while the electrons are analogous to the raisins embedded in it. In perhaps a more vivid analogy, the massless positive charge is the red flesh of a watermelon and the black seeds are analogous to the electrons. A picture of Thomson's model is shown below in Figure 5.8. Thomson received the 1906 Nobel Prize in Physics for his discovery of the electron. His work in determining the charge to mass ratio of electrons is considered to be one of the greatest experiments of physics.

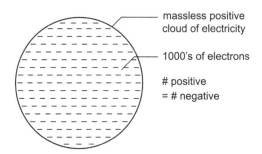

Figure 5.8 A diagram of Thomson's plum-pudding model of the atom.

Thomson did not stop with his discovery of the electron. Starting in 1907 or so, he turned his attention to what were known as "canal rays". These were originally discovered by Eugen Goldstein in 1886. Goldstein had built a cathode ray tube in which the cathode was perforated with small holes. A diagram of this tube is shown in Figure 5.9. As in a normal cathode ray tube, the cathode rays (electrons) move from right to left from the negative cathode toward the positive anode. However, another set of rays moved from left to right through the holes in the cathode and on to a detector built into the right side of the tube. Since this second set of rays moved towards the cathode, in the opposite direction from the cathode rays, it was concluded that they were positively charged. Since they moved through the holes (or "canals" – albeit very short canals) they were dubbed "canal rays" (or, as Goldstein called them, Kanalstrahlen, "channel rays").

Thomson proceeded to deflect the canal rays with magnetic and electric fields. The charge-to-mass ratios of these rays were much smaller than had been observed for cathode rays. He interpreted this result to be because these positive particles were much more massive. With greater mass, they had greater momentum and therefore were more difficult to deflect with magnetic and electric fields. Furthermore, the e/m ratio and amount of deflection varied with the gases used in the tubes. He was producing positively charged cations, He^+ if helium was the gas in the tube, Ne^+ if the gas was neon, *etc.* When he carefully studied neon, he found two types of particles, one a little more massive than the other, were present. This was one of the first indications that atoms of a given element could have different masses, that is,

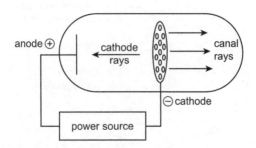

Figure 5.9 A diagram of a canal ray tube.

Electricity and the Atom 135

that there were "isotopes". Isotopes are varieties of the atoms of a given element, differing only in their masses. Frederick Soddy, in 1913, formally proposed the idea of isotopes. (See Chapter 6, pp. 168–169 for more on Soddy and isotopes.)

5.5.2 Travel Sites in Cambridge Related to J. J. Thomson

5.5.2.1 The Old Cavendish Laboratory. (52.203391°, 0.119955°), Free School Lane, Cambridge. This was described earlier when discussing Clerk Maxwell sites. It's a part of the "New Museums Site" of Cambridge University. Recall that the laboratory was named to commemorate the British chemist and physicist Henry Cavendish (See Chapter 3) and his relative William Cavendish, the 7th Duke of Devonshire, who was the Chancellor of the University and gave the money for the construction of the laboratory. There is a slate plaque on Free School Lane that records that this was the site of Cavendish Laboratory for 100 years from 1874 to 1974. Another oval plaque on the lane notes this is the place where J. J. Thomson discovered the electron.

5.5.2.2 Museum at the Cavendish Laboratory. (52.209240°, 0.091759°). Again, this was described earlier. Note that it is, fittingly, located on J. J. Thomson Avenue. Thomson's beautiful, original tube for carrying out his experiment is on display here. Also, see Chapter 6 for other important equipment housed in this museum that is related to the experimental work done by Ernest Rutherford, James Chadwick, and Francis Aston.

5.5.3 Travel Sites in London Related to J. J. Thomson

5.5.3.1 Westminster Abbey. See Chapter 7 on the graves and memorials of scientists in Westminster Abbey. The ashes of Joseph John (J. J.) Thomson and his wife Rose are buried here in the altar area.

5.6 SUMMARY

At the new Royal Institution (Ri) in London in 1806–1807, Humphry Davy used Volta's chemical battery to discover six elements but, some would argue, his most important "discovery" may well have been Michael Faraday. In 1812, Faraday attended Davy's "Farewell Lectures" at the Ri, elaborately bound his notes and sent them to Davy. As a result, Davy hired him an assistant and had him serve as both scientific assistant and valet while Davy and his new wife set out on a grand European tour to Paris and Italy. Among others, they met with Gay-Lussac, Berthollet, and Ampère in France, and Volta in Italy. Upon his return, Faraday settled in at the Ri as a chemist. However, after Denmark's Christian Ørsted discovered electromagnetism in 1820, Faraday decided that this was going to be his new field. He developed the ideas of "lines of force" for magnetic and electric fields and produced results that led directly to modern electric motors, transformers, and generators. His electrochemical research led to a set of laws that govern electrolysis. An atomic agnostic, he nevertheless showed that ions have charges that are integral multiples of a single fundamental unit of charge. In London, we can visit Faraday's birthplace, his grave, various plaques and statues, as well as the great Royal Institution with its lecture hall and Faraday magnetic laboratory.

James Clerk Maxwell provided the mathematical basis of Faraday's ideas on the relationship between light, electricity, and magnetism, providing the basis of what have become known as Maxwell's Equations. Clerk Maxwell also worked on the mathematics of the kinetic molecular theory of gases, which pictures atoms or molecules of a gas moving at great speeds and colliding with the walls of their containers, thereby producing pressure. He also determined that the additive primary colors were red, green, and blue. His primary contribution, however, was to convert Faraday's lines of force into what is now known as the electromagnetic field, and showed that light is electromagnetic radiation. Clerk Maxwell was chosen to be the first Professor of Experimental Physics at the new Cavendish Laboratory in Cambridge and, as such, designed and supervised the building of that facility. In Edinburgh, we can visit his birthplace (which now houses the James Clerk Maxwell Foundation), the place where he lived while attending the Edinburgh Academy and the University of Edinburgh, and the newly erected statue in his honor. We can also visit "Glenlair", the refurbished Clerk Maxwell estate, his grave in Parton, and the

Electricity and the Atom 137

church in Corsock with its great stained-glass window erected in Clerk Maxwell's honor. In Cambridge, we can visit the Old Cavendish Laboratory with its Maxwell Lecture Theatre, and the Museum at the new Cavendish Laboratory.

J. J. Thomson was the third Cavendish Professor of Experimental Physics. He continued the works of others to investigate the nature of cathode rays. In his Crookes tube, he discovered what became known as the electron, the first subatomic particle. He went on to determine the charge-to-mass ratio of the electron and concluded that it was exceedingly light, much lighter than even the lightest atom. He proposed a model of the atom, called the plum-pudding model, composed of a cloud of massless positive electricity embedded with thousands of electrons. Finally, he investigated "canal rays" that were the positive particles left over after one or more electrons had been stripped from their atoms. He also presented the first evidence for the existence of isotopes. At the Museum at the Cavendish Laboratory we can see a Thomson's tube for carrying the determination of the charge-to-mass ratio of the electron. Both Clerk Maxwell and Thomson are honored in Westminster Abbey.

ADDITIONAL READING

- F. A. J. L. James, Displaying Science in Context at the Royal Institution of Great Britain, in *Science History: A Traveler's Guide*, ed. M. V. Orna, American Chemical Society, Washington, DC, 2014.

On Humphry Davy:

- J. Z. Fullmer, *Young Humphry Davy: The Making of an Experimental Chemist*, American Philosophical Society, Philadelphia, 2000.
- D. Knight, *Humphry Davy: Science and Power*, Cambridge University Press, Cambridge, 1992.
- Memoirs of the Life of Sir Humphry Davy, in *The Collected Works of Sir Humphry Davy*, ed. J. Davy, Smith, Elder and co., London, 1839, vol. I.

On Michael Faraday:

- Faraday's London, Royal Institution, http://www.rigb.org/docs/faradays_london_2.pdf, accessed March 2019.
- S. K. Bolton, *Lives of Poor Boys who Became Famous*, Thomas Y. Crowell Company, 1885.

On James Clerk Maxwell:

- B. Mahon, *The Man Who Changed Everything: the Life of James Clerk Maxwell*, John Wiley & Sons Ltd., Chichester, England, 2003.
- I. Falconer, Cambridge and Building the Cavendish Laboratory, in *James Clerk Maxwell: Perspectives on his Life and Work*, ed. R. Flood, M. McCartney and A. Whitaker, Oxford University Press, 2014.
- J. G. Crowther, *The Cavendish Laboratory 1874 – 1974*, Science History Publications, New York, 1974.
- J. W. Arthur and J. Donald, *Brilliant Lives: The Clerk Maxwells and the Scottish Enlightenment*, imprint of Birlinn Ltd, Edinburgh, 2016.

On J. J. Thomson:

- B. Pippard, The Whipple Museum and Cavendish Laboratory, in *The Physical Tourist: A Science Guide for the Traveler*, ed. J. S. Rigden and R. H. Stewarart, Birkhäuser, Basel, 2009.
- B. Pippard, Thomson, Rutherford and atomic physics at the Cavendish, in *Cambridge Scientific Minds*, ed. P. Harman and S. Mitton, Cambridge University Press, Cambridge, 2002.
- J. G. Crowther, *The Cavendish Laboratory 1874–1974*, Science History Publications, New York, 1974.
- For an unusual description of life at the Cavendish during Thomson's time, take a look at "JJ and the Cavendish" at, http://www.phy.cam.ac.uk/history/years/jjandcav, accessed March 2019.

REFERENCES

1. T. K. Kenyon, Science and Celebrity, Humphry Davy's Rising Star, *Chem. Heritage*, 2009, **26**, 30–35.
2. A. Hirshfeld, *The Electric Life of Michael Faraday*, Walker & Company, New York, 2006.
3. Portrait of Sir Humphry Davy, Available at, https://wellcomecollection.org/works/f48jat3q, accessed June 2019.
4. N. Forbes and B. Mahon, Faraday, *Maxwell, and the Electromagnetic Field: How Two Men Revolutionized Physics*, Prometheus Books, Amherst, NY, 2014.
5. J. J. O'Connor and E. F. Robertson, History topic: A visit to James Clerk Maxwell's house, November 1997, http://www-history.mcs.st-andrews.ac.uk/HistTopics/Maxwell_House.html, accessed March 2019.
6. The History of the Cavendish Laboratory, http://www.phy.cam.ac.uk/alumni/files/Cavendish_History_Alumni.pdf, accessed June 2019.
7. James Clerk Maxwell, Introductory Lecture as first Cavendish Professor of Experimental Physics, 15 October 1871.
8. J. Satterly, The Postprandial Proceedings of the Cavendish Society, *Am. Phys. Teach.*, 1939, 7, 179–185.

CHAPTER 6

The Brits, Led by the "Crocodile" and His Boys, Take the Atom Apart: Ernest Rutherford (England, Scotland, Ireland, New Zealand, and Montreal)

6.1 A QUICK LOOK AT PLACES TO VISIT "TRAVELING WITH THE ATOM" IN ENGLAND, SCOTLAND, IRELAND, NEW ZEALAND, AND MONTREAL

In New Zealand: the Lord Rutherford Memorial Reserve ✹✹ in Brightwater, the Rutherford-Pickering Memorial ✹✹ and the Havelock School ✹ in Havelock, the Lord Rutherford Memorial Hall ✹ (formerly the Foxhill School) in Foxhill, Nelson College ✹ in Nelson, the Rutherford Gallery ✹ in the Historic Cape Light and Museum in Taranaki, and Rutherford's Den in Christchurch ✹✹✹; in Montreal, Canada: the Rutherford Museum ✹✹✹ and the Rutherford-Frederick Soddy plaque ✹✹ in the former Macdonald Physics Building ✹ and the Rutherford residence ✹; in Manchester, England: the University of Manchester's Rutherford Building ✹✹ where the alpha-particle, gold-foil experiments were performed that led Rutherford to propose the nuclear atom and, inside the building,

---

Traveling with the Atom: A Scientific Guide to Europe and Beyond
By Glen E. Rodgers
© Glen E. Rodgers 2020
Published by the Royal Society of Chemistry, www.rsc.org

the plaque ❂ in the room where Rutherford carried out the first nuclear transformation reaction; the Schuster Laboratory ❂❂ with its Rutherford, Bragg, Brackett, and Moseley Lecture Halls; in Cambridge, England: the Old Cavendish Laboratory entrance ❂❂, the plaque ❂ about its 100 year history, the Crocodile Carving ❂❂ honoring Rutherford, the Whipple Museum of the History of Science ❂❂, the Rutherford residence at the Newnham Cottage ❂, and the Cavendish Museum ❂❂❂❂❂ exhibits including equipment used by Rutherford to characterize alpha particles and carry out the first nuclear transformation, the Chadwick neutron chamber, Francis Aston's mass spectrograph, and the rectifier tube from the 1930 Cockcroft–Walton machine; in London: Rutherford's grave site at Westminster Abbey ❂❂❂❂❂; in Eastbourne, England: Soddy's birthplace and boarding school ❂; in Glasgow, Scotland: two Soddy plaques ❂; in Oxford: two Moseley plaques ❂, the University Museum of Natural History ❂❂, Trinity College ❂❂, and the History of Science Museum in Old Ashmolean ❂❂; in Dungarvan, Ireland Walton Causeway Park ❂❂; in Waterford, Ireland ETS Walton Building and plaque ❂; in Dublin, Ireland Walton's Grave ❂ in Deansgrange Cemetery.

> **The Crocodile and "His Boys" Take the Atom Apart**
> - After earning three baccalaureate degrees from Canterbury College in Christchurch, New Zealand, Ernest Rutherford, in 1895, receives an "Exhibition of 1851 Scholarship" and travels to England to study at "the Cavendish" at Cambridge University.
> - Working with J. J. Thomson, he characterizes alpha and beta rays emanating from the radioactive elements newly discovered in Paris by Pierre and Marie Curie.
> - In 1902, Rutherford and Frederick Soddy, working together at McGill University in Montreal, characterize nuclear transformations and define the half-life of a radioactive element.
> - Moving to Manchester, Rutherford, Ernest Marsden and Hans Geiger carry out the famous 1909 alpha-particle gold-foil experiments.
> - Rutherford formulates his "nuclear atom" in 1911 and carries out the first artificial nuclear transformation in 1917.
> - Henry Moseley (1887–1915) discovers atomic numbers in 1913.
> - Working at the University of Glasgow, Soddy defines isotopes in 1913.
> - Francis William Aston (1877–1945), starting in 1919, fully characterizes various isotopes using his mass spectrometer.

(continued)

- In 1928 John Cockcroft and Ernest Walton devise their "Cockcroft–Walton Machine" that uses strong electrical fields to accelerate and slam protons and alpha particles into nuclei resulting in a variety of nuclear transformations. Their results involve the first experimental verification of Einstein's $E = mc^2$ equation.
- The neutron is finally discovered by James Chadwick in 1932. The "Crocodile" carving, commissioned in honor of his friend Rutherford by Peter Kapitza, is unveiled at the Mond Laboratory at the Cavendish in 1933.

6.2 ERNEST RUTHERFORD (1871–1937)

"For Mike's sake, Soddy, don't call it transmutation. They'll have our heads off as alchemists …. Make it transformation." Ernest Rutherford

"… the atom itself is not the smallest unit of matter but is a complicated structure made up of a number of small bodies." Ernest Rutherford

Sir Ernest Rutherford, 1st Baron of Nelson, is one of the most fascinating characters in the history of the atomic concept. Fortunately, there are a number of quality "traveling with the atom" sites associated with Rutherford including his birthplace and those related to his primary, through undergraduate education in New Zealand, his graduate education at the Cavendish Laboratory in Cambridge, England, his important early work on nuclear transformations at McGill University in Montreal, Canada, his discovery of the nucleus and early nuclear reactions work at the University of Manchester, and finally his work as the Director of the Cavendish Laboratory back in Cambridge. Rutherford's nickname was the "Crocodile" and he is permanently honored with a carving of a crocodile at the "Old Cavendish" Laboratory at Cambridge University.

6.3 RUTHERFORD IN BRIGHTWATER, FOXHILL, HAVELOCK, NELSON, AND CHRISTCHURCH, NEW ZEALAND

Rutherford's family emigrated from Scotland to New Zealand in 1842. Ernest was born in Brightwater, near Nelson on the South Island in 1871, the fourth of 12 children. He was a dedicated and worthy student who occasionally came in only second as he progressed from high school to university. At Canterbury College in Christchurch he earned three baccalaureate degrees – BA, MA (double first-class honors), and BS – in mathematics, physics and physical science but, more importantly, carried out pioneering research into the magnetization of iron by high-frequency electrical discharges.

In 1851, "The Great Exhibition" had been held in London and was a great success, so much so that it generated all the funds necessary to build the Albert Hall, the Victoria and Albert Museum, and The Science Museum. The profits were also used to fund the "Exhibition of 1851 Scholarships", which were intended to sponsor talented science students from all over the United Kingdom to study in England. In 1895, there were two entrees in the New Zealand competition for these scholarships (New Zealand used to be under British rule). The result was both bad news and good news for young Ernest. The bad news was that, unfortunately, though Rutherford's work was well done and his application in good order, he did not win the competition. The good news was that, in a fantastic piece of luck for Rutherford and physics, the winner unexpectedly declined the prize because, being recently married, he could not give up his well-paid position to take the scholarship. When word came to Rutherford that he was to receive the award and would be an "1851 man", he was reportedly digging potatoes on his father's flax farm. He tossed his spade aside and said, "That's the last potato I'll dig". As it turned out, Rutherford had marriage in mind as well, but he and his intended, Mary ("May") Newton, Ern's landlady's daughter, decided to delay their marriage and Ernest was off to Cambridge to study under J. J. Thomson ("J. J.") at "The Cavendish". 1895 was the first year that Cambridge University accepted 1851 scholars. Figure 6.1 shows Rutherford as a young man, about 21 years old.

6.4 TRAVEL SITES IN NEW ZEALAND RELATED TO ERNEST RUTHERFORD

See Chart 6.1 for the location of these sites.

6.4.1 Lord Rutherford Memorial Reserve

※※※, Brightwater, New Zealand (−41.377297°, 173.101790°). This birthplace memorial in a garden setting is highlighted by a bronze statue of a young child carrying a "Mathematics Primer"

Figure 6.1 Ernest Rutherford, possibly taken for his 21st birthday in 1892. Reproduced from ref. 1.

heading out to the future. There are 14 picturesque and informative display panels and six sound stations that cascade down from the statue. It is located, not surprisingly, on Lord Rutherford Road at the junction of the Brightwater Deviation (State Highway 6). The prominent "RUTHERFORD BIRTHPLACE" stone marker on Highway 6 proudly announces one's arrival at the site. The memorial is floodlit at night and never closed. The actual house where Rutherford was born was demolished in 1921.

6.4.2 Lord Rutherford Memorial Hall

658 Wakefield-Kohatu Highway, Foxhill, New Zealand (−41.43999°, 172.97169°). Rutherford's primary school (the Foxhill School) has been converted into a meeting house. There is an informative panel located in front of the building. The Rutherford Farm where "Ern" lived from age 5 to 11 (1877–1883) was located directly across the street. During this time his father was a flax-mill operator and a wheelwright. Ern earned some money picking the nearby hops.

6.4.3 Rutherford-Pickering Memorial

(−41.27925°, 173.76668°) and the nearby Havelock School (−41.17924°, 173.76683°), Havelock, New Zealand. The memorial,

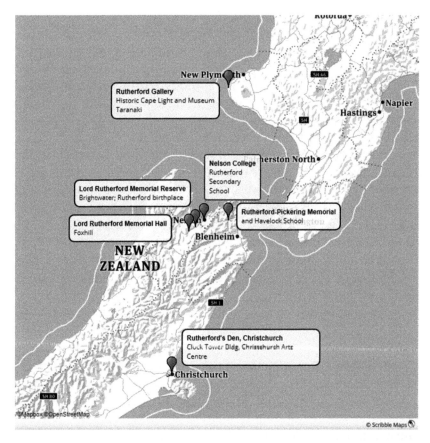

Chart 6.1 Travel sites on South Island, New Zealand related to Ernest Rutherford including his birthplace in Brightwater (the Lord Rutherford Memorial Reserve), the Rutherford-Pickering Memorial and the nearby Havelock School (now the Youth Hostel) in Havelock, Nelson College and the Rutherford Boarding Hostel in Nelson, the Foxhill School (now the Lord Rutherford Memorial Hall) in Foxhill, Rutherford's Den in the Clock Tower Building of the Arts Centre at Christchurch and, on the North Island, the Rutherford Gallery at the Historic Cape Light and Museum in Taranaki.

built around the pre-existing white memorial to the shipwreck of the Schooner *Ronga*, commemorates Ernest Rutherford and William Hayward Pickering (1910–2004), who both attended school in this pleasant coastal village. Pickering was a space-age pioneer who directed the Jet Propulsion Laboratory (JPL) at Pasadena, California for 22 years. The eight information panels outline the work and life of the two scientists in addition to providing background on Havelock, geological history, and the local Maori settlement. At the front are Lord Rutherford's crest on the left and, on the right, a roundel representing the JPL, the National Aeronautics and Space Administration (NASA) and California Institute of Technology (from which Pickering graduated). At the bottom of Rutherford's crest are the words from Lucretius, '*Primordia quaerere rerum*', meaning "to seek the first principles of things", which is the motto he chose when he became 1st Baron Rutherford of Nelson. (May wished that he had chosen Havelock for his title, rather than Nelson.) Just a block south of the memorial is Havelock Youth Hostel, 46 Main Rd, which was formerly the Havelock School that both Rutherford and Pickering attended; Rutherford from age 12 to 15 (1883 to 1886). Rutherford said that he had a "keen remembrance" of his days at the school. Just inside the hostel for "Rutherford Backpackers" there is a large bulletin board about the history of the Havelock School.

6.4.4 Nelson College

46 Waimea Road, Nelson (−41.28548°, 173.27831°). On his second attempt, Ern won the sole Marlborough Scholarship to attend secondary school at Nelson College (1887–1889). The original wooden building that existed during Rutherford's days burned down in 1904. Its replacement was severely damaged by an earthquake in 1929 and was also replaced. Soon afterward, the nearby Rutherford House was built as the secondary school boarding house. After a five-year renovation, it is now fully available for a new generation of Nelson College students. A large outdoor display about the history of the college includes two photographs showing Rutherford with his classmates.

6.4.5 The Rutherford Gallery

at the Historic Cape Light and Museum, 377–379 Cape Rd, Taranaki, New Zealand (−39.217728°, 173.793879°). In 1888, the Rutherford family moved here from Havelock and set up another flax mill. The gallery is on the ground floor of this replica of the nearby original Cape Egmont Lighthouse, located in the shadow of the dormant volcano, Mt. Taranaki, or Mt. Egmont, in nearby Egmont National Park.

6.4.6 Rutherford's Den, Christchurch

(−43.531450°, 172.629017°). Clock Tower Building 2, Christchurch Arts Centre, 2 Worcester Boulevard, Central City, Christchurch 8001, New Zealand. This is a multi-media exhibition space housed where Rutherford studied, attended lectures, and carried out his "electrical researches" at Canterbury College in Christchurch. Rutherford and another student received permission to work in a former "gown room" under the lecture hall. This "den" had a concrete floor suitable for carrying out experiments adversely sensitive to vibrations. In 2010 and 2011, the devastating Canterbury earthquakes caused severe damage to 22 of the 23 buildings at the Arts Centre. Scientific travelers are fortunate that it has been fully restored and transformed into one of New Zealand's most significant heritage sites celebrating the life and legacy of her greatest scientist. Interactive displays suitable for both scientists and non-scientists including children, a fully restored and modernized lecture theatre with its original straight-backed benches, an elaborate timeline of Rutherford's life spiraling up a staircase, and the actual den where Rutherford worked, featuring a full wall-length video screen depicting his experiments and actual sound recordings of his and other voices. It is a special occasion for the scientific/historical traveler to sit here in this dark and low-ceilinged little room where Ern started his scientific journey to international fame. The Arts Centre also has the Bunsen café where you can sit back with fellow travelers and contemplate where Sir Ernest Rutherford started his journey to become one of the greatest figures in the discovery of the atom. Alternatively, you could ride the Christchurch Tram to colorful New Regents Street for a fine local craft beer. Ern and May would disapprove, however, particularly May, who followed her mother's example as a lifelong prohibition advocate.

6.5 RUTHERFORD'S EARLY YEARS AT CAMBRIDGE UNIVERSITY AND THE MOVE TO MONTREAL

Rutherford intended to continue his electromagnetic work at Cambridge. He had brought his equipment with him all the way from Christchurch. The boys from elsewhere in the UK ("the provincials") were sometimes ridiculed by the sophisticated young gentlemen who more usually attended a place like Cambridge.

Moreover, Rutherford himself was a large, rough-hewn bloke who spoke loudly with a strange New Zealand accent. He was, however, a quick student and a hard-worker, and soon made his mark. Early on he gave a lecture on his research and accompanied his talk with some experiments, which made a hit. However, as it turned out, that work was not the wave of the future in physics in the late 1890s. In 1895, Wilhelm Röntgen discovered X-rays at Strassburg, Germany. (See Chapter 13 for an account of this amazing discovery and the sites to visit related to Röntgen.) Then only a few months later, in early 1896, Henri Becquerel discovered what Marie and Pierre Curie soon termed "radioactivity". See Chapter 14 for that account and the related places to visit in Paris. In 1897, J. J. announced his discovery of what soon became known as electrons (See Chapter 5, pp. 132–133, for that account.) The paradigm of physics was radically altered by these discoveries and so was the fate of the young Ernest Rutherford. J. J. invited him to work on characterizing "Röntgen's rays" (still called that in Germany to this day), a project which was quickly expanded to include "Becquerel's rays" as well. By 1898, Rutherford, working in Cambridge, and the Curies in Paris had discovered that Becquerel's rays were composed of two types of radiation, which Rutherford, logically enough, called alpha and beta rays. (Gamma rays were discovered several years later.)

About this time, the Physics Laboratory at McGill University in Montreal needed to hire a new Professor of Experimental Physics. McGill had a great benefactor in Sir William Macdonald who, among other projects, financed the construction of the chemistry, engineering, and physics buildings on that campus. These were first-class facilities. For example, no iron or steel was used in the construction of the physics building to minimize magnetic interferences. Only copper, bronze, and brass were used for nails and other fixtures, including the radiators. The Macdonald Physics Building was completed in 1893. Macdonald also sponsored faculty chairs at McGill, including the new professorship in experimental physics. In a visit to the Cavendish, the McGill representative asked J. J. Thomson who they should hire, and he recommended the young (then just 26) Rutherford. Thomson wrote that "I have never had a student with more enthusiasm or ability for original research than Mr Rutherford." Even though he was labeled as a bit of a foreigner there, Ernest was a bit reluctant to leave one of the major centers of physics – he'd only just arrived there himself for heaven's sake – but there were some major advantages to taking the position. First, McGill, with the help of William Macdonald, was establishing itself as an up-and-coming place. Second, the emphasis on experimental physics played to Rutherford's interest and greatest strength. Third, and perhaps most importantly, the salary was sufficiently large that he and his fiancée May Newton could now finally get married! He accepted the position at McGill in 1898 and in 1900 returned to Christchurch to marry and bring his bride back to Montreal *via* Hawaii and the Canadian Rockies.

Rutherford's years in Montreal (1898–1907) were most productive. He published 69 papers (an amazing 49 of which were single-authored) during his eight years here – a phenomenal and difficult-to-imagine rate of production. As noted above, his work in Cambridge had revealed that two types of radiation were "emanating from" (to use Rutherford's expression) radioactive elements like uranium, thorium, and the newly discovered radium. The two different types of radioactivity, which he had dubbed "alpha" and "beta" rays, were characterized by their different abilities to penetrate sheets of metal of various thicknesses. (The paper with these results established in Cambridge was actually published after he had arrived at McGill.) Beta radiation, which penetrated through various sheets relatively well, was composed of negatively charged particles whose straight-line trajectories were fairly easily bent by applying a relatively weak magnetic field. Therefore, he concluded, beta rays were best described as beta particles. Alpha rays, which did not penetrate through much of anything, could not at first be deviated by magnetic fields, so their nature was uncertain. Using a strong magnet borrowed from the McGill electrical engineering department, Rutherford finally bent the path of the alpha radiation (in the opposite direction from that of beta particles) and concluded that they were positively charged particles, that is, "α-particles". He characterized these alpha particles derived from a variety of radioactive sources. They were always the same no matter what their source.

So, fellow atomic travelers, it was becoming apparent to scientists such as Thomson in Cambridge and Rutherford in Montreal that certain types of atoms spontaneously broke apart and spit out little charged particles that could penetrate through various materials (including, not incidentally, human flesh). So, for example, if a uranium atom spit out an alpha particle, what was left over? It was pretty clear that the product was not uranium anymore. But what was it? How much energy did the emitted particles possess? And how long did it take for the radioactive element to decompose or "decay" in this manner? Working with a young 23-year-old chemist named Frederick Soddy, Rutherford (himself still only 29) defined the term "half-life" as the time it takes for half a given material to decay and measured that value for radium and other radioactive elements. Rutherford and Soddy published 9 important papers together in only 18 months, including "The cause and nature of radioactivity", published in two parts in 1902. One of the trickiest parts of their work was that when a radioactive element, take thorium for example this time, decayed by ejecting an alpha particle, the atomic fragment produced, dubbed "thorium-X", was also radioactive. *Via a succession of transformations involving a series of both alpha and beta decays, a variety of elements of varying half-lives were produced.* Ultimately, this and other transformations ended in various forms (later designated "isotopes") of lead. Incidentally, knowing the half-life of the most common form of uranium (4.47 billion years) enabled Rutherford to theorize on the age of the earth. The dizzying number of radioactive products with varying half-lives was most complicated. It must have been an amazing challenge to Rutherford and Soddy.

6.6 TRAVEL SITES IN MONTREAL RELATED TO ERNEST RUTHERFORD AND FREDERICK SODDY

See Chart 6.2 for the location of these sites.

6.6.1 The Rutherford Museum ※※※※, Ernest Rutherford Physics Building, McGill University

3600 rue University (45.507063°, −73.578410°). This building was built in 1977 and the museum is located in Room 110 near the big lecture hall. There is bust of Rutherford as you approach

The Brits, Led by the "Crocodile" and His Boys 151

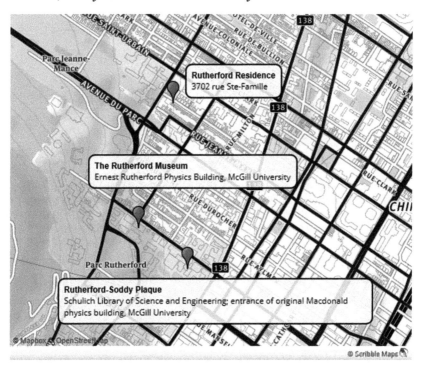

Chart 6.2 Travel sites in Montreal related to Ernest Rutherford and Frederick Soddy. These include the Rutherford Museum in the Rutherford Physics Building and Rutherford-Soddy Plaque in the Schulich Library of Science and Engineering (at the entrance of the original Macdonald physics building) and the Rutherford residence at 3702 rue Sainte-Famille.

the lecture hall next to the museum. The museum contains an impressive collection of the actual apparatus used by Rutherford and his co-workers, including Soddy. We scientific travelers are fortunate to be able to view such a collection because the practice of the day was to dismantle such equipment after it had served its experimental purpose. Fortunately, his colleagues recognized that Rutherford was doing pioneering work in radioactivity and his equipment should not be "cannibalized", as the saying went. The equipment was put aside and not disturbed for several decades after Rutherford left McGill. It is now displayed in five beautiful hand-made cabinets, one of which is shown in Figure 6.2. Adjoining the room housing the museum, is the Anna MacPherson Collection of physics equipment and also the desk

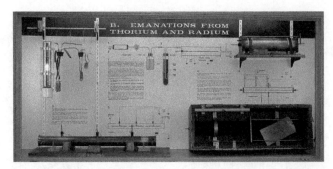

Figure 6.2 One of the display cases in the Rutherford Museum, Ernest Rutherford Physics Building, McGill University. This shows the emanations from thorium and radium. Used with permission from The Rutherford Museum, McGill University.

that Rutherford used there at the university. To make an appointment to see the museum and collection, contact curator@physics.mcgill.ca

6.6.2 Rutherford-Soddy Plaque, Schulich Library of Science and Engineering

809 rue Sherbrooke Ouest (45.505291°, −73.575368°). On November 30, 2009, as part of its Historic Sites initiative, the American Physical Society presented a pair of plaques (one English, one French) to McGill University to honor the achievements of Ernest Rutherford and Frederick Soddy. Placed at the entrance to the original Macdonald physics building and the new library, the English plaque reads: "At this location, Ernest Rutherford and Frederick Soddy, during 1901–1903, correctly explained radioactivity as emission of particles from the nucleus and established the laws of the spontaneous transmutation of the elements."

6.6.3 Rutherford's Residence

3702 rue Ste-Famille (45.512431°, −73.576442°). The former Rutherford residence is now a private dwelling that, unfortunately, has no plaque installed to mark it as Rutherford's home. However, it is still worth standing in front of the house and thinking about the good fortune and hard work it took to bring Ernest and May together here in Montreal. Eileen Mary Rutherford was born in March of 1901 and was six when the Rutherfords moved back to England. The desk used by Rutherford in this residence is housed by the window of the Rutherford Museum in the new physics building.

6.7 RUTHERFORD MOVES TO MANCHESTER, ENGLAND

Rutherford was restless in Montreal and longed to move back to England. In 1906, he was given the opportunity to move to Manchester, once the home of John Dalton and James Prescott Joule. (See Chapter 4 for a description of the places to visit in Manchester relative to John Dalton, who proposed and defended the first concrete atomic theory.) Rutherford accepted the position as the Langworthy Professor of Physics at the Victoria University of Manchester (now the University of Manchester) and moved there in the summer of 1907. As at McGill, Rutherford was to work in another new, state-of-the-art physics building called the Coupland Building, built in 1900. (It was renamed the Rutherford Building in 2006.) It was designed by Arthur Schuster who had recruited Rutherford to succeed him at Manchester. Here again, Rutherford was the beneficiary of another sizable increase in salary. The days when he had to leave May behind in New Zealand due to insufficient funds were but a distant memory by then. Somewhat ironically, on the basis of the work he did at McGill as the experimental professor of physics, the following year (1908) he was awarded the Nobel Prize in Chemistry for "researches on the disintegration of the elements and the chemistry of radioactive matter". Rutherford remarked with considerable (and most likely loud) irony that his instantaneous transformation from a physicist to a chemist was one of the most amazing he had ever witnessed. (See Chapter 19 for a discussion of the politics that resulted in Rutherford's transformation.)

Rutherford kept working on his alpha particles. Working with a young German postdoctoral fellow whose name most all of us now readily recognize, Hans Geiger, they proved that an alpha particle was a helium atom stripped of its two electrons, that is, it could be represented as He^{2+}. This was done by collecting the particles in a tube that was sealed and subjected to an electric charge. When the tube was analyzed spectroscopically, the spectrum was the same as that of helium, first observed on the sun. (See Chapter 11 for more on the development of spectroscopy in Heidelberg, Germany.) Note that a beam of alpha particles or helium cations, was an example of the "canal rays" that Thomson was working on at Cambridge. (See Chapter 5, p. 134.) Later, in 1914, Rutherford would suggest that the lightest of these canal rays would occur when a hydrogen atom was stripped of its one and only electron to produce a hydrogen cation, H^+, or what would soon be termed a proton.

For now, however, first in 1906 while still at McGill and then when he got settled in at Manchester, Rutherford, working with a new supply of radium that produced alpha particles, started shooting them at thin sheets of metal, some light, like aluminum but others heavy, like gold. The alpha particle–gold foil experiments are among the most famous ever conducted. At this time, Rutherford and Geiger's working model of the atom was Thomson's "plum pudding model" (Chapter 5, p. 133), composed of thousands of very light electrons embedded in a cloud of massless positive electricity. Since there was nothing at all massive in such a model, it was expected that the alpha particles would pass right through the thin metal foils to a detector set up behind the strip of metal. The detector was a thin layer of zinc sulfide that would light up (or "phosphoresce") when a charged particle hit it. For the most part, this is exactly what happened.

During this time, a 19-year-old, third-year undergraduate named Ernest Marsden approached Geiger about the possibility of starting a small research project. The hierarchy of the British system being what it was, it is doubtful that Marsden would have normally approached Rutherford directly with his request. Richard Reeves, however, in his book *A Force of Nature, the Frontier Genius of Ernest Rutherford*,[1] reports that Rutherford had encountered Marsden while in a hurry looking for a prism, mistakenly accused the young man of absconding with it and then, discovering that someone else had borrowed it instead, had a good discussion with Marsden about his "life and ambitions". Even so, it was most likely left to Geiger to approach Rutherford, a Nobel laureate after all, about the possibility of young Marsden getting started in a research project. Rutherford, perhaps after sitting back and puffing on his ever-present pipe, proposed that

Marsden try building a new section of the zinc sulfide detector, so he could look for alpha particles that bounced nearly straight back from the gold foils. This possibility was viewed as a distinctly unlikely result but, in the meantime, Rutherford might have surmised, Marsden would gain a good deal of experimental where-with-all that he could put to good use when he started doing research that would produce a significant result. Now when alpha particles hit the zinc sulfide detector the little flashes of light (called "scintillations") that were produced were very difficult to see. The room needed to be severely darkened and the observer, using a microscope, needed to let his eyes adjust for at least 20 minutes before starting to work. There were also reports that some observers would ingest a drug called "belladonna" that would dilate the pupils and make the scintillations easier to see. (The name "belladonna" is derived from the Italian for "beautiful lady" because this herb was used by women to dilate the pupils of their eyes to make them more seductive. Belladonna was also used to dilate the pupils of painters' models.) In any case, in 1909, Marsden dutifully modified the apparatus, set up the source of alpha particles, let them impinge on the gold foil and settled in with his microscope, perhaps with a little belladonna, who knows, to see if any alpha particles bounced directly back toward him. Quite amazingly, they did! When Marsden and Geiger reported this to Rutherford, the rest became history, "atomic history". Rutherford is said to have remarked that

> "It was quite the most incredible event that has ever happened to me. It was almost as incredible as if you fired a 15-inch shell at a piece of tissue paper and it came back and hit you."

So why did Rutherford deem this result to be so incredible? Rutherford and his colleagues immediately knew that in order for an alpha particle to be deflected directly back at the observer, there had to be something of relatively great mass and/or highly positively charged in the atom – that is, sufficiently massive or positive to at least match the mass and positive charge of the alpha particle. Since only a very few alpha particles were bounced back, this massive, positively charged structure had to be exceptionally small. In any case, Marsden's observations that the alpha particles bounced directly back at him were totally inconsistent with Thomson's model of the atom. Rutherford stewed on this result and its significance. He was sure that a new model of the interior of the atom was needed. By late 1910, the idea of what we now call the atomic "nucleus" was percolating in his mind. In 1911, he announced that he "knew what the atom looked like". In his nuclear model, the atom is mostly empty space. At its very center is the incredibly small, point-like nucleus containing all the positive charge and most of the mass of the atom. The tiny, extremely light, negatively charged electrons would orbit around the nucleus. When alpha particles come in, they most always miss the nucleus or just glance off it and are unaffected if they happen to encounter an electron. This model would explain why most of the alpha particles pass directly through the thin metal foils. On rare occasions (1 out of about 8000!), however, the mostly positive and relatively heavy alpha particle hits the massive, positively charged, teeny nucleus and bounces straight or nearly straight back. The situation is depicted in Figure 6.3.

It's difficult to visualize Rutherford's nuclear model of the atom. One useful analogy is that the atom is much like a pinhead

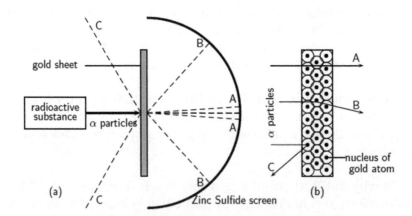

Figure 6.3 A schematic diagram of the Rutherford-Geiger-Marsden alpha-particle gold-foil experiment. Most of the α-particles (A) pass through unaffected. A few (B) glance off a nucleus and are diverted to some extent. A very few (C) collide directly with the nucleus of the gold atom and bounce nearly straight back at the observer. Reproduced from ref. 3 [https://cnx.org/contents/pYpuVhcS@1/Derived-copy-of-The-Atom-Grade-10-CAPS] under the terms of a CC BY 4.0 license [https://creativecommons.org/licenses/by/4.0/deed.en], Copyright © 1999–2019, Rice University.

in the middle of St. Paul's Cathedral in London ●●. Almost all of the mass is in the "pinhead" and the rest of the atom (represented by the massive dome of the cathedral) is virtually empty. The innermost electron is on the inner edge of the dome. Other similar expressions include "the fly in a cathedral" which is the title of Brian Cathcart's 2004 book.[2] Rutherford himself referred to a "gnat in the Albert Hall". Travelers with the atom might like the St. Paul's Cathedral analogy best. When you visit St. Paul's, take a moment to find the memorial to Robert Hooke (See Chapter 3, p. 41) in the crypt and then go back upstairs to sit down and think about Rutherford's nuclear model of the atom. Literally or figuratively, hold up a pinhead to represent the nucleus. Now look way up at the dome. Whoops, there goes an electron.

Ernest Rutherford was knighted in 1914. World War I interrupted much of the basic scientific research being carried out in Britain. Sir Ernest Rutherford secretly carried out experiments designing underwater listening devices to be used for anti-submarine warfare. As a result, he invented the so-called Anti-Submarine Detection Investigation Committee (ASDIC) devices which were later renamed "sonar" (standing for "sound navigation and ranging") by American researchers. The war split up the teams at Manchester. Hans Geiger returned to Germany to serve in the artillery. His name will always be associated with his invention of a counter for measuring levels of radiation. In 1915, Ernest Marsden, a fellow Kiwi himself, became, on Rutherford's recommendation, a professor of physics at the University of New Zealand in Wellington and was considered the country's *second* most distinguished scientist – after Rutherford himself of course.

In 1915, Rutherford and his colleagues, including Ernest Marsden before he left for New Zealand, had started to use alpha particles to bombard a sample of nitrogen gas contained in a cylinder. Although Marsden soon left, Rutherford continued to work on these experiments on his own during the war. Rather amazingly, hydrogen atoms were detected in the resulting product. It turned out that the alpha particles were knocking hydrogen nuclei out of the nitrogen atoms and that Rutherford had actually carried out a nuclear transformation: helium (as the doubly charged ion, He^{2+} or an alpha particle) hits a nitrogen atom, resulting in hydrogen (as the hydrogen ion, H^+) and oxygen. Some said he had "split the atom" but he had really just knocked a chip off it. During this time,

Rutherford started referring to the hydrogen ions produced in this transformation as "protons". If an ordinary hydrogen atom is composed of just one proton and one electron, then the hydrogen nucleus is just a proton. When Rutherford took up his position as Director and Fourth Professor of Physics of the Cavendish Laboratory at the University of Cambridge in 1919, he took the small brass tube used to investigate this nuclear transformation with him. So, it turns out, it is actually on display at the Museum at the Cavendish Laboratory in Cambridge.

6.8 TRAVEL SITES IN MANCHESTER RELATED TO ERNEST RUTHERFORD

See Chart 6.3 for the location of these sites.

6.8.1 Rutherford Building, University of Manchester

(53.465990°, −2.234723°). This is located just off Oxford Road, sandwiched between Bridgeford and Coupland Streets, and was originally known as the Coupland Building. This is where Rutherford and his colleagues worked during his time at the university. (It was renamed the Rutherford Building in 2006.) A Royal Society of Chemistry blue plaque on the side of the building says that Rutherford led this laboratory between 1907 and 1919 and "Herein discovered the nuclear atom, split the atom, and initiated the field of nuclear physics". Marsden and Geiger did the alpha-particle, gold-foil experiment here in 1909. The Danish grand theorist Niels Bohr (See Chapter 12) worked here for a time in the 1910s, combining Rutherford's model with Max Planck's quantum theory (See Chapter 15) to propose the orbital model of atomic structure. So too did James Chadwick, who discovered the neutron. It was also here that Henry Moseley started his X-ray diffraction work that established the relationship between nuclear charge and atomic number. As noted above, one of Rutherford's final achievements at Manchester was to demonstrate the artificial disintegration of nitrogen by alpha-particle bombardment, an achievement often referred to as the "first splitting of the atom." There is no formal access to the interior of this building, but if you ask a porter at the entrance of the building

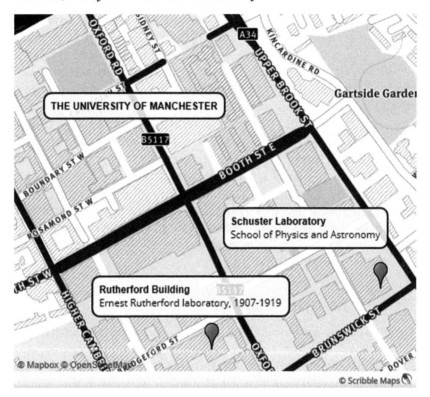

Chart 6.3 Travel sites at the University of Manchester, Manchester, England, related to Ernest Rutherford. These include the Rutherford Building on Coupland Street and the Schuster Laboratory (or the Schuster Building) on Brunswick Street. The reader might also consult Chart 4.2, p. 93, which shows the extensive number of Dalton sites in Manchester.

he will most likely be willing to take you to the room where Rutherford "split the atom". There is small plaque there to denote the occasion.

6.8.2 Schuster Laboratory (or the Schuster Building)

Brunswick Street, University of Manchester (53.467210°, −2.230593°) houses the present-day School of Physics and Astronomy at the university. The building houses four large lecture rooms around the foyer on the ground floor, named after physicists who taught or carried out research in the department including Rutherford, Bragg, Blackett, and Moseley. Lawrence Bragg shared a Nobel Prize with his father for inventing X-ray crystallography. Patrick Blackett won a Nobel Prize for his work on cosmic rays. Moseley, as we know, discovered the atomic number which also denotes the total positive charge (and thus the number of protons) in the nucleus of an atom. Rutherford Hall is the largest of the lecture theatres. There is a meeting room on the roof of one of the wings, called the Niels Bohr Common Room. The building also houses a small cafe on the ground floor, named the "Error Bar". In the hallway outside the lecture halls the author found a picture of the "Staff and Research Students of Manchester University Physics Department 1910". It shows Rutherford in the front row, Geiger in the second row and, immediately above him, Marsden in the third row. Look for it or ask if it is still to be found.

6.9 RUTHERFORD MOVES BACK TO "THE CAVENDISH"

In 1919, after an absence of 21 years, Rutherford became the Director of the Cavendish Laboratory at the University of Cambridge, succeeding his mentor, J. J. Thomson. Thomson was most pleased with this development, writing to Rutherford to say that "Nothing would give me more pleasure as to have my successor be my most distinguished pupil". "The Provincial" had returned to the Cavendish. He brought with him James Chadwick. Chadwick, a Manchester native by birth, had been a promising researcher at the university. In 1913, Rutherford had arranged for him to have a position with Geiger, who had returned to Berlin. When the war broke out, Chadwick was stranded there as a designated "enemy alien", where his health declined due to malnourishment and depression. After the war, with Chadwick nearly penniless, Rutherford welcomed him back to the Cavendish, where he remained for 16 years. He received his PhD in 1921 and was made assistant director of the lab.

A word about what it was like to work at the Cavendish in these days. The Cavendish had suffered during the war, losing both researchers and funding. However, Rutherford attracted both and the laboratory was soon flourishing once again. So, what was a typical workday like? First, although the starting time was generally agreed to be a gentlemanly 10AM, the closing time was a rigidly enforced 6 PM. It did not matter the state of your experimental work, you were expected to be headed home by 6 PM. Time to stop lab work and go home and think. At 4 PM there was British "tea", provided by the director's wife. Mary, or Lady Rutherford, agreed to carry on this tradition started by Mrs Thomson. This was a short break, where the scientists gathered for tea and sticky buns, a little gossip, and lots of shop talk. Harking all the way back to Clerk Maxwell's time, the Cavendish Physical Society met every second Wednesday in the main lecture hall (still standing). Founded by the "visiting" Russian physicist (who visited for 13 years or so, from 1921 to 1934) Peter Kapitza, the Kapitza Club met regularly, often in his rooms at Trinity College. This was an informal gathering, where the normal hierarchical relationships between professors, post docs, graduate students down to undergraduates were largely broken down. Even the newest and youngest workers were expected to speak up. Presentations were limited to "chalk talks", where only a portable blackboard and white chalk were allowed. It was expected that you would be interrupted early and often.

At the annual Christmas-time Cavendish dinners, started during Thomson's tenure, the lively tradition of light and irreverent verse and good fun continued. Here's another example:

> ... that handsome, hearty British lord
> We knew as Ernest Rutherford.
> New Zealand farmer's son by birth,
> He never lost the touch of earth;
> His booming voice and jolly roar
> Could penetrate the thickest door,
> But if to anger he inclined
> You should have heard him speak his mind
> In living language of the land
> That anyone could understand!
>
> From ref. 3, pp. 221–222

Rutherford and Thomson in his day sometimes wound up, as Brian Cathcart notes, "on their chairs and singing at the top of their voices". This must have been quite a sight.

The early 1920s was a troubling time for atomic scientists, particularly for Rutherford and his colleague Chadwick at the Cavendish. By that time, atoms seemed to be composed of two fundamental particles: the proton with a +1 charge and the electron with a −1 charge. For an atom to be electrically neutral it must contain an equal number of protons and electrons. Accordingly, a hydrogen atom would logically be made up of one proton and one electron and therefore was electrically neutral. It was assigned a mass of 1 and that seemed to be due almost entirely to the weight of the proton. (An electron had been calculated to weigh about one 2000th the weight of a proton.) But helium, since it had an assigned weight of 4, would have to contain four protons and four electrons. However, given the existence of alpha particles (known by then to be helium nuclei, He^{2+}) there seemed to be only two orbiting electrons that could be removed to produce an alpha particle. So, this accounts for two protons and two electrons. What about the other two protons and two electrons? Rutherford was of the opinion that a proton and an electron could be fused together to produce another fundamental but neutral particle, which Rutherford, logically enough, called a "neutron". Under this theory, there were two types of electrons; orbiting electrons and "hidden electrons", the latter being fused together with protons to produce neutrons. Neutrons and protons must co-exist in the nucleus. Alarmingly, the number of hidden electrons could become overbearing. In ordinary uranium, for example, it was posited that there would be 238 protons, 92 orbiting electrons and 146 "hidden electrons" that were fused with 146 protons, thereby comprising 146 neutrons in the nucleus. Good Lord, what a mess.

In one of the saddest incidents of these troubling times, in 1930, the Rutherfords' daughter Eileen, 29 years old and the mother of three, gave birth to a fourth child, but one week later died of a lethal blood clot. Rutherford was devastated and some said he was never quite the same, but his face "always lit up when he spoke of" his four grandchildren. Two weeks after Eileen died, he was created First Baron Rutherford of Nelson, New Zealand, and Cambridge. Lord Rutherford of Nelson designed his own coat of arms that included a kiwi bird, a Maori warrior, and the motto *Primordia Quaerere Rerum*, meaning "to seek the first principles of things".

To compound the difficulties, it was getting more problematic to study the internal structure of the atoms using the alpha particles derived from a radioactive source like radium. These particles were not numerous or powerful enough to yield further information. To boost these energies, some extraordinary efforts were taken by two workers at the Cavendish: John Cockcroft and Ernest Walton. Cockcroft was an electrical engineer who went on to receive his PhD from Cambridge in 1928. With his background, he was well equipped to figure out how to accelerate particles in an electrical field. Walton, who, like Rutherford was another "1851 Man" – Walton was from Ireland – had initiated the idea to carry out this acceleration. They saw that, instead of always accelerating alpha particles into nuclei, they might have better success with protons. They used their high-voltage "Cockcroft–Walton Machine" to do just that. Cockcroft and Walton received the 1951 Nobel Prize in Physics. Cockcroft was knighted in 1948.

Two experiments conducted by Cockcroft and Walton are of particular note. First, in 1932, they accelerated a proton into a lithium nucleus and detected two alpha particles as the result. Using Aston's mass spectrograph (see below) they determined the masses of the reactants (the proton and the lithium nucleus) and products (the two alpha particles). They noted that a little mass had disappeared during this process. When they measured the energy of the alpha particles they found that the mass had been converted to energy as predicted by Einstein's $E = mc^2$ equation. Although Einstein formulated his famous equation in 1905, this was the first time, according to Einstein himself, that it was experimentally verified.

Second, also in 1932, the Cockcroft–Walton machine was used to accelerate alpha particles into beryllium nuclei. This produced carbon atoms and an intense radiation that had more energy than the bombarding alpha particles. At first Frédéric and Irène Joliot-Curie in Paris, using alpha particles derived from the decay of Po-210 (See Chapter 14), thought this radiation was gamma rays but when Chadwick repeated their experiments he disagreed and finally figured out that this was a beam of neutrons. It did

164   *Chapter 6*

not take long to disprove the idea that the neutron was a proton–electron pair. It was, in fact, a third atomic constituent, joining the proton and electron. Chadwick received the 1935 Nobel Prize in Physics and was knighted in 1945.

6.10 TRAVEL SITES IN ENGLAND RELATED TO ERNEST RUTHERFORD AND "HIS BOYS"

See Chart 6.4 for the location of these sites.

6.10.1 Cambridge

6.10.1.1 Museum at the Cavendish Laboratory. (52.209240°, 0.091759°) The museum, as described previously, is housed in the New Cavendish Laboratory, fittingly located on J. J. Thomson Avenue, off Madingley Road in West Cambridge. In Chapter 5, it was noted that this museum houses material related to James Clerk Maxwell and J. J. Thomson. Here we note several

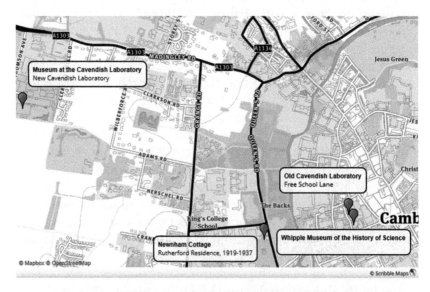

Chart 6.4 Travel sites in Cambridge, England related to Ernest Rutherford and Francis Aston. These would include the Museum at the Cavendish Laboratory in the New Cavendish Laboratory, the Crocodile Carving just inside the entrance and to the right at the Old Cavendish Laboratory, the Whipple Museum of the History of Science, and the Rutherford residence (1919–1937) at Newnham Cottage on Queens Road. A larger scale chart of the Free School Lane area is given in Chart 5.4, p. 126.

The Brits, Led by the "Crocodile" and His Boys 165

other pieces of equipment used (1) to prove spectroscopically that α-particles were helium nuclei, (2) to show how a beam of α-particles could be deflected by a strong magnetic field, and (3) to measure the charge on the α-particle. Also on display is (1) the small brass tube with some glass valves and brass extensions that Rutherford used to carry out the first nuclear transformation, (2) Chadwick's neutron chamber, (3) Francis Aston's mass spectrograph and (4) the rectifier tube from the 1930 Cockcroft–Walton Machine. There are also several photographs of the 1930, 1932, and 1936 machines. A brief word about the role of these rectifier tubes. A rectifier converts a.c. (alternating current) to d.c. (direct current). Direct current was required in order to propel protons or alpha particles toward target nuclei. No rectifiers of the right capacity were available, so Cockcroft and Walton had to design and make their own. At the high voltages they were using, these experiments were very dangerous. The rectifiers were tubes of glass that had to be just the right size and shape and able to withstand high vacuums.

6.10.1.2 The Old Cavendish Laboratory. (52.203369°, 0.118818°) Free School Lane, Cambridge. This is part of the "New Museums Site" of Cambridge University. Above the door of the entrance is, on the left, the badge of the house of Devonshire and, on the right, the seal of the University of Cambridge. Recall that the laboratory was named to commemorate the British chemist and physicist Henry Cavendish (See Chapter 3, pp. 52–53) and his relative William Cavendish, the 7th Duke of Devonshire, who was the Chancellor of the University and gave the money for the construction of the laboratory. There is a slate plaque on Free School Lane that records that this was the site of Cavendish Laboratory for 100 years, from 1874 to 1974. When the results of Cockcroft's and Walton's experiments were announced in May of 1932, the laboratory on Free School Lane was mobbed by reporters and photographers from all over the world. You might lean against the wall by this gray plaque and think about what that might have been like.

6.10.1.3 The Crocodile Carving. (52.203380°, 0.119112°). Also in the "New Museums Site" of Cambridge University. On the side of the round front of the Mond Laboratory on the site of the original Cavendish Laboratory in Cambridge, there is an engraving in Rutherford's memory in the form of a crocodile, this being the nickname given to him by its commissioner, his colleague Peter Kapitza (see Figure 6.4). In Russia, crocodiles are a symbol of raw power. Kapitza had worked for fourteen years at the Cavendish with Rutherford. In his letters home, Kapitza wrote that the crocodile "is ... regarded with awe and admiration because it has a stiff neck and cannot turn back. It just goes straight forward with gaping jaws – like science, like Rutherford". Sadly, the highly respected and charismatic Russian had only a couple of years to enjoy the carving. In 1934, at the end of a holiday back in the Soviet Union, Kapitza was barred from leaving the country and never set foot in Cambridge again. (Joseph Stalin had personally ordered that he was never to leave the Soviet Union.) Before returning to the Soviet Union, Kapitza had worked at developing ultrahigh magnetic fields but since the equipment for this research had to stay in Britain, he turned to low temperature work. As part of this effort, he discovered the "superfluidity" of liquid helium. In 1978, Kapitza won the Nobel Prize in Physics "for his basic inventions and discoveries in the area of low-temperature physics".

6.10.1.4 Whipple Museum of the History of Science. (52.202778°, 0.119199°). As noted in Chapter 5, this museum, just down Free School Lane from the old Cavendish and within the New Museums Site, houses a number of important

Figure 6.4 The Crocodile carving honoring Ernest Rutherford. On the wall of the Mond Building in the Old Cavendish Laboratory, Cambridge University. Photo by Glen Rodgers.

collections. Most pertinent to our interests is the Cavendish Laboratory Collection, which includes many of the instruments used in the laboratory including optical and electrical instruments, X-ray tubes, vacuum tubes, and thermometers. The door of the Whipple Museum has a sign over it that says, "Laboratory of Physical Chemistry".

6.10.1.5 Newnham Cottage. (52.202164°, 0.111865°) Queen's Road. Ernest Rutherford lived here from 1919 to his death in 1937. It is a short walk from the Cavendish Laboratory on the other side of the River Cam. It was the Rutherford's practice on Sunday to invite his close associates and students for afternoon tea at the cottage. No alcohol was served and Mary always brought the sessions to a definite close. Rutherford's death in 1937 was caused by falling from a tree he was pruning in the garden near the cottage. Emergency surgery to repair a "slight umbilical hernia" was followed by a massive infection, which caused his death.

6.10.1.6 Eagle Public House. (52.203998°, 0.118198°). This pub was founded in 1667 and was a favorite place for the Cavendish Laboratory scientists to hang out. It also contains the picturesque "Airman's Bar" frequented by RAF and American pilots during the Second World War. In February 1953, Francis Crick and James Watson came here to announce that they had determined the structure of DNA. A blue plaque outside the pub notes this event. You could duck down here after visiting the Old Cavendish and have a pint of DNA Ale. (From the Old Cavendish, walk north on Free School Lane to where it intersects with Bene't Street. The Eagle Pub is off to your left at 8 Bene't St.)

6.10.2 London

6.10.2.1 Westminster Abbey. See Chapter 7 for a guide to the graves and memorials of scientists in Westminster Cathedral. After his untimely death in 1937, Rutherford was honored by the fact that his ashes were buried near Sir Isaac Newton's tomb in Westminster Abbey. The diamond marking the interment is in the nave and is often roped off in an altar area right next to the diamond honoring J. J. Thomson. Sometimes, if one is admitted to this area (see a verger, that is, an attendant, for permission) the elaborate cloths covering the altar table have to be lifted to see the two diamonds honoring Rutherford and J. J. Thomson.

6.11 SOME OTHER PLACES TO VISIT RELATED TO RUTHERFORD'S "BOYS"

6.11.1 Frederick Soddy (1877–1956)

6.11.1.1 Eastbourne, England. Frederick Soddy was born and educated in Eastbourne in southern England. He was born at 6 Bolton Road (50.767768°, 0.285733°). There has been a small bronze plaque at this location but, unfortunately, it was the victim of metal thieves in 2013. The plaque noted that "Here was born Frederick Soddy Nobel Laureate in Chemistry, foremost radiochemist of his time. He discovered the transmutation of matter and the concept of isotopes. His life was dedicated to human welfare through the benefits of science." It has been difficult to prevent the theft of this plaque. If you are in the area, take a look and see if one is currently in place.

Soddy was educated at Eastbourne College where he was a "home boarder" or day boy. In 1998, a blue plaque ❋ was placed on the roadside wall of what is now the Design & Technology (D&T) Building, Eastbourne College on Blackwater Road (50.764063°, 0.279885°). This notes that he was "Educated at Eastbourne College, Nobel Laureate 1921, for his fundamental contributions to the understanding of radioactivity."

6.11.1.2 Glasgow, Scotland. At the University of Glasgow from 1904 to 1914, Soddy studied how radioactive changes moved materials to the left and right in the periodic table (what he called his "Displacement Law"). (For example, when U-235 emits an alpha particle, the uranium atom (U) becomes a thorium (Th) atom which is two spaces to the left of uranium in the periodic table. When that thorium atom subsequently emits a beta particle, it becomes a protactinium (Pa) atom which is one space to the right of thorium in the table.) He also formulated the idea of and coined the term "isotopes" that were chemically identical elements with different atomic weights. (Today, we say that isotopes have the same atomic number but a different mass number. The atomic number is the number of protons while the mass number is the total of the number of protons and neutrons, collectively known as "nucleons".) A plaque, located at the "Soddy Landing" in the chemistry department (55.872049°, −4.293205°) at the University of Glasgow, commemorates Soddy's work. There is also a Soddy plaque at the George Service House at 11 University Gardens (55.872980°, −4.291058°) which, rather amazingly, reads "At a dinner party held in this house in 1913, Frederick Soddy (1877–1956) introduced the concept of "ISOTOPES". He was awarded the Nobel Prize in 1921 for his work on radioactivity". Not many plaques start with "At a dinner party ...!"

6.11.2 Henry Gwyn Jeffreys Moseley (1887–1915)

6.11.2.1 Oxford, England. Rutherford's Manchester days produced many startling and significant results. Another we should mention is the work of Henry Gwyn Jeffreys (Harry) Moseley. Moseley received his degree in physics from Trinity College, Oxford in 1910 and wrote to Rutherford to ask to work with him in Manchester. He joined the laboratory in 1910 and quickly made his mark. Using X-rays that are emitted from elements, he established the concept of the atomic number, which is the positive charge in the nucleus of an atom and also the number of protons in that nucleus. In his own words, he said that "there is a fundamental quality of the atom which increases by regular steps as we pass from one element to the next. This quality can only be the charge on the central positive nucleus." As part of this work he noted that there were some elements missing from those known at the time. Specifically, he did not see evidence for metals of atomic number 43, 61, and 75. These elements are now known to be Technetium (Tc), Promethium (Pm), and Rhenium (Re), respectively. Toward the end of 1913, Moseley decided to move back to Oxford. Reports vary about the reasons for this move. Some say that he did not have a high regard for Rutherford, noting that he was "the son of a New Zealand flax farmer who possessed neither languages or culture".

There were also reports that he felt that Rutherford had no more to teach him. Others say that he wanted to be closer to his widowed mother, while continuing his experiments. In any case he declined a fellowship that Rutherford offered him and, in November 1913, he moved back to Oxford where he was provided with laboratory facilities to continue his X-ray studies but no financial support. He took up his work at the Electrical Laboratory, now the Townsend Building of the Clarendon Laboratory. When World War I broke out, Moseley immediately joined the Royal Engineers of the British Army and was killed by a sniper's bullet in June of 1915 in Gallipoli, Turkey. Most observers agree that Moseley would have been a prime candidate to receive the 1916 Nobel Prize in Physics for his work had he survived.

6.11.3 Travel Sites in Oxford Related to Harry Moseley

See Chart 6.5 for the location of these sites.

6.11.3.1 Trinity College, Oxford University. (51.754579°, −1.255835°). Harry Moseley was in the class of 1910. Starting at the Porter's Lodge (where one pays a small entrance fee), walk toward the Trinity College Chapel, where he matriculated in 1906. Proceed to "the Hall" where Moseley, who held the only science scholarship, dined with fellow scholars at the table closest to the large, open fireplace. From here, proceed to the Garden Quadrangle where Moseley lived for three years. Finally proceed to the War Memorial Library where Moseley is listed on the "War Memorial Board" as one of 154 men who died in World War I. The names are arranged by date of admission with Harry's name (1906) in the middle column near the top.

6.11.3.2 Moseley Family Home Plaque, 48 Woodstock Road, Oxford. (51.761288°, −1.261725°) A blue plaque on the backside of the house honors Moseley, "who established the atomic numbers of the chemical elements. Killed in action at Gallipoli."

6.11.3.3 Clarendon Laboratory Plaque, Sherrington Road, Oxford. (51.75937°, −1.25653°). This National Chemical Landmark of the Royal Society of Chemistry plaque notes that Moseley completed his "pioneering studies on the frequencies of X-rays emitted from the elements. His work established the concept of atomic number and helped reveal the structure of the atom".

6.11.3.4 History of Science Museum in Old Ashmolean, Oxford. (51.754427°, −1.254748°). The apparatus used by Moseley is on display here. Also on display is the so-called

The Brits, Led by the "Crocodile" and His Boys 171

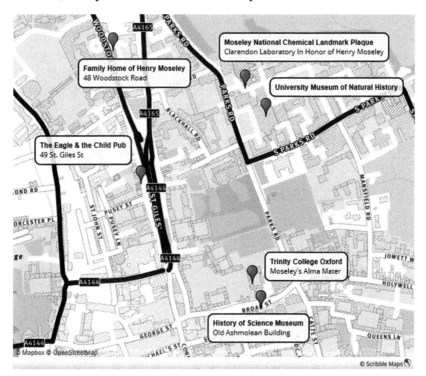

Chart 6.5 Travel sites in Oxford, England, relative to Henry Moseley. This chart includes the two plaques at Woodstock Road and the Clarendon Laboratory, the University Museum of Natural History, Trinity College, the History of Science Museum in the Old Ashmolean building and "The Eagle and the Child" pub.

"Einstein's Blackboard" that Albert Einstein used on 16 May 1931 in his lectures while visiting the University of Oxford. His lecture explained how his theory of relativity could help to explain the expansion of the universe as indicated by the red-shift of spectral data from faraway galaxies. The blackboard was rescued at the end of the lecture.

While you are in this part of Oxford, you might want to find the University Museum of Natural History on Parks Road (51.758700°, −1.255456°). This is briefly described in Chapter 3, p. 37. Here, you can walk around the perimeter of the courtyard to view statues of some of the greatest scientists of all time, including Isaac Newton, Hans Ørsted, Humphry Davy, and Joseph Priestley. Had Harry Moseley not been cut down at Gallipolli at the age of 28, his statue would almost certainly have been included here. To rest and contemplate all this, you might now head to the nearby Eagle and Child Public House, 49 St. Giles. It was here that J. R. R. Tolkien, C. S. Lewis and other writers dubbed the "Inklings" read their works to each other and shared a pint. They nicknamed the pub "the bird and baby".

6.11.4 Ernest Walton (1903–1995)

6.11.4.1 Travel Sites in Ireland Related to Ernest Walton

See Chart 6.6 for the location of these sites.

6.11.4.1.1 Walton Causeway Park. (52.092728°, −7.618778°) in Abbeyside, Dungarvan, County Waterford, was dedicated in his honor with Walton himself attending the ceremony in 1989. Dungarvan, a coastal town located at the foot of the Comeragh Mountains, is situated at the mouth of the Colligan River, which divides the town into

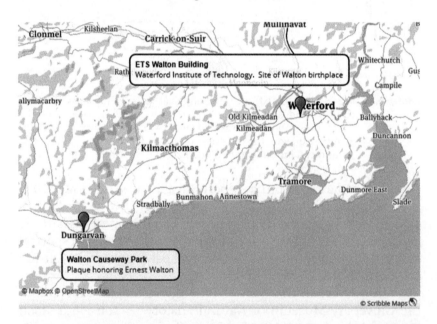

Chart 6.6 Travel sites in Dungarvan and Waterford, Ireland, relative to Ernest Walton. This chart includes the Walton Causeway Park in Abbeyside, Dungarvan and the ETS Walton Building on the campus of Waterford Institute of Technology on Cork Road in Waterford. For other sites in this area of Ireland, see Chart 2.1, p.14.

two parishes – that of Dungarvan to the west and Abbeyside to the east. A plaque located in this park says that Walton received the 1951 Nobel Prize in Physics jointly with John Cockcroft for their work in 1932 on splitting the atomic nucleus.

6.11.4.1.2 ETS Walton Building. Waterford Institute of Technology, Cork Road, Waterford. (52.245710°, −7.137317°) A plaque has been placed at the site of his birthplace. The Walton Building is located on the Institute's main Cork Road campus close to the award-winning Institute library, the Luke Wadding Library.

6.11.4.1.3 Earnest Walton's Grave. (53.278169°, −6.163671°) is located in Deansgrange Cemetery, south County Dublin at the site designated 90 A–St. Nessan. Walton is the only Irishman to have won a Nobel Prize in Physics.

6.12 SUMMARY

During the late nineteenth and early twentieth centuries, Ernest Rutherford and his many distinguished associates (whom he often characterized as "his boys") started to describe radioactivity and then proceeded to take the atom apart piece by piece. Stimulated by Röntgen's 1895 discovery of X-rays and Becquerel's 1896 discovery of radioactivity, Rutherford defined and named alpha and beta particles. Working with Frederick Soddy in Montreal in 1902, he described complicated sets of nuclear transformations of radioactive elements and defined and measured the half-life of these elements. Soddy went on, from 1904 to 1914, to define and study isotopes and showed how radioactive changes moved materials to the left and right in the periodic table. Now in Manchester, Rutherford and Hans Geiger proved that an alpha particle was a helium nucleus, He^{2+}. Rutherford, Geiger, and Ernest Marsden in 1909 carried out the alpha-particle gold-foil experiments that were inconsistent with then-prevalent "plum-pudding model" of the atom devised by J. J. Thomson. Rutherford proposed his nuclear model of the atom in 1911 and this "fly in the cathedral" model became the accepted way to visualize the internal structure of the atom. Just before World War I, Henry Moseley used X-rays to characterize the atomic number of an atom.

Starting in 1915, Rutherford carried out an artificial nuclear transformation in which an alpha particle knocked a proton (hydrogen nucleus) off a nitrogen atom, transforming it into an oxygen atom. Back at the Cavendish, Rutherford proposed the idea that protons and electrons might co-exist as a coupled pair in the nucleus. In 1928, John Cockcroft and Ernest Walton constructed their Cockcroft–Walton Machine, which could accelerate alpha particles and protons into nuclei, producing a variety of nuclear reactions. In 1932, they fired protons at lithium to produce alpha particles. The mass data from this experiment was analyzed using Francis Aston's mass spectrometer and this provided the first experimental verification of Einstein's $E = mc^2$ equation, first proposed in 1905. In another experiment, alpha particles were fired at beryllium, producing an intense radiation that James Chadwick correctly identified as a beam of neutrons.

In this chapter we have identified many sites to visit. In New Zealand, there are the birthplace and early education sites of Ernest Rutherford in Brightwater, Havelock, Nelson, and Christchurch on the South Island. At McGill University in Montreal there is the amazing "Rutherford Museum" that shows off an impressive collection of the original equipment used by Rutherford and his associate Frederick Soddy to characterize nuclear transformations and measure half-lives. A plaque honoring the work of Soddy and Rutherford is proudly displayed at the entrance of the original Macdonald physics building. The nearby Rutherford residence is also identified. In Manchester, England, there is the Rutherford Building where the nucleus was discovered and the first artificial nuclear transformation was carried out. The new Schuster Laboratory is also described as worthy of a visit. Back in Cambridge, the Museum at the Cavendish Laboratory is one of the best "travels with the atom" sites; it is not to be missed. The Old Cavendish Laboratory site with its entrance portal, plaques noting what occurred here, and most of all, the "Crocodile" carving on the exterior wall of the Mond Building are worth exploring. The Newnham Cottage, Rutherford's residence on Queen's Road, is noted. As we will detail in the next chapter, Westminster Abbey is where Rutherford's ashes were interred. The Abbey is definitely worth visiting for all the graves and memorials to a variety of scientists. Frederick Soddy sites in Eastbourne, England and Glasgow, Scotland are noted as are Henry Moseley sites in Oxford and Ernest Walton sites in Dungarvan, Waterford, and Dublin, Ireland.

ADDITIONAL READING

- J. Campbell, *Rutherford, Scientist Supreme*, AAS Publications, Christchurch, New Zealand, 1999.
- B. Pippard, The Whipple Museum and Cavendish Laboratory, in *The Physical Tourist: A Science Guide for the Traveler*, ed. J. S. Rigden and R. H. Stuewer, Birkhäuser, Basel, 2009.
- B. Pippard, Thomson, Rutherford and atomic physics at the Cavendish, in *Cambridge Scientific Minds*, ed. P. Harman and S. Mitton, Cambridge University Press, Cambridge, 2002.
- J. G. Crowther, *The Cavendish Laboratory 1874–1974*, Science History Publications, New York, 1974.
- G. Gamow, *Biography of Physics*, Harper & Row Publishers, New York, 1961.

REFERENCES

1. R. Reeves, *A Force of Nature, the Frontier Genius of Ernest Rutherford*, Atlas Books, W. W. Norton & Company, New York, 2008.
2. B. Cathcart, *The Fly in the Cathedral, How a Group of Cambridge Scientists Won the International Race to Split the Atom*, Farrar, Straus and Giroux, New York, 2004.
3. G. Gamow, *Biography of Physics*, Harper & Row Publishers, New York, 1961.

CHAPTER 7

Scientists at the Heart of Westminster Abbey

7.1　A QUICK LOOK AT PLACES TO VISIT "TRAVELING WITH THE ATOM" IN WESTMINSTER ABBEY AND GREATER LONDON

Atomic scientists buried and/or honored in Westminster Abbey (in the order encountered on the prescribed walking tour) include the memorials to Sir Humphry Davy and John Wm Strutt, 3rd Baron of Rayleigh in St. Andrew's Chapel, the memorial stone to honor Robert Hooke in the northeast corner of the crossing, or "the lantern"; in the nave: the magnificent Newton monument in the Gothic Revival choir screen and the gravesite of Sir Isaac Newton; the side-by-side memorial stones in honor of Michael Faraday and James Clerk Maxwell; the burial sites of Sir Joseph John Thomson and Ernest, 1st Baron Rutherford under the altar table; the memorial to Paul Dirac; in the North Choir Aisle the wall memorial for Sir William Ramsay and a memorial bust of James Watt. Other scientists mentioned here that are memorialized (m) and/or buried (b) in the Abbey include: Thomas Young (m), Lord Kelvin (m, b), George Green (m), Sir William Herschel (m), Sir John Herschel (b), Charles Darwin (m, b), Edmond Halley (m), James Prescott Joule (m),

---

Traveling with the Atom: A Scientific Guide to Europe and Beyond
By Glen E. Rodgers
© Glen E. Rodgers 2020
Published by the Royal Society of Chemistry, www.rsc.org

and Stephen Hawking (b). In addition, special note is made of the Hunterian Museum at Glasgow University in Scotland (Kelvin), the Herschel Museum of Astronomy in Bath, the William Ramsay collection at University College London, the Royal Observatory in Greenwich, and Darwin's House in Downe. The latter two places are among the best scientific-historical sites in the world.

> **Atomic Scientists in Westminster Abbey**
>
> - Visiting Westminster Abbey gives us the opportunity to see memorials and actual gravesites of many atomic scientists, most of whom have been described in earlier chapters covering the United Kingdom.
> - From the entrance at the north door leading us to the North Transept, we follow the normal touring route around the cruciform-shaped Abbey. We find the memorials to Humphry Davy, Lord Rayleigh, and Thomas Young in St. Andrew's Chapel, and the memorial to Robert Hooke in the "crossing".
> - In the nave, we stand on the memorial to William Herschel and look into the roped-off altar area to view the grave and magnificent monument for Isaac Newton. Then we turn to the adjacent memorials to Michael Faraday and James Clerk Maxwell. We can have a verger take us to see where the ashes of J. J. Thomson and Ernest Rutherford are interred under the altar table.
> - We also discuss the other scientists honored in this restricted area including Lord Kelvin, George Green, Paul Dirac, and Stephen Hawking.
> - Outside the altar area we see the gravesites of John Herschel and Charles Darwin, and then go into the North Choir Aisle to see memorials to Darwin, James Prescott Joule, Sir William Ramsay, and James Watt.
> - As we move from one memorial or gravesite to another, we briefly recall each scientist's contribution to the development of the atomic concept and the varied traveling-with-the-atom sites available for us to visit all over the United Kingdom. These sites have been described in detail in previous chapters.

So far, our traveling-with-the-atom adventure has involved visiting castles, laboratories, museums, libraries, schools, and universities. We have been to birthplaces, residences, and graveyards related to scientists involved in developing the atomic concept. We've found plaques, streets-named-for-scientists, pubs, landmarks, and statues. Sometimes we've been in remote villages and small towns, but other times in some of the most glamorous cities of Europe. Sometimes our travels have taken us to obscure places far off the traditional paths that tourists normally frequent. We have even found ourselves in chapels, churches, meeting houses, and even cathedrals in pursuit of these men and women. Some of these houses of worship have been small, like the Mill Hill Chapel in Leeds (Priestley), the Quaker Meeting House in Pardshaw Hall (Dalton), the Corsock and Kirkpatrick-Durham Church in Corsock, Scotland (Clerk Maxwell), and St. Mary's Collegiate Church in Youghal, Ireland (Boyle). Others have been large and elaborate, like St. Patrick's Cathedral in Dublin (Boyle), the Corpus Christi Roman Catholic Church in London (Boyle and Godfrey), and St. Paul's Cathedral in London (Hooke). So far, we have made only passing references to visiting one of the most prominent houses of worship in all of Europe, the magnificent Westminster Abbey in London. As scientific-historical travelers, we might pinch ourselves as we queue up to get into this great structure in the heart of London. If we are Europeans, travelers like us occasionally see scientists honored in a variety of places, but if we are Americans, we are not used to seeing scientists honored *anyplace*, never mind a great Christian Church like the Abbey.

Several million people per year visit this gothic abbey church located on the northern bank of the River Thames. Located in the City of Westminster near the Houses of Parliament and Big Ben, the Collegiate Church of St. Peter at Westminster, known more familiarly as Westminster Abbey, attracts tourists who want to see where English and British monarchs have been married, crowned, and buried for centuries. (Construction of the present church was begun in 1245 by Henry III, who selected the site for his own burial.) Technically, Westminster Abbey once had the status of a cathedral, but it is now a "royal peculiar" and, as such, is a place of worship directly responsible only to the British sovereign. Ergo, royalty are married and buried here. Those married here include Queen Elizabeth II, her children Princess Margaret, Princess Anne, and Prince Andrew, and most, recently, her grandson, Prince William who married Kate Middleton in 2011. (Prince Charles was married to Lady Diana Spencer in 1981 in St. Paul's Cathedral. Her funeral was held at the Abbey in 1997.)

Those buried or commemorated here include not only monarchs, ranging from Henry III to Queen Elizabeth I but soldiers, sailors, statesmen, musicians, poets, writers, actors, and, as it somewhat unexpectedly turns out, some of the greatest scientists in British history.

Admittance to the Abbey is not gained through the West Front, that is, through the Great West Doorway under the iconic towers at the western end of the Abbey. (This entrance leading directly to the nave is only used for ceremonial occasions.) Instead, as has been tradition for centuries, folks line up at the north entrance, under an impressive gothic façade, which opens into the North Transept. Recall that churches like the Abbey are shaped like a cross, that is, they are cruciform in shape, with the vertical section of the cross usually oriented in the east-west direction and the horizontal sections (the transepts) in the north-south direction. In the line leading to the entrance we scientific/historical travelers, like others, talk excitedly among ourselves. But once we get inside, our shoes clack on the stone floors, our eyes are drawn to the great vaulted ceiling, and our voices, somewhat muted now, echo off the centuries old walls covered by plaques, inscriptions, and statues. Visitors are expected to follow the velvet ropes to the left and up and around the ornate apse. Following this one-way, prescribed route, you are led by the tombs of many kings and queens, the Coronation Chair, and finally down into the South Transept, where the so-called "Poets' Corner" is located. From there, visitors normally walk around the cloisters and come into the nave on the south side. Throughout the tour they admire the vaulted ceilings and the stained-glass windows.

We travelers-with-the-atom, like everyone else, will have to follow the prescribed route and we will certainly recognize and appreciate the memorials, graves, statues, and other reminders of the poets, musicians, statesmen, royalty, and others honored here. However, primarily we will be on the lookout for burial sites and memorials for scientists, particularly those involved with the history of the atomic concept. So, let's get started.

As we enter the North Transept, we find St. Andrew's Chapel on the left. In the "Official Guide to Westminster Abbey" this is described as "structurally the eastern aisle of the north transept". This can be a little difficult to find so feel free to consult a verger to get oriented. Although you can't get too far into the chapel without assistance, in the foreground you will see the marker to Sir Humphry Davy (1778–1829) and immediately behind that the memorial to John Wm Strutt, 3rd Baron of Rayleigh (1842–1910).

Recall from Chapter 3 (pp. 66–68) that Humphry Davy was from "tinner" territory in Penzance (near Lands End at the southwestern tip of England), where his Victorian era statue is on Market Jew Street. Davy learned his chemistry by reading the works of Lavoisier and Priestley and, in 1798, became the director of the Thomas Beddoes' Pneumatic Institution in Bristol. The buildings of this institute still stand today. Davy's enticing, intoxicating and altogether dazzling experiments on nitrous oxide ("laughing gas"), developed as part of an "inhalation therapy", honed his skills as a chemist. The data he produced on the mass ratios of the various nitrogen oxides as well as ammonia and other compounds were used by John Dalton in formulating the first concrete atomic theory in 1803. In 1801, Davy was invited to be a lecturer at the newly formed Royal Institution. In Chapter 5 (pp. 100–103), we describe how this charismatic, handsome, flamboyant, almost foolhardy pneumatic chemist turned into a daredevil (and still foolhardy) electrochemist. Using his giant voltaic pile, he discovered six elements in three years (1807–1809). He retired from the Ri in 1812, married the rather acerbic Jane Apreece and set out on his grand European tour with Michael Faraday as his valet and assistant. The Davy exhibit and famous lecture hall in the Royal Institution are among the best traveling-with-the-atom sites we have described. Here in Westminster Abbey, the memorial in St. Andrew's Chapel says that Davy was "distinguished throughout the world by his discoveries in chemical science". He is not buried here, but instead in Geneva at the *Cimetière des Rois* (French for "Cemetery of Kings") or *Cimetière de Plainpalais*, (46.202589°, 6.137349°) outside the city walls. John Calvin, the Protestant reformer, is also buried there.

John Wm Strutt, 3rd Baron of Rayleigh or "Lord Rayleigh" was the second Cavendish Professor of Experimental Physics at Cambridge University. The initial director of the "Cavendish" was James Clerk Maxwell. Like Clerk Maxwell, Rayleigh worked on electrodynamics, electromagnetism, optics, and color vision. His name is associated with the "Rayleigh scattering" of visible light, which is the common explanation for the blue color of the sky and the red colors of sunsets. He also played a role in the mathematics of "blackbody radiation", the explanation of which led to the development of quantum mechanics by Max Planck in 1900. (See the section on Planck in Chapter 15.) Here in Westminster Abbey, the memorial in St. Andrew's Chapel says that Rayleigh was "An unerring leader in the advancement of natural knowledge." In 1904, Rayleigh received the Nobel Prize in Physics and Sir William Ramsay was awarded the Nobel Prize in Chemistry for their joint discovery of the inert gas argon. Ramsay, who went on to isolate four more inert gases (helium, neon, krypton, and xenon) in the span of only four years, is honored in Musician's Aisle in the Abbey. We will describe that memorial at the end of this chapter.

Also in St. Andrew's Chapel is a medallion bust to Thomas Young (1773–1829). We have not encountered Young before in this book, but he was one of the first to establish the wave theory of light. His memorial is difficult to see as it is behind the large Norris Memorial to the right of the memorials to Rayleigh and Davy.

The "lantern" area of the Abbey is just a short walk from St. Andrew's Chapel and the North Transept. The lantern is atop the "crossing" between the choir and the sanctuary. It is called the lantern because it is under a small square of roof containing eight windows that admit light to the altar area. This was the only place in the Abbey where, in 1941, a German bomb came down into the church and exploded. One can still see pock marks in the southwest columns about 10 to 15 feet up. It is possible that you will need to seek the help of a verger to be admitted to this area, which is close by the Sanctuary and High Altar. In the floor in front of the steps leading up to the Sanctuary we can see a black marble diamond installed in 2005 to honor Robert Hooke (1635–1703). Hooke, in fact, had been responsible for the laying of this floor. It simply says, "Robert Hooke 1703". Facing the altar, the diamond is in the northeast corner, that is, to the left immediately in front of the steps near to the pulpit. This is a place of great honor. In Chapter 3 we relate how Robert Boyle hired Robert Hooke as his private assistant. Together, they set up a laboratory in Oxford, a fact now denoted by a marker on the outer wall of University College. Hooke was instrumental in devising Boyle's air pump (see Figure 3.1), which was used to gather the data leading to Boyle's Law. Boyle imagined air pressure as being due to "corpuscles" (what we call atoms or molecules) of the air slamming against the inside walls of the container. This model, now fully mathematically fleshed out, is still used today and is called the kinetic-molecular theory of gases. James Clerk Maxwell (see Chapter 5) and Ludwig Boltzmann (see Chapter 10) contributed significantly to this mathematics. Hooke also helped Boyle and his sister Lady Katherine Ranelagh design a laboratory in her house in London. Hooke was, to say the least, a controversial figure. Isaac Asimov described him as "a nasty, argumentative individual, antisocial, miserly, and quarrelsome". He had a particularly intense and acrimonious rivalry with Isaac Newton. They disagreed about the nature of light (Newton describing it as particulate; Hooke as waves), and whom should get credit for the law of inverse squares as applied to gravitation forces. He was active in all manner of scientific fields including physics, astronomy, and biology. In Chapter 5 we describe the markers at the Hooke birthplace ※※ on the picturesque Isle of Wight, his grave ※ at St. Helen Churchyard at Bishopsgate, London, and two others, one ※ in St. Paul's Cathedral and the other ※※ at the Monument to the Great Fire of London. Sadly, and ironically, there is no memorial in Westminster Abbey to honor Hooke's mentor, Robert Boyle. Isaac Newton, Hooke's great rival, however, is honored in grand style just on the other side of the Choir.

From these memorials, we need to follow the crowd up the North Ambulatory and to the Chapel of Henry VII to see the tombs of British royalty, as well as the famous Coronation Chair. Coming down the South Ambulatory we come to the South Transept, where Poets' Corner is located. Take a moment here to find your favorite poet, be it Chaucer, Spenser, Milton, Shakespeare, Keats, or Shelley, or the memorials to Humphry Davy's friends Samuel Taylor Coleridge, Robert Southey, and William Wordsworth, all found next to each other (see Chapter 3, pp. 69–70). (Recall also that Davy was a poet, although Southey said his work was "tedious and feeble".) From here one walks around the cloister and comes into the nave at the South Door. Now the fun really begins for the scientific/historical traveler.

The South Door is midway along the South Aisle. In front of you to the right is the altar at the top of the nave. It is generally elaborately roped off with thick royal purple velvet ropes. This

Scientists at the Heart of Westminster Abbey

is one of the most sacred areas in the Abbey. Quite amazingly, it is within or near the altar that we find memorials and burial sites for some of the greatest scientists in British history, many associated with the development of the atom. Occasionally, it has been referred to as "Science Corner" but it never gets the press accorded to the more famous Poet's Corner. To get oriented, go around the roped off area to the North Aisle on the other side of the nave and get close to the magnificent Gothic Revival choir screen where there is the stunning monument in honor of Isaac Newton. If you stand on the diamond memorializing the great astronomer, William Herschel (1738–1822), you are ready to take in these memorials and burial places. See Chart 7.2 to help get situated.

Stand on the William Herschel diamond and peer over the velvet ropes into the altar area. Immediately on the other side of the rope you will see the two diamonds memorializing Michael Faraday (1791–1867) on the right and James Clerk Maxwell (1831–1879) on the left. Just beyond these diamonds is the large rectangular stone under which Sir Isaac Newton (1642–1727) is buried. To the left of Newton's grave in the choir screen is the monument to Newton. It is with Newton that we should certainly start.

Rupert Hall, in his book "*The Abbey Scientists (The Memorials of Westminster Abbey)*,[1] says that "Newton was buried before the Choir Screen in the Abbey after one of the grandest funerals ever accorded to a commoner in England, with the Royal Society in attendance." After lying in state, his coffin was followed to the grave by most of the members of the Royal Society, of which he was president for the last 24 years of his life. The dates of birth and death are according to the Julian calendar which was in effect during Newton's lifetime. In 1752, Britain adopted the Gregorian calendar, using which the dates are different. These are shown in Table 7.1. Note that by the old calendar, Newton was born on "Christmas Day". This is no longer the case using the modern calendar. His gravestone is inscribed *Hic Depositum Est Quod Mortale Fuit Isaaci Newtoni*, which translates to "Here lies that which was mortal of Isaac Newton". The white marble monument in the choir screen is certainly one of the most impressive in the Abbey. Newton is shown reclining, with his elbow on a stack of four of his great books including *Opticks* and

Table 7.1 The dates of birth and death for Isaac Newton using the Julian and Gregorian calendars. England switched from the Julian to the Gregorian calendar in 1752.

	Julian calendar	Gregorian calendar
Date of Birth	25 Dec 1642	4 Jan 1643
Date of Death	20 Mar 1726	31 Mar 1727

the *Principia*. Above him is a Celestial Globe, upon which reclines a female statue representing the Genius of Astronomy. Under Newton, in the black sarcophagus, groups of children carved in white marble illustrate some of Newton discoveries. One carries a prism and another a reflecting telescope. The inscription in Latin under the monument relates that Newton was the first to demonstrate the laws that determine the forms and orbits of the planets and the nature of light and color. It ends with the admonition that we mortals should rejoice that "there has existed such and so great an ornament of the human race".

In Chapter 3 (p. 43) we described Woolsthorpe Manor in Grantham, where Newton was born in 1642 and to which he returned in 1666–1667 for his *annus mirabilis*. Here he wondered about the forces that caused the apple to fall from the tree in the garden outside the manor and here too he used a prism to split white light into its component colors. It was in his *Opticks* that he stated his ideas on the nature of atoms as "solid, massy, hard, impenetrable, moveable Particles". Newton was also an avid alchemist and we have noted the location of his alchemical laboratory near Trinity College in Cambridge. In London, one should visit the Royal Society and see some of the plaques put in place near his residences.

Newton split light into its component colors, but the question remains, what is light anyway? This turns out to be a simple question with a complex answer. A substantial part of the answer came from the work of Michael Faraday, one of the greatest experimental scientists of his time. As we described in Chapter 5 (pp. 103–111), his initial work was in chemistry, but he soon turned to the study of electromagnetism, which led directly to the development of electric motors, generators, and transformers. James Clerk Maxwell (see Chapter 5, pp. 111–127) took it from there and used Faraday's ideas to formulate the mathematical relationship between light, electricity, and magnetism. Today, one of the best depictions of light available to us is as an electromagnetic wave radiating through space, that is electromagnetic radiation, often abbreviated as EMR.

Michael Faraday's memorial here in the Abbey is most fitting. (It is noteworthy that Faraday had consciously turned down being buried in the Abbey.) In Chapter 5, we have detailed numerous Faraday sites in London, ranging from the memorial marking his birthplace, the plaque where he worked as a bookbinder's apprentice, his statue at Savoy Place, the Faraday House on Hampton Court Road, and his grave at Highgate West Cemetery . Most famous of all is the Royal Institution where we find paintings of him, his wonderful statue at the base of the grand staircase, the famous lecture hall where he and his predecessor and mentor, Humphry Davy, entertained and educated so many people, and the Faraday Museum on the ground floor.

It is also fitting that the James Clerk Maxwell memorial sits right next to Faraday's and directly adjacent to Newton's grave here in the Abbey. The Faraday and Clerk Maxwell plaques were unveiled on the same day. In Chapter 5, we described Maxwell's efforts in electrodynamics, the mathematics of the kinetic-molecular theory of gases, the physics of primary colors and color vision and even the composition of the rings of Saturn. His Maxwell's Equations firmly established the mathematics of electromagnetic fields and his picture of light as perpendicular oscillating electric and magnetic fields propagating through space revolutionized physics in a way that is arguably just as important as Newton's gravitational theory. It is with his work that the role of electricity in the structure of matter was firmly established. In that chapter, we went on to describe the many traveling-with-the-atom sites related to Clerk Maxwell. These include his birthplace in Edinburgh, now the home of the James Clerk Maxwell Foundation , his residence on Heriot Row , the school he attended , and a magnificent new statue recently erected in his honor. In addition, there are the sites in Corsock and Parton in Scotland. These include the Clerk Maxwell homestead Glenlair , now under the care of the Maxwell at Glenlair Trust, his grave at Parton Kirk , and the James Clerk Maxwell Window at the Corsock and Kirkpatrick-Durham Church.

Clerk Maxwell was also the first Professor of Experimental Physics at the new Cavendish Laboratory of Physics. The Museum of the Cavendish Laboratory ★★★★★ contains exhibits related to Clerk Maxwell.

There is more to see just beyond the Newton, Faraday, and Clerk Maxwell stones but you will need the assistance of a verger to view them. Under the altar table are the graves of Sir Joseph John Thomson (1856–1940) ★★ and Ernest Rutherford, 1st Baron Rutherford (1871–1937) ★★. These are not just memorials to these pre-eminent atomic scientists but, in fact, their ashes are buried under these stones. Rose Thomson, who helped make the atmosphere at the Cavendish so welcoming, is also buried with her husband here. A verger will take you into the restricted area and help you lift the heavy tapestries that cover the altar table so you can see the inscriptions on the tablets. In Chapter 5 (pp. 128–135), we described the contribution of "J. J." Thomson and Chapter 6 was devoted almost entirely to the work of Lord Rutherford. For travelers-with-the-atom, these two gravestones are as close to the holy grail as we are going to get today.

Thomson, the 3rd Cavendish Professor of Physics, had a hand in a number of discoveries that firmly established the electrical nature of the atom. In 1897, Thomson discovered the electron, determined its charge-to-mass ratio and estimated its mass. This was the first sub-atomic particle and in 1906 he received the Nobel Prize in Physics for its discovery. In 1904, based on his discovery he advocated the Plum-Pudding model of the atom. Starting in 1907 he investigated "canal rays", a line of research that led to isotopes and the development of the mass spectrometer. In Chapter 5, we noted that a traveler-with-the-atom had to visit the Old Cavendish Laboratory ★★ and the Museum at the Cavendish Laboratory ★★★★★, where one can see a beautiful replica of Thomson's apparatus for carrying out these experiments. Before or after you view J. J. and Rose's gravestone, you might enjoy rereading the short poem on pp. 129–130 of Chapter 5.

Rutherford, the 4th Cavendish Professor of Physics, was born in New Zealand, but emigrated to England where he took up with Thomson's group investigating radioactivity. He characterized and named alpha and beta particles before moving to Montreal where, working with Frederick Soddy, he further described nuclear transformations and defined what we know today as the half-life of a radioactive isotope. Now working at the University of Manchester, Rutherford and Hans Geiger identified alpha-particles as helium nuclei. In 1911, based on the results of the alpha-particle, gold-foil experiments of Geiger and Ernest Marsden, he announced his discovery of the nuclear atom (sometimes referred to as the fly-in-the-cathedral model), which immediately replaced Thomson's plum-pudding model. Just before leaving Manchester, Rutherford carried out a nuclear transformation by bombarding nitrogen atoms with alpha particles. Now taking up as the director at the Cavendish, he and James Chadwick tried to determine the exact role of electrons and protons in the nucleus. This endeavor ended when Chadwick discovered the neutron in 1932. During Rutherford's time as director, John Cockcroft and Ernest Walton devised their machine, using which they carried out several nuclear transformations, one of which provided the data that was the first confirmation of Einstein's $E = mc^2$ equation. A second transformation led to Cavendish's discovery of the neutron.

The travels-with-the-atom sites connected to Rutherford are numerous. We have detailed them in Chapter 6. Starting on the South Island, New Zealand, his birthplace at Brightwater, a memorial in Havelock, and his "den" in Christchurch have been described. In Montreal, the Rutherford Museum is one of the best places to see the original "little science" equipment that Rutherford used in his early work. The Rutherford-Soddy plaque in the original Macdonald Physics Building and his nearby residence are worth visiting. In Manchester, we can visit the Rutherford Building where Rutherford and his colleagues did their seminal work, and then go over to the Schuster Laboratory with its named lecture halls and pictures of the physics department during Rutherford's days. Back in Cambridge, there is the Museum at the Cavendish, the Old Cavendish Laboratory with its Crocodile Carving that Kapitza had carved on the wall of his laboratory, the Whipple Museum of the History of Science and Rutherford's residence at Newnham Cottage. Before or after you visit Rutherford's gravestone in the Abbey, sit back and recall his many contributions to the development of the atomic concept, the travels-with-the-atom sites associated with him that you have already visited, and perhaps reread the short poem about "that handsome, hearty British lord" found on p. 161 of Chapter 6.

If we return to our vantage point on William Herschel's memorial stone, several other sites within the sacred altar area are apparent and should be briefly described. First is the burial site of William Thomson, Lord Kelvin (1824–1907). His elaborate funeral in 1907, before the memorial stones for Faraday, Clerk Maxwell, Thomson or Rutherford had been put in place or even imagined, established this area as the "Science Corner". Kelvin estimated the age of the earth (but did so before the discovery of radioactivity and therefore his estimate was much lower than Rutherford's, see Chapter 6, p. 150), was responsible for the Kelvin temperature scale, introduced the term "kinetic energy" (meaning energy of motion) and posited that it was zero at the absolute zero of temperature (-273 °C or 0 °K), but was most famous for his work improving trans-Atlantic telegraph cables. He was made a baron in 1892 (becoming the first "Science Lord") and took the name "Kelvin" from the River Kelvin that runs near the University of Glasgow, where he was a professor for more than half a century. In addition to his grave in the Abbey, a memorial window in the Musician's Aisle, as noted in Charts 7.1 and 7.2, is located nearby. The Hunterian Museum at Glasgow University, Scotland has a permanent display called "Lord Kelvin: Revolutionary Scientist". It is located on the balcony level of the Hunterian Museum main hall and, although Kelvin was not directly involved with the history of the atomic concept, it is worth a visit.

Also in the altar area just beyond Newton's rectangular gravesite and next to the diamond over Lord Kelvin's grave is a memorial diamond honoring George Green (1793–1841). Green was a mathematical physicist who wrote an early but famous essay applying mathematics to the theories of electricity and magnetism. His ideas set the stage for the work of James Clerk Maxwell and others.

Near the upper right-hand corner (as we face it from William Herschel's memorial stone) of Isaac Newton's grave is a memorial

Scientists at the Heart of Westminster Abbey

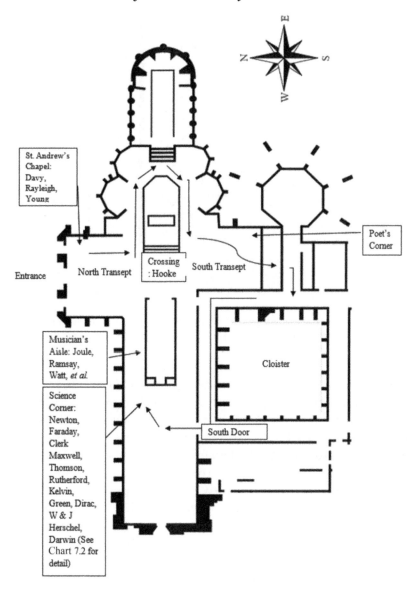

Chart 7.1 Memorials and graves of scientists in Westminster Abbey. Adapted from [https://commons.wikimedia.org/wiki/File:ABBY1.jpg] by Mavin_101, under the terms of a CC BY-SA 3.0 license [https://creativecommons.org/licenses/by-sa/3.0/]. This adaptation is licensed under the terms of a CC BY-SA 3.0 license.

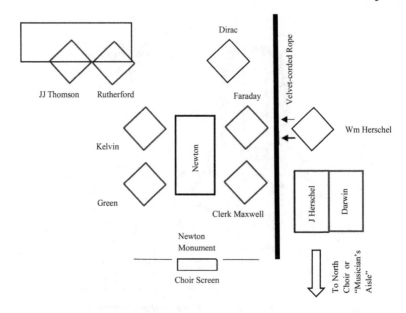

Chart 7.2 Location of memorials and graves in the science corner of Westminster Abbey. In June 2018, Stephen Hawking was also buried in this area.

stone honoring Paul Adrien Maurice Dirac (1902–1984) . Another mathematical physicist, Dirac worked on wave mechanics and shared the 1933 Nobel Prize in physics with Erwin Schrödinger (See Chapter 16). Remarkably, his relativistic wave equation is actually inscribed on his stone. He also was the first to suggest the existence of the positron, a particle of the same mass of an electron but with a positive rather than negative charge. The positron was discovered two years after Dirac proposed it. It was the first example of anti-matter. We will have the opportunity to discuss quantum mechanics in Chapters 16 and 17.

We have been standing on the diamond-shaped memorial stone dedicated to the German-English astronomer Sir William Herschel (1738–1822) . In 1781, with their beautiful handmade telescope set up in their garden in Bath, England, William and his sister Caroline accidentally discovered a small bluish-green planet as it drifted in a nearly circular orbit far beyond that of Saturn, then the outermost known planet in our solar system. The Latin inscription ("*Coelorum perrupit claustra*") on William's stone translates to "He broke through the confines of heaven" but modern observers note that it should say "they" not just "he" because Caroline deserves equal credit for the discovery. In Bath, England, the Herschel Museum of Astronomy (51.382532°, −2.366956°), while not related to the development of the atomic concept, is very much worth visiting. It is located at 19 New King Street in Bath.

Right next to the Sir William Herschel memorial stone is the large rectangular marker for the grave of William's son, Sir John Frederick Herschel (1792–1871). John Herschel was a mathematician, chemist, astronomer, and philosopher. From 1833 to 1838 he set up an observatory in South Africa in order to survey the southern skies. In 1836, the *HMS Beagle,* with Charles Darwin as its naturalist visited Cape Town toward the end of its historic 5 year tour around the world. Charles and John became good friends.

The Herschels are not the only astronomers honored and/or buried in the Abbey. For example, in the south cloister there is a tablet in memory of Edmond Halley (1656–1742). In 1721, he was appointed Astronomer Royal and was first to predict the return of the periodic comet which now bears his name. This slate tablet is in the shape of a stylized comet and was placed here in 1986 to mark the return of "Halley's comet", which he had himself had observed in 1682. He concluded that this same comet had also been observed in 1456, 1531, and 1607, that it appeared every 75 or 76 years, and was due to return in 1758. Incidentally, if you find the history of astronomy attractive, a must place to visit is the Royal Observatory in nearby Greenwich. This is the location of the prime meridian of the world and the home of Greenwich Mean Time. The Royal Observatory (51.476880°, −0.000511°) is one of the best scientific-traveling sites in the world and should not be missed. Take the opportunity to place one foot in the western hemisphere and the other in the eastern. It's a great spot to consider what fun it is to be a scientific-traveler.

As noted above, Charles Darwin (1809–1882), author of *The Origin of the Species,* setting out the theory of evolution by natural selection, is buried in the north aisle, not far from Sir Isaac Newton and directly adjacent to the grave of his friend, Sir John Herschel. Mr Darwin lived with his family in his home in Downe, in Kent, not far from London. The Down House (spelled without the "e") (51.330371°, 0.052767°) is today another of those fascinating scientific-traveling sites that should not be missed. Charles died there and expected to be buried in St. Mary's churchyard in the village. A simple coffin was built by the village carpenter and Charles was placed in it. However, as was the case with Newton, members of the Royal Society exerted their influence. Telegraphs to the family, petitions to members of Parliament, articles in newspapers noting Darwin as the greatest Englishman since Newton all had their effect. The simple coffin was returned and replaced with one that the village carpenter said, "you could to see to shave in" and Charles Darwin, a professed agnostic most of his life, was interred in Westminster Abbey.

This brings us to the north choir aisle, sometimes known as "musician's aisle". (See Chart 7.2) Being located near the North Transept, this is the last part of the Abbey that a visitor is able to visit. As indicated by its nickname, a number of important musicians are honored here. These include Sir Edward Elgar (memorial), Ralph Vaughan Williams (ashes), Benjamin Britten (memorial), Henry Purcell (burial), and Orlando Gibbons (memorial bust), among others. But here too are found some prominent scientists, many of whom were contemporaries. To find these, first locate the elaborate memorial to Lord John Thynne (1798–1881) who had been Canon of Westminster for 49 years. He died just before Darwin and his memorial occupies a prime spot. Perhaps Darwin would have had this spot if not for the order of their deaths. The monument is of white and red veined marble and shows a reclining, life-sized white marble effigy. Looking above Thynne's memorial and moving from right to left, you will find a series of five plaquettes or wall monuments. First is a bronze life-sized relief bust of Charles Darwin (1809–1882). Then there are three white marble roundels in honor of the naturalist Alfred Russel Wallace (1823–1913) (who independently formulated a theory of evolution), Lord Joseph Lister (1827–1912) (surgeon and pioneer of antiseptic treatment), and John Couch Adams (1819–1892) (mathematician and astronomer). Finally, there is bronze roundel in honor of Sir George Stokes (1819–1903) (physicist and mathematician). Below Stokes is a white marble square memorial to James Prescott Joule (1818–1889) and below that a rectangular memorial to Sir William Ramsay (1852–1916).

We mentioned James Prescott Joule in Chapter 4 (p. 96). He was briefly a student of John Dalton (1766–1844) and beautiful white statues of Joule and Dalton face each other in the foyer of the Manchester Town Hall . Joule was not an academic but a brewer and was mostly self-educated, although he did have some instruction from Dalton. He was not involved in the formulation of the atomic concept. As his tablet in the Abbey notes, he is best known for the formulation of the Law of Conservation of Energy (the first law of thermodynamics) and establishing the mechanical equivalent of heat. The International System of Units (SI) primary unit of energy is the Joule (pronounced "jool" or "jowle") in his honor.

As noted earlier, Sir William Ramsay received the 1904 Nobel Prize in Chemistry for his part in the discovery of argon and the other inert (now called "noble") gases. In that same year, John William Strutt, the third Lord Rayleigh received the Nobel Prize in Physics for his work with Ramsay in their joint discovery of argon. Ramsay went on to isolate four more inert gases (helium, neon, krypton, and xenon). He isolated helium from clevite, a uranium-containing mineral, in 1895. (Recall that Rutherford and Geiger also determined that alpha-particles were just helium nuclei. See Chapter 6.) In 1898 Ramsay's group isolated krypton, neon, and xenon from liquid air. In 1908 he isolated and determined the density of radon which had been discovered by Dorn in 1900. Thus, he had a prominent role in isolating all six of the inert gases, a group of the periodic table that was heretofore unknown. Playing a role in the discovery of six previously unknown atoms and adding a group to the periodic table is a daunting accomplishment, matched only by Davy's discovery of six elements (potassium, sodium, magnesium, strontium, barium and calcium) from 1807 to 1808 (see Chapter 5). For all of his accomplishments, his memorial in the Abbey simply says "William Ramsay, Chemist". Ramsay spent much of his professional life at University College London (UCL) , where they hold many items, including his Nobel Prize and numerous examples of his discharge tubes. There is a portrait of Ramsay in the lobby of the Christopher Ingold Building at 20 Gordon Street (51.525237°, −0.132509°). The best way to inquire about visiting UCL to see the Ramsay materials is to search for "UCL Science Collections". The Science Museum also holds materials related to Sir William Ramsay. There is blue Royal Society of Chemistry Chemical Heritage plaque honoring Sir William Ramsay at the Slade School of Art on Gower Street (51.524934°, −0.134387°), not far from the Christopher Ingold Building. This plaque marks where Ramsay's laboratory was once located.

James Watt (1736–1819), who most famously made the steam engine economically feasible, is honored farther up the north aisle, where a bronzed plaster bust of him sits on a window-ledge. It is noteworthy that a large marble statue of a seated Watt carved by Francis Chantrey had originally been placed in the Abbey in 1825. However, it was so heavy that it broke through the floor and revealed "rows upon rows of gilded coffins" underneath. In 1960, it was moved out of the Abbey and, after several stops, was moved in 1996 to the Heriot-Watt University in Edinburgh, Scotland. Chantrey also carved the statue of John Dalton that resides in the Manchester Town Hall and, incidentally, the statue of George Washington in the Massachusetts State House in Boston. Recall that Gregory Watt, James' son, suffered from consumption and moved in 1797 to Penzance, where he stayed with Humphry Davy for the winter. They became good friends. It was Gregory Watt who recommended Davy to Thomas Beddoes, who was trying to find a director for his Pneumatic Institute in Bristol. It was here that Davy made his mark, became an excellent chemist and soon moved to the Royal Institution. If it hadn't been for Gregory and James Watt, Humphry Davy might never have achieved chemical stardom. James Watt also devised a machine to produce the large amounts of gases needed at the institute.

7.2 SUMMARY

During a visit to Westminster Abbey, travelers expect to find memorials, statues, and graves of kings and queens, great clergymen, statesmen, and yes, even poets. But great atomic scientists, the men who had discovered the fundamental building blocks of matter – to see their markers and those for other scientists here in this great abbey is a little unexpected. Following the prescribed visitor route through the Abbey, we have found in St. Andrew's Chapel markers for Humphry Davy, John William Strutt, Lord Rayleigh, and Thomas Young. Nearby, in the lantern area we find a recently placed marker to Robert Hooke.

Moving to the nave, we have peered over velvet ropes to see Isaac Newton's gravesite and the stunning white marble monument in the choir screen erected in his honor, memorial markers for Michael Faraday and James Clerk Maxwell, the stones under which the ashes of J. J. Thomson and Ernest Rutherford are buried, the gravesite of Lord Kelvin, and markers to George Green and Paul Dirac. Just outside the altar area we have used the memorial stone honoring William Herschel to help us locate the graves of his son John Herschel and the latter's friend, Charles Darwin. In the North Aisle we have found markers honoring Darwin, Alfred Russel Wallace, Joseph Lister, John Couch Adams, George Stokes, James Prescott Joule, William Ramsay, and James Watt.

For many of these (most notably Davy, Rayleigh, Hooke, Newton, Faraday, Clerk Maxwell, Thomson, and Rutherford) we have discussed in earlier chapters a number of significant places to visit including birthplaces, statues, residences, laboratories, and museums. In this chapter, we have noted several other places to visit that are not directly tied to the history of the atom. These include the Hunterian Museum at Glasgow University in Scotland (Kelvin), the Herschel Museum of Astronomy in Bath, the Royal Observatory in Greenwich, and Darwin's House in Downe. The latter two places are among the best scientific-historical sites in the world.

ADDITIONAL READING

- J. Physick, Men of Science in Westminster Abbey, *Interdiscip. Sci. Rev.*, 1990, **15**, 373.
- I. Asimov, *Asimov's Biographical Encyclopedia of Science and Technology,* Doubleday & Company, Garden City NY, Second revised edition, 1982.

REFERENCE

1. A. Rupert Hall, *The Abbey Scientists (The Memorials of Westminster Abbey)*, Westerham Press, Kent (1966).

CHAPTER 8

The New French Chemistry and Atomism: Franklin, Lavoisier, Berthollet, Gay-Lussac, Ampère (Paris I)

8.1 A QUICK LOOK AT PLACES TO VISIT "TRAVELING WITH THE ATOM" IN PARIS

Benjamin Franklin House ✦✦ *in London and Statue* ✦✦ *in Paris; Antoine Lavoisier: rue Pacquet birthplace* ✦*, St. Merri christening site* ✦*, International Historic Chemical Landmark at College Marazin* ✦✦*, La Petit Arsenal laboratory plaque* ✦✦*, Laboratory at the Musée des Arts et Métiers* ✦✦✦*, La Concierge cell* ✦✦*, Place de la Concorde guillotine site* ✦✦*, Errancis Cemetery marker* ✦*, City Hall statue* ✦*, Eiffel Tour name* ✦*; Claude Berthollet: statue* ✦✦ *in Jardin de l'Europe, Annecy, Arcueil house plaque* ✦*; Joseph Louis Gay-Lussac: Eiffel Tower name* ✦*, Cimitiere du Pere Lachaise grave* ✦✦*, Gay-Lussac Café* ✦ *and marker for Louis Pasteur Laboratory* ✦*; Louis Pasteur: laboratory markers rue d'Ulm* ✦*, Pasteur Museum in Pasteur Institute* ✦✦✦✦*; André-Marie Ampère: Eiffel Tower name* ✦*, rue Monge house plaque* ✦*, Montmartre Cemetery grave* ✦✦*.*

Traveling with the Atom: A Scientific Guide to Europe and Beyond
By Glen E. Rodgers
© Glen E. Rodgers 2020
Published by the Royal Society of Chemistry, www.rsc.org

Antoine Lavoisier, the New French Chemistry, and the Advance of Atomism

- Benjamin Franklin (1706–1790), who had encouraged Priestley in his investigations of electricity and "airs", also was a close friend to Antoine Lavoisier and his wife, Marie-Anne.
- Lavoisier (1743–1794) placed a large emphasis on the construction of accurate balances. Using these instruments, he measured the masses of reactants and products before and after a reaction and formulated the Law of Conservation of Mass.
- Lavoisier was an opponent of the phlogiston theory and successfully replaced it with his theory of combustion. He renamed Joseph Priestley's "dephlogisticated air" oxygen and Cavendish's "inflammable air" hydrogen. Under his new nomenclature of chemistry, metals were "oxidized", that is, they reacted with oxygen to produce "oxides".
- In his 1789 seminal work, *Elementary Treatise on Chemistry* Lavoisier summarized these ideas and listed 33 elements which he defined as substances that chemical analyses had failed to break down into simpler entities. Chemists could now write equations to represent chemical reactions. The fundamental building blocks of these equations were atoms.
- Claude Berthollet, although he ultimately accepted Dalton's Law of Multiple Proportions and Proust's Law of Definite Proportion, remained skeptical of Dalton's atomism.
- Joseph Gay-Lussac discovered the Law of Combining Volumes which, with André-Marie Ampère and Amadeo Avogadro's Hypothesis that equal volumes of gases contain equal numbers of particles, demonstrated that water was H_2O and hydrogen and oxygen gaseous molecules were diatomic. Gay-Lussac gradually became a more vocal advocate for atomism, particularly so after Claude Berthollet died in 1822. These advances, however, were not recognized as valid until 1860.

8.2 BENJAMIN FRANKLIN (1706–1790)

8.2.1 Paris (Franklin)

Recall that in Chapter 3 on "Pneumatists Set the Atomic Stage (Western England and Northumberland, Pennsylvania)" we described how Ben Franklin had arranged for Joseph Priestley to join the Royal Society and encouraged his young friend to pursue an interest in electricity which Franklin had so avidly and ably investigated. With Franklin's encouragement, Priestley published his *History and Present State of Electricity*, which established his reputation in the sciences. Priestley went on to discover a number of gases ("airs") including his "dephlogisticated air", which Lavoisier later renamed oxygen. Priestley's scientific career would most likely not have been possible without the encouragement of Ben Franklin.

In London you may have had a chance to visit the Franklin House 36 Craven St, London (51.507619°, −0.124903°) where he lived from 1757 to 1775. In 1765, the young Joseph Priestley (then at Warrington) came to London to attend a meeting of "the Club of Honest Whigs", which Franklin had helped to establish. They met in the London Coffee House and it was during this time that Franklin and his friends greatly encouraged Priestley to work on his book about electricity.

Now here in Paris you can visit the magnificent statue of Franklin close to the Eiffel Tower near the Pont de Bir-Hakeim, Place du Trocardaro, 38 rue de Benjamin Franklin (48.861781°, 2.286470°). There are some magnificent views of the Eiffel Tower in this area. You can stand under this statue and recall Franklin, not only as one of the founding fathers of the United States and its first ambassador to France, but also as an important scientist and mentor for other scientists who made significant contributions to the development of the atomic concept. Franklin was a frequent visitor to Antoine Lavoisier's private residence (the Arsenal) in Paris while he was the United States ambassador to France (1776–1785). Madame Lavoisier, who ably illustrated Lavoisier works, painted a portrait of Benjamin Franklin and presented it to him in 1788. Her portrait is shown in Figure 8.1.

8.3 ANTOINE LAVOISIER (1743–1794)

> "The importance of the end in view prompted me to undertake all this work, which seemed to me destined to bring about a revolution in...chemistry. An immense series of experiments remains to be made." Antoine-Laurent Lavoisier, 1773.

The New French Chemistry and Atomism

Figure 8.1 Marie Anne Pierrette Paulze Lavoisier's portrait of Benjamin Franklin. Courtesy of the Phillips Museum of Art at Franklin & Marshall College.

Back in Chapter 3, "Pneumatists Set the Atomic Stage" in the subsection entitled "Scottish and English Pneumatic Chemistry: 'Fixed', 'Phlogisticated' and 'Inflammable' Airs (1756–1772)" (p. 47) we left what many readers would view as unanswered questions. For example: what did Priestley think he had discovered? Why did he call it "dephlogisticated air"? What did Daniel Rutherford discover and why did he call it "phlogisticated air"? What was the true nature of Cavendish's "inflammable air"? Why did Lavoisier call Priestley's gas oxygen and Cavendish's gas hydrogen? In this section, we answer these questions and describe the role that these airs played in Lavoisier's new way of looking at the world, what is often called the "chemical revolution" or the new French Chemistry.

To get the possible answer to these questions, let's back up just a bit. We already have an appreciation of the fact that the transition from alchemy to modern chemistry was a gradual one, replete with haphazard advances and retreats. We have briefly described Robert Boyle's contribution to the initial stages of that transition and his place as a "chymist" positioned between the medieval alchemist and the modern chemist (See Chapter 3). One of the last bastions of alchemy remaining in Priestley and

Lavoisier's day was the concept of phlogiston. Both started their careers as "phlogistonists". This mystical fire-like element can be traced back to the German alchemist Johann Becher (1635–1682), who accepted the Greek elements (fire, water, air, and earth) but articulated three different types of earths: *terra fluida* that gave metals their shininess and heaviness, *terra pinguis*, an oily earth that made things combustible, and *terra lapidea* that made things solid. Terra pinguis was given the new name "phlogiston" by Becher's student Georg Stahl (1659–1734). In 1700, he proposed that phlogiston could be used to explain a number of processes and therefore was important in the natural philosopher's arsenal of ideas to explain the nature of matter. Even as late as 1796 (nearly a century later), Joseph Priestley heralded phlogiston as the "greatest discovery that had ever been made in science". Here, basically, is how it worked.

When a candle burns, it is easy to imagine that something quite intangible, even mystical, is given off. Burning substances, particularly where flames are involved, have always excited the human imagination. We all have wondered: "what constitutes the fire?" What supports the flames? Stahl proposed that during burning, the "imponderable" element phlogiston (from the Greek *phlogistón*, n. of *phlogistós* burnt up, inflammable) swirls off into the air. To phlogistonists it was almost as if they could see it leap away from the substance. If the candle is placed in an enclosed container, it burns for a while but then stops. Stahl claimed that this was because the air became too rich in phlogiston and could no longer accept any more. With no place for the phlogiston to go, the flame of the burning candle is snuffed out. In a stroke of insight, Stahl correctly recognized that the rusting of metals was analogous to the burning of wood or a candle. Again, he maintained that a metal possessed phlogiston but its brittle rust or powdery ash (or what he called the 'calx') did not. When a metal rusts it gives off phlogiston and all that is left is the calx. He also maintained that charcoal was a great source of phlogiston, so that combining a calx and charcoal in a fire could reconstitute the metal. Even life, including Mary's mouse (see Chapter 3, p. 62), was dependent on phlogiston being evolved during respiration. If the mouse was placed in a closed bell jar, the air was soon saturated with phlogiston and its life imminently endangered. Burning wood or metal, the winning of metals from their ores and respiration all were explained for *more than a century* using phlogiston. No wonder Priestley and Lavoisier, as well as Black, Rutherford, and Cavendish, accepted its central role.

Still, there were chinks in the phlogistic armor. Even early on, it appeared that when some metals burned the resulting calx weighed *more* than the original metal. Phlogistonists explained this by assuming that in these cases phlogiston had a *negative mass*. Today we find such an assertion to be absurd but it's important to realize that the concepts of mass and weight were not such prevalent ideas even late into the eighteenth century. "Negative mass" is a term a modern chemist would not understand but back then there were other ways to understand it. The French philosopher Condorcet said that phlogiston was "impelled by forces that give it a direction contrary to that of gravity". Putting it more poetically, others declared that phlogiston "gave wings to early molecules". "Substances" such as Newton's "aether", Franklin's "electrical fluid" and the fluid of heat called "caloric" were not readily described in terms of their masses. So, phlogiston's explanatory appeal survived its apparently negative mass. It survived also the fact that every once in a while a calx, such as that of mercury, could be turned back to the metal merely by heating without using phlogiston-rich charcoal. This too was a mystery if phlogiston was so central to the process of winning metals from their ores.

Gradually, the importance of mass and its measurement became paramount. While Black, Cavendish, and others took pains to measure the masses of their "airs", Lavoisier took these measurements to a considerably higher level. Independently wealthy, he had very accurate, advanced balances constructed for his personal use and he used them to their full potential. The results were at least two-fold. First was Lavoisier's formal declaration of the Law of Conservation of Mass, that is, that matter is neither created nor destroyed, but is always conserved, that is to say, the total mass is unchanged. This was an ever-present assumption throughout his investigations and, as we saw in Chapter 4 on John Dalton, it proved to be a critical tool in the development of the atomic concept. Second, by using huge vessels containing large volumes of air that Lavoisier could mass very accurately, he was able to demonstrate that in all cases of combustion where an increase in weight was observed, air was absorbed and he could measure its mass. His advanced balances, like those shown in Figure 8.3, quantitatively showed that a component of the air reacted with the metal to produce the calx. Lavoisier believed that Priestley's "dephlogisticated air" was this component of ordinary air.

But why, you ask, did Lavoisier, while not credited with its discovery, rename Priestley's "dephlogisticated air" oxygen? His choice of this name can be traced back to his work heating phosphorus and sulfur over a fire while accurately measuring the mass of the reactants and products. Again, the residues weighed more than the elements before they were heated and the total mass (including the air in which the elements were heated) was found to be conserved. But, in a separate experiment, he also found that the residues absorbed water to produce acids. Lavoisier, a little impulsively perhaps, extrapolated these results to imply that Priestley's "dephlogisticated air" was a central component of *all* acids. For this reason, he chose to rename Priestley's air "oxygen". The name comes from the Greek *oxys* "sharp, acid" combined with the French *-gène*, for "something that produces" (from the Greek *-genes* "formation, creation"). So, the name "oxygen" means "acid-producer". As logical as this assumption was relative to phosphorus and sulfur, where it produced what today we call phosphoric and sulfuric acids, it soon proved generally incorrect. Hydrochloric acid and hydrocyanic acids were produced soon afterwards and contained no oxygen. In a twist of fate, Priestley had discovered the gas that modern chemists call hydrogen chloride in 1772. Hydrogen chloride, as its modern name implies, does not contain oxygen. When dissolved in water it produces hydrochloric acid and this fact quickly helped to disprove Lavoisier's assumption.

Oxygen was essentially the opposite of phlogiston. It was *absorbed* (not given off) when something burned, a metal was heated in air, or Mary's mouse took a breath. Lavoisier's oxygen-based combustion theory was the basis of the chemical revolution. Not only was it a revolutionary new way to think of combustion and respiration, but it redefined the language of chemistry forever. Combustion was soon known as oxidation. Calxes were compounds of oxygen called oxides. For example, Priestley's mercury calx was mercuric oxide and it liberated oxygen and the element mercury when heated. Black's "fixed air" was carbon dioxide, Daniel Rutherford's "phlogisticated air", which did not support combustion, eventually was called nitrogen. Cavendish's inflammable air was renamed hydrogen by Lavoisier, because when it combined with oxygen it produced water. (Both Cavendish and Priestley had demonstrated that the combination of their two airs produced water.) The name hydrogen is from the Greek *hydr-*, stem of *hydros* "water" plus the French *gène* meaning "producing". So "hydrogen" means "water producing".

Lavoisier preferred the word gas over "air". Common air, no longer viewed as a single substance, was recognized as a mixture of gases such as nitrogen, oxygen, carbon dioxide, and so forth. For leading the chemical revolution Lavoisier is often called the "Father of Chemistry". The transition from alchemy to modern chemistry was completed by Lavoisier and his colleagues in the late eighteenth century. Arguably, the chemical revolution was just as important as the American or the on-going French Revolution. These were truly revolutionary days. Lavoisier was the father of one, but the victim of another.

Antoine Lavoisier was born, married, prospered, and died (*via* the guillotine) in Paris. His mother died young and he was raised by his aunt. He attended Mazarin College, where he excelled at classics and literature. He then attended the law school at the University of Paris, but had time to study sciences such as astronomy, mathematics, botany, geology, and chemistry. In 1766 (at the age of 21) he was awarded a gold medal from the Royal Academy of Sciences for his work in investigating the best way to light the streets of a large town. He was elected a member of the Royal Academy at the age of 25. Royal Academy members (unlike Royal Society members in England) received a salary from the king and worked on scientific and technological projects. For Lavoisier, this involved the purity of cider, the nature of meteorites, the best way to raise cabbages, the mineralogy of the Pyrenees, and many other projects. Everything was going well. He had inherited some money, had a steady income, and liked what he was doing. But to increase his income, he made a fatal (literally, as it turned out) mistake. In 1768, still only aged 27, he invested in the Ferme générale, which was a private "tax farming system" engaged by the government to collect taxes on land, earnings, salts, wine, tobacco, and a variety of other goods. To Lavoisier this was a sound investment and he used the resulting considerable income to finance his scientific work. He even found his wife through his associations with the "Tax Farm". Marie-Anne-Pierette Paulze, the beautiful and intelligent 14 year-old daughter of a tax farm executive, Jacques Paulze, married Lavoisier in 1771. Marie-Anne loved the sciences as well, and became a great aid to her husband. She took lessons in English, chemistry, and drawing and was active in recording experimental procedures and results, translating texts, illustrating the resulting books, and making sketches of equipment and even the general scenes of laboratory life. Her sketch of the Lavoisier lab during an experiment to measure human respiration is shown in Figure 8.2.

In 1774, the new king, Louis XVI, in an effort to clean up corruption, set up the Gunpowder Commission to improve the supply and quality of gunpowder. Lavoisier was appointed as one of four commissioners and, therefore, in 1776, set up a state-of-the-art laboratory at the Petit Arsenal, which served as a gunpowder warehouse for the nearby Bastille. Here, Antoine and Marie-Anne had a private apartment and a large library as well as an immense and beautifully equipped laboratory in the attic. Marie-Anne, then in her early twenties, was a vivacious and engaging hostess. The Lavoisiers entertained regularly both

Figure 8.2 A view of Lavoisier's laboratory in the Paris Arsenal, *c.* 1790 (A drawing by Madame Lavoisier depicting human respiration. She is shown taking notes on the right side of the drawing.) Reproduced from ref. 3 with permission from Wellcome Collection, under the terms of a CC BY 4.0 license [https://creativecommons.org/licenses/by/4.0].

their Parisian friends and colleagues but also visitors from all over the world including Joseph Priestley, Benjamin Franklin, and even Thomas Jefferson.

As their chemical work progressed, the tax farm prospered. As well might be imagined, it was not popular. Many of the tax farmers were dishonest and collected more taxes than were actually owed. In 1784, it started to build a wall around Paris to make sure that tolls were paid on all the appropriate goods entering the city. In the latter half of the 1780s, things started coming to a head. First, Lavoisier and his colleague, Claude Berthollet (see below), and several others published *Methods of Chemical Nomenclature*, in which they established the new language of chemistry that replaced the many trivial names (like dephlogisticated and inflammable air) for chemical elements and compounds with systematic names (oxygen and hydrogen). By 1789, the production and quality of gunpowder in France had become the ideal of the world. France was even able to export gunpowder abroad, including to the American rebels still fighting their war of independence against England. Also in 1789, after 15 years of research, Lavoisier published *Elementary Treatise on Chemistry*. Many historians of science maintain that it was this book that marked the birth of modern chemistry. But 1789 also marked the Fall of the Bastille, an event that many historians mark as the beginning of the French Revolution. The Bastille was close to the Petit Arsenal. Things were getting dangerous.

By 1792, the French Revolution and the "reign of terror" were in full swing. The monarchy fell and was replaced by the Republic. The hated Tax Farmers were arrested. Lavoisier said he was a natural philosopher (again, we would say "scientist") and should not be arrested but the revolutionists famously stated, "The Republic has no need of scientists" and beside that, he *was* a tax farmer. Quick and unfair trials resulted in convictions. In a few short years, Lavoisier went from a preeminent natural philosopher and valuable citizen to making the trip to the guillotine. Lavoisier and 27 other tax farmers (including Anne-Marie's father) were executed on May 8, 1794.

But Antoine Lavoisier had changed chemistry forever. The phlogiston theory was replaced by his combustion theory. The modern language of chemistry was quickly proposed and accepted.

As part of his "Elementary Treatise on Chemistry" he listed 33 elements, which he defined (harking back to Robert Boyle's definition) as substances that chemical analyses had failed to break down into simpler entities. His elements included the gases oxygen, nitrogen, and hydrogen; the metals antimony, arsenic, bismuth, cobalt, copper, gold, iron, lead, manganese, mercury, molybdenum, nickel, platinum, silver, tin, tungsten, and zinc; and the non-metals sulfur, phosphorus, and carbon. (Lavoisier also erroneously included light and heat as elements.) Many of those elements reacted with oxygen to produced oxides and these reactions could be represented with chemical equations. Equations showed how reactants were converted into products at the molecular level and these changes could be followed quantitatively. The fundamental building blocks, the alphabet if you will, of these equations were atoms.

8.3.1 Travel Sites in Paris Related to Antoine Lavoisier

See Chart 8.1 for the location of these sites.

8.3.1.1 Rue Pacquet. (48.8598°, 02.3560°) Lavoisier was born and spent the first five years of his life at rue Pacquet. As James L. Marshall says in his article "Rediscovery of the Elements: Phlogiston and Lavoisier",[1] this short street was originally a cul-de-sac. It was rebuilt in 1954 and the northern end now opens up onto rue de Rambuteau. The narrow streets and buildings (some of which seem literally to bulge out toward you) are much the same as they were in Lavoisier's day.

8.3.1.2 Saint Merri Church. (48.8592°, 02.3503°) Lavoisier was christened here at l'Eglise Saint Merri at 78, rue Saint Martin, which is within walking distance from his birthplace. The church can also be seen from the funky automated Fontaine Stravinsky beside the Centre Pompideau Art Museum. There is a plaza here where you can enjoy a coffee or a cup of tea, watch the fountain in action, look back at the church, and talk to your friends about the amazing Antoine Lavoisier.

8.3.1.3 College Marazin. Lavoisier attended the College Marazin from 1754 to 1761. This is now the Academy of Sciences of the Institute of France, 23 Quai de Conti (48.857453°, 2.337770°). In 1999, the academy and the American Chemical Society placed

The New French Chemistry and Atomism 207

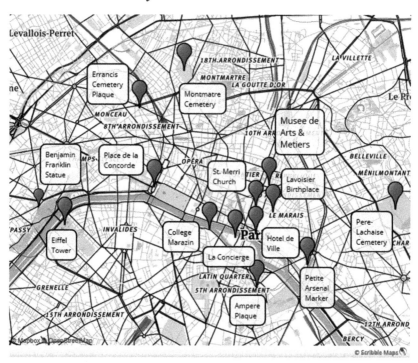

Chart 8.1 Travel sites in Paris related to Benjamin Franklin and Antoine Lavoisier. For Lavoisier, these include his birthplace, christening site, college, residence and laboratory, burial sites, and memorials.

an International Historic Chemical Landmark here. Here is the English translation of the citation on the landmark:

> "In these buildings, then 'Collège Mazarin' or 'des Quatre-Nations', Antoine-Laurent Lavoisier (1743–1794) studied from 1754 to 1761. He was elected to the Royal Academy of Sciences in 1768, where he presented his important studies on oxygen in chemistry. These began with a 'pli cacheté' of Nov. 2, 1772, and, after he experimentally proved the chemical composition of water by the quantitative method, culminated in his abandoning of the phlogistic theory in 1785. In 1787, he proposed the principles of a new Méthode de Nomenclature Chimique, in collaboration with the chemists Guyton de Morveau, Berthollet, and Fourcroy and with the help of the mathematicians Monge and Laplace. The publication of his Traité Elémentaire de Chimie two years later convinced French

and foreign chemists of his theories. His papers, stored in the Archives of the Academy of Sciences, bear witness to the conception and maturing of his revolutionary ideas, which are at the foundations of modern chemistry."

8.3.1.4 La Petit Arsenal. The Lavoisiers lived and experimented at La Petit Arsenal from 1775 to 1792. At the corner of Boulevard Bourdon and rue Bassompierre, there is a plaque ※※ (48.8508°, 2.3670°) that translates to "Here was located the Administration Office of Gunpowder, where Antoine-Laurent Lavoisier worked and lived 1776–1792, manager of gunpowder and saltpeter, who here constructed his chemical laboratory." Le Petit Arsenal was actually located at the end of rue de la Ceraisie, 100 meters north along the Boulevard Bourdon. Continuing north on Boulevard Bourdon we find the Place de la Bastille where the famous prison once stood.

8.3.1.5 Musée des Arts et Métiers. ※※※ At the Musée des Arts et Métiers, 60 rue Réaumur (48.866061°, 2.355467°), there is a reconstructed "Lavoisier's Laboratory" prominently located at the foot of the main staircase. This exhibit, part of which is shown in Figure 8.3, displays his pneumatic troughs, huge brass vessels and two-pan balances for measuring the masses of gases, his calorimeters and three of his specialized and beautiful cabinet balances.

8.3.1.6 La Concierge. ※※ La Concierge (48.856025°, 2.345474°) is located on the Île de la Cité on the Seine River near the Pont Au Change. What is now called the Concierge

Figure 8.3 Lavoisier's laboratory in Musée des Arts et Metiers. © Musée des Arts et métiers-Cnam, Paris/photo M. Favareille.

was part of the former royal palace, the seat of the medieval kings of France, called the Palais de la Cité, which also included the Palais de Justice and the Sainte-Chapelle. In 1358, King Charles V moved the palace to the Louvre and appointed a concierge to take care of the old palace. To this day, a concierge is a care-taker or a "keeper". In 1391, part of the building was converted for use as a prison and took on the name of the concierge. This former royal palace is where Lavoisier and many others, including Queen Marie Antoinette were held before their execution during the French revolution. Nearly 3000 prisoners were tried and convicted in this building. Here, the condemned prisoners had their hair cut off from the back of their necks before being placed in groups of 12 to await a cart that carried them off to the scaffold. The imprisonment and pre-execution procedures are explained through re-creations of the cells. There is a small exhibit that includes Lavoisier. These reconstructions have been reported to be somewhat difficult to find. They are located near the museum shop.

8.3.1.7 *Place de la Concorde.* (48.865647°, 2.321236°). It was here that Lavoisier faced the guillotine. During the French Revolution, the area was renamed the Place de la Révolution. King Louis XVI was executed here on January 21, 1793 and was followed by Queen Marie Antoinette on October 16, 1793. Lavoisier was tried and convicted at La Concierge and guillotined here on May 8, 1794 at the age of 50, along with 27 co-defendants, all of whom were tax collectors for the Ferme Générale. This octagonal square is magnificent place to visit in its own right, but we scientific/historical travelers can wander around and realize that it was here that one of the greatest chemists of all time was beheaded. Joseph-Louis Lagrange, the great French-Italian mathematician, was quoted to the effect that it took "Only a moment to cut off that head and a hundred years may not give us another like it".

8.3.1.8 *Errancis Cemetery.* This cemetery (literally the "cemetery of the wandering") was only open for four years (1793–1797). It was used to dispose of the corpses of 1119 victims of the guillotine. Lavoisier was interred here, but all the bodies were moved to the Catacombs of Paris around 1848. There is memorial plaque marking the location of the cemetery at coordinates

(48.880774°, 2.316823°). It is on rue de Monceau between number 97 and the corner with rue de Rocher. The marker is directly above the rue de Monceau signpost.

8.3.1.9 City Hall. The Hôtel de Ville, Place de l'Hôtel de Ville (48.856648°, 2.351553°). There is a statue of Lavoisier here . If you are visiting Notre Dame, head north on rue d'Arcole, cross the River Seine and find the magnificent city hall. The square in front of the hotel is considered one of the best in the world. The statue is located on the façade of the city hall.

8.3.1.10 Eiffel Tower. Lavoisier is one of the 72 names of eminent French scientists, engineers and mathematicians inscribed on the Eiffel Tower which was built for the 1889 World's Fair. His name is located on the northwest side right beside that of Ampère. Other atomists on the tower include Gay-Lussac and Becquerel on the southwest side.

8.4 LAVOISIER'S SUCCESSORS

> "La chimie est une science francaise; elle fut constitute par Lavoisier d'immortelle mémoire." ("Chemistry is a French Science. It was founded by Lavoisier of immortal memory.") French chemist, Adolphe Wurtz, 1868.

It is not difficult to see why Wurtz maintained that chemistry was a French science. Lavoisier had changed chemistry forever, but he did not live long enough to see those changes come to full fruition. It was left to Claude Berthollet, Joseph Gay-Lussac, André Marie Ampére, and even the temporary Frenchman, Amadeo Avogadro to carry on the legacy. When John Dalton visited Paris in 1822, he couldn't meet with Lavoisier, but he did meet with his successors.

8.4.1   Claude Louis Berthollet (1748–1822)

Claude Berthollet had been one of the first to accept Lavoisier's new theories. He and his colleagues and students continued with the new gravimetric- and oxygen-based chemistry first advocated by their friend. Berthollet was a co-author with Lavoisier of the

Methods of Chemical Nomenclature (1787). He and Lavoisier also helped to establish the metric system of units.

Berthollet was born in Talloires, near Annecy, in eastern France and received medical degrees from both the University of Turin and the University of Paris. He arrived in Paris in 1772 and by 1780 had been elected to the Royal Academy of Sciences. He bought a house in the then-Paris-suburb of Arcueil and had a magnificent laboratory installed there. From 1807 to 1813 this house served as the regular meeting place of what became known as the Society of Arcueil (Sociéte d'Arcueil). Under this Society's auspices, a variety of young scientists (including Jean Biot, Alexander von Humboldt, and Joseph Gay-Lussac) experimented, read, and subsequently published their work.

Although he had reservations at first, Berthollet ultimately accepted Dalton's Law of Multiple Proportions and Proust's Law of Definite Proportion. However, he always remained skeptical of Dalton's atomism. Unlike the somewhat unlucky Lavoisier, Berthollet was a major player in French politics. He met Napoleon in Egypt in 1798 and wound up teaching the emperor chemistry. Under the aegis of the emperor, Berthollet became a senator, a count, and a grand officer of the Legion of Honor. Berthollet was active in many areas of chemistry, including bleaching, dyeing, iron and steel making, and munitions. He also helped establish the Polytechnic School (the École Polytechnique) in 1794. Some have called him the Doyen of French Chemistry during the first two decades of the nineteenth century.

Napoleon Bonaparte (Napoleon I, Emperor of the French from 1804 until 1814, and again in 1815) had a great interest in science, engineering, and mathematics. Tanford and Reynolds, in their book *The Scientific Traveler*,[2] state that "No other ruler of any modern country has been anywhere near Napoleon's equal as a patron of science". The Napoleonic Age is often taken as 1795 to 1815, but even before that, as a young Army Officer, Napoleon helped establish the École Polytechnique (1794) in Paris, which soon became a leader in the education of engineers. The word "polytechnic" is still found in the names of American schools like Rensselaer Polytechnic Institute, as well as in related terms at institutes of technology like MIT, Georgia Tech and Cal Tech. He was a great supporter of many scientists, including Berthollet. Oftentimes, scientists were invited to discuss and demonstrate their discoveries with the enthusiastic emperor. One prominent example is when Alessandro Volta (1745–1827) was invited to Paris in 1801 to demonstrate his "voltaic piles" to Napoleon. Voltaic piles were the first chemical batteries. One is on display at the Musée des Arts et Métiers. For more on Volta and his voltaic piles, see pp. 229–231 in Chapter 9. Humphry Davy used a giant voltaic pile located in the basement of the Royal Institution in London to isolate six elements in two years (1807–1808). See Chapter 5, p. 101 for more on these experiments.

8.4.2 Travel Sites Related to Claude Louis Berthollet

8.4.2.1 Annecy. Annecy is located in eastern France, near the borders with both Switzerland and Italy. It is located about 20 miles south of Geneva, Switzerland. Here there is a statue of Berthollet in the Jardins de l'Europe (45.899287°, 6.132247°). On the top of the pedestal there are four small scenes done in relief. One shows Berthollet with Napoleon Bonaparte at the pyramids of Egypt (1798–1801).

8.4.2.2 Arcueil. The house where Berthollet established the Societe d'Arcueil stood at 16–18 rue Berthollet in Arcueil, just south of Paris (48.803667°, 2.330667°). It no longer stands and has been replaced by La Caisse des Dépôts et Consignations. There is a plaque there to the memory of Berthollet.

8.4.3 Joseph Louis Gay-Lussac (1778–1850)

As a chemistry professor or student, you may have thought that Gay-Lussac was a bit of a strange name. It turns out that Antoine Gay, Joseph's father, acquired a piece of property, a "croft", named Lussac and then added this place name to his surname to produce the new surname of Gay-Lussac. Joseph was born in 1778 and was home-schooled until safely after the French Revolution ended. In 1795, Joseph was sent to Paris, where after several years he enrolled at the new Polytechnic School (the École Polytechnique) under Berthollet, who readily befriended him.

Young Joseph immediately made his mark. In 1802, he discovered the law that states that the volume of a gas increases with temperature. (This is often called Charles's Law, but it is more properly called Gay-Lussac's Law. Although Jacques Charles discovered the law, he did not think it important enough to publish his results.) In 1804 Gay-Lussac and Jean-Baptiste Biot started taking balloon flights to test the composition of the air at high altitudes. In a subsequent solo flight, he reached a height of four miles, higher than the tallest peak of the Alps!

One amazing story about Gay-Lussac was how he met his wife. Upon entering a draper's shop in Paris, he encountered Geneviève Rojot, a charming girl of 17. Gay-Lussac noticed that she was reading a book in between waiting on customers. To Gay-Lussac's amazement, the book was a treatise on chemistry. One could say that this book served as a catalyst. Geneviève's education had been interrupted by the revolution, but Gay-Lussac arranged for her to resume her studies at his expense. In 1808, they were married and had a long, happy marriage resulting in five children. Geneviève read German and English, painted, and was a musician.

Also in 1808 he was appointed professor of physics at the Sorbonne, a post he held for 24 years. In his first year he discovered the element boron by reacting potassium (just discovered by Davy in 1807, see Chapter 5, p. 104) with boron oxide. Most importantly for our purposes, as we investigate the development of the atomic concept, Gay-Lussac made a discovery in 1809 that had significant implications about the nature of atoms and molecules. Unfortunately, his discovery put him in conflict with his friend and mentor, Berthollet.

Gay-Lussac's second law regarding gases is called the Law of Combining Volumes. Using data from other investigators, as well as his own, he discovered that gases combine in proportions by volume that could be expressed in ratios of small whole numbers. For example, two volumes (say 2 liters, *i.e.*, 2 L) of hydrogen combined with 1 volume (say 1 L) of oxygen to form 2 L gaseous water. What, if any, it was asked, are the implications of this result for atomic theory? How does one make sense of this new Law of Combining Volumes? Since the volumes are in ratios of whole numbers and Dalton's just-published atomic theory said that atoms combine in whole number ratios to make molecules, doesn't it make sense that the volume ratios must be reflective of the atomic ratios. Doesn't it make sense that equal volumes of gases contain equal numbers of atoms? [Sometimes this assumption is abbreviated as EVEN; Equal Volumes (of gas) contain Equal Number of atoms, molecules, or particles in general.] This would mean that if 2 volumes of hydrogen combine with 1 volume of oxygen to make 2 volumes of water vapor then it follows that 2 atoms of hydrogen combine with 1 atom of oxygen to make 2 molecules of water. And, just by the way, since it takes 2 atoms of hydrogen but only one atom of oxygen to make water, then a water molecule must contain two atoms of hydrogen and one of oxygen. Its formula must be H_2O.

Dalton was certainly not going to buy any of this. Following his rule of greatest simplicity, Dalton adamantly maintained that water was HO not H_2O. (See Chapter 4, p. 87, for more on this.) And furthermore, regardless of which formula of water is correct, if two atoms of hydrogen combine with one atom of oxygen to make water, that would mean that we will have to split the oxygen atom in half to make the two molecules of water. Dalton was *certainly* not going to buy that. Atoms cannot be split. So, what is the answer to this set of dilemmas? The answer seemed to be that the fundamental particles of hydrogen and oxygen gases were not atoms; they were molecules. Specifically (using the simplest assumption possible), they were diatomic molecules, that is, each molecule of hydrogen contained two atoms of hydrogen. The same with oxygen – each of its molecules contain two atoms of oxygen. The way to represent these molecules is by writing H_2 and O_2. Furthermore, since Gay-Lussac had found that

> 2 volumes of hydrogen + 1 volume of oxygen produces 2 volumes of water,

we can use the assumption that equal volumes of gases contain equal numbers of particles (molecules in this case), to write that

> 2 molecules of hydrogen + 1 molecule of oxygen produces 2 molecules of water

Or, using the formulas for the hydrogen, oxygen and water molecules,

$$2H_2 + 1O_2 \text{ produces } 2H_2O$$

Or, finally, using modern chemical notation, $2H_2(g) + O_2(g) \rightarrow 2H_2O(g)$. The (g) indicates that the reactants and products are all gases. The number 1 in front of the O_2 is assumed.

All of this makes good sense to us in the twenty-first century, but at the time it was most controversial. Berthollet did not believe in atomic theory. Dalton certainly did, but would not accept the Law of Combining Volumes, the idea of H_2O, or the fact that the fundamental particles of gaseous oxygen and hydrogen are diatomic molecules. As we will see in a moment, Amadeo Avogadro's hypothesis came along within two years to explain Gay-Lussac's law, but it was ignored for half a century. In the meantime, Gay-Lussac was a young man caught in somewhat of a bind. Initially a closet atomist, he gradually became a more vocal advocate, particularly so after Claude Berthollet died in 1822.

8.4.4 Travel Sites in Paris Related to Joseph Louis Gay-Lussac

8.4.4.1 Eiffel Tower. As we have noted, Gay-Lussac's name is one of the 72 French scientists, engineers, and mathematicians inscribed on the Eiffel Tower. His name is located on the southwest side. (Lavoisier and Ampère are on the northwest side.)

8.4.4.2 Cimetière du Pere Lachaise. 6 rue du Repos (48.861407°, 2.393328°). This is an amazing graveyard to visit if you have time. The gravestones and small chapels of this beautiful, large (70 000 tombs) and famous cemetery are squeezed right on top of each other. This is the final resting place of some of the most famous people in the world. The big names prominently touted are almost always artists, musicians, authors, and actors like Chopin, Molière, Sarah Bernhardt, Jim Morrison, Isadora Duncan, Marcel Proust, Gertrude Stein, and Oscar Wilde. Most people are not scientific/historical travelers, of course, but those of us who are know that this is also where Joseph Gay-Lussac is buried in Division 26. Get a chart and see if you can find him and/or use these coordinates: 48.861944°, 2.394167°. Interestingly, at least on one chart he is listed as "GAY-LUSSAC Louis savant (phys. et chimie)". Joseph Guillotin is listed as a famous interree but not Joseph Gay-Lussac. Other scientists buried here include the mathematician and astronomer Pierre Laplace and the chemist Henri Moissan. Moissan received the 1906 Nobel Prize in Chemistry for his isolation of the element fluorine. Dmitri Mendeleev was also a nominee that year, but Moissan edged him out by one vote. Mendeleev died the next year and never received a Nobel Prize.

8.4.4.3   Gay-Lussac Café.   15 rue Royer-Collard (48.846032°, 2.340891°). Although this café has no connection to the scientist, it is on the corner of rue Royer Collard and rue Gay-Lussac. Across the street is a plaque marking the site of the laboratory of Pasteur. You can enjoy a good meal and a pint and talk about Joseph Gay-Lussac if you can find anyone who will listen.

8.4.5   Amedeo Avogadro (1776–1856) and André-Marie Ampère (1775–1836)

Turin, where the Italian chemist Amedeo Avogadro started his professional career as an ecclesiastic lawyer, was under the control of the French who, led by Napoleon Bonaparte, had recently overrun the northern Italian states. Avogadro started his work in physics and chemistry in 1800 and published his early papers in French. He was familiar with Dalton's atomic theory (1808) and Gay-Lussac's Law of Combining Volumes (1809). In a relatively obscure journal in 1811 he published a "landmark article" that he evidently regarded as an extension of Gay-Lussac's work. His paper supports the idea (EVEN) that equal volumes of gases, at the same temperature and pressure, contain equal numbers of atoms, molecules, or other particles. This statement is often termed Avogadro's hypothesis. He also suggested that elementary molecules might be diatomic. Avogadro spent his entire life in the Turin area and never had direct personal access to Parisian chemists such as Gay-Lussac or Berthollet. He was teaching at the Royal College of Vercelli in 1811 when he published his paper. There is a marker in Vercelli that we will describe in Chapter 9, p. 237.

In 1814 André-Marie Ampère, using a different methodology, published a similar theory in a more prestigious journal that commanded a bit more attention. (It was evidently in the form of a long letter to Berthollet.) He too advocated the EVEN theory and the possibility of diatomic molecules of elementary substances. Despite its logic, the so-called Avogadro-Ampère gas theory seems to have been disregarded for some 50 years. Only in 1860 at the Karlsruhe Congress was its significance finally recognized through the work of Stanislao Cannizzaro. This work and the travel sites related to it will be described in the next chapter. We will see how this dilemma was finally resolved.

Ampère went on to describe the science of electric currents in motion, which he called electrodynamics. It is now referred to as classical electromagnetism. He differentiated this area from electrostatics, the study of stationary electric charges, in which Ben Franklin's (and Joseph Priestley's) work had been so pivotal. The unit of electric current, the ampere (or amp for short) is named in his honor. Officially, an ampere is a measure of the amount of electric charge passing a given point in an electric circuit per unit of time.

8.4.6 Travel Sites Related to André-Marie Ampère

8.4.6.1 Eiffel Tower. As we have noted, Ampère's name is one of the 72 French scientists, engineers, and mathematicians inscribed on the Eiffel Tower, which was built for the 1889 World's Fair. His name is located on the northwest side along with Lavoisier. (Gay-Lussac is on the southwest side.)

8.4.6.2 André-Marie Ampère Plaque. The plaque is located above a branch of HSBC bank at 29bis, rue Monge (48.846466°, 2.352014°). He lived at this location from 1818 until 1820.

8.4.6.3 Montmartre Cemetery. (48.887929°, 2.329884°) 20 avenue Rachel, Île-de-France, Paris, Division 30 (48.8868°, 2.3300°). Ampère's life was short and marred by several tragedies. His father was beheaded in the French Revolution and his wife died at an early age. He died at 50 in Marseille, but is buried here at the Cimetière de Montmartre beside his brother, Jean-Jacques, an important language scholar, who was named after André's father. This world-renowned scientist did not have an easy life; the epitaph on his gravestone reflects that difficult lot as it reads "Tandem felix" ("Happy, at last"). The cemetery with its striking sculptures and ornate gravestones is only a short distance from Sacré-Cœur Basilica (Sacred Heart). It is known for the graves of other prominent figures including Hector Berlioz, Edgar Degas, Alexandre Dumas, Stendahl, and Émile Zola. The latter was moved to the Pantheon in 1908, but his family's gravestone is still there with Émile's name on it. Also buried here is Adolph Sax, inventor of the saxophone. This beautiful cemetery was built on an abandoned gypsum quarry used as a mass grave during the French revolution.

8.4.7 Louis Pasteur (1822–1895)

Although Louis Pasteur was not an atomist *per se*, a scientific/historical traveler should not leave Paris without a visit to the Pasteur Museum at the Pasteur Institute. We can tie Pasteur into a discussion of atomism in that John Dalton did not speculate on the shape of his molecules, even though his pictographs look nicely symmetrical and seem to imply a shape. It appears that Pasteur, in 1848, was one of the first to seriously consider the shapes of molecules and their interactions with light. (He actually discovered that molecules can come in left- and right-handed forms and that each "hand" interacts with polarized light in a different manner). Later the Dutch chemist Jacobus van't Hoff (1852–1911) and the French chemist, Joseph Le Bel (1847–1930) continued this early consideration of the three-dimensional structures of molecules. Pasteur, of course, was more famous for his discoveries that prevented various diseases such as rabies, syphilis, and anthrax and for the process which bears his name, pasteurization, which he applied to milk, wine, and beer.

8.4.8 Travel Sites Related to Louis Pasteur

8.4.8.1 Commemorative Plaque. 45 Rue d'Ulm (48.842312°, 2.344311°). This plaque ※ marks the site of Pasteur's Laboratory on the side of a building of the École normale superieure.

8.4.8.2 Pasteur Museum, Pasteur Institute. 25 rue du Doctor Roux (48.840376°, 2.311249°). This outstanding scientific/historical travel site ※※※※ includes: (1) the apartment where Pasteur and his family spent the last seven years of his life; (2) a scientific souvenir room that traces his major discoveries based on the original scientific instruments he used; and (3) a magnificent Byzantine-style crypt where Louis Pasteur is buried alongside his wife. In the apartment, look for a small stained-glass window showing a chemist at work. The souvenir room features his copper equipment shined up to a high luster and the intricate glassware used for sterilization. The crypt is walled with colorful mosaic tiles depicting his major works. Be sure not to miss the ceiling where at the four corners are inscribed the words Foi (faith), Esperance (hope), Charite (charity), and ... wait for it ...

not Amour (love) but rather Science (science). Sometimes love is just not enough!

8.5 SUMMARY

Benjamin Franklin was a mentor to Joseph Priestley and a friend to Antoine Lavoisier. After visiting the Franklin House in London, we now have the opportunity to stand near Franklin's statue near the Eiffel Tower. Priestley was an avowed phlogistonist and called the gas he had isolated from the calx of mercury "dephlogisticated air". For nearly a century, the phlogiston theory had explained processes such as burning, rusting, and breathing in terms of negatively massed phlogiston being released into the air. Lavoisier, on the other hand, believed that a component of the air is actively consumed during these processes. Lavoisier maintained that Priestley's gas was that component. Lavoisier renamed dephlogisticated air oxygen and maintained that it was a component of all acids. Lavoisier, being independently wealthy from the profits he made from being a "tax farmer", commissioned the building of extremely accurate balances that he used in his research. Using these balances, he established the Law of Conservation of Mass. His *Elementary Treatise on Chemistry* and *Methods of Chemical Nomenclature* (the latter co-authored with Berthollet) are the basis of what is often called the "chemical revolution" or the new French chemistry. Unfortunately, Lavoisier's affiliation with the Ferme générale (the tax farm) ultimately resulted in his losing his life at the guillotine in the French Revolution. There are travels-with-the-atom sites all over Paris that mark Lavoisier's birthplace, christening, college, laboratory, and places of imprisonment, execution, and unidentified burial. Of greatest interest is the portion of his laboratory that is preserved in the Musée des Arts et Métiers. He is honored at the Place de l'Hôtel de Ville and at the Eiffel Tower.

Lavoisier's successors include the atomic skeptic Claude Louis Berthollet and the atomist Joseph Louis Gay-Lussac. Berthollet had been one of the first to accept Lavoisier's new theories. After his contemporary's death, Berthollet and his colleagues and their students continued with Lavoisier's new gravimetric- and oxygen-based chemistry. Berthollet ultimately accepted Dalton's

Law of Multiple Proportions and Proust's Law of Definite Proportion, but was skeptical of Dalton's atomism. Gay-Lussac, on the other hand, was an atomist. His Gay-Lussac's Law (also sometimes known as Charles' Law) established the relationship between the temperature and the volume of a gas. His Law of Combining Volumes, together with the work of Andre-Marie Ampère and Amadeo Avogadro (often collectively referred to as Avogadro's hypothesis) supported the atomic theory and put it on a firm foundation. Furthermore, the work of Gay-Lussac, Ampère, and Avogadro supported the possibility of diatomic molecules and the formula of water being H_2O rather than Dalton's HO. Unfortunately, these results were not accepted by the wider chemical community until 1860. Gay-Lussac and Ampère are honored at several sites in the City of Light, including the Eiffel Tower and their gravesites in the Cimetière du Pere Lachaise and Montmartre Cemetery, respectively. No scientific/historical traveler should leave Paris with having visited the Pasteur Museum at the Pasteur Institute.

ADDITIONAL READING

- C. Djerassi and R. Hoffmann, *Oxygen, A Play in 2 Acts*, Wiley-VCH, New York, A video recording of the University Theatre's production of this play was produced by Educational Innovations, Inc. but may no longer be available.
- S. Johnson, *The Invention of Air, A Story of Science, Faith, Revolution, and the Birth of America*, Riverhead Books, Published by the Penguin Group, New York, 2008.
- J. Jackson, *A World on Fire, A Heretic, An Aristocrat and the Race to Discover Oxygen*, Viking published by the Penguin Group, New York, 2005.
- M. S. Bell, *Lavoisier in the Year One, The Birth of a New Science in Age of Revolution*, Atlas Books, W. W. Norton & Company, New York, 2005.
- International Historic Chemical Landmarks, Antoine-Laurent Lavoisier: The Chemical Revolution, American Chemical Society, 1999, http://www.acs.org/content/acs/en/education/whatischemistry/landmarks/lavoisier.html, accessed September 2019.

- S. W. Weller, Napoleon Bonaparte, French Scientists, Chemical Equilibrium, and Mass Action, *Bull. Hist. Chem.*, 1999, **24**, 61.
- P. J. T. Morris, *The Matter Factory, A History of the Chemistry Laboratory*, Reaktion Books, Ltd., London, 2015.

REFERENCES

1. J. L. Marshall and V. R. Marshall, *Rediscovery of the Elements: Phlogiston and Lavoisier*, The Hexagon, 2005.
2. C. Tanford and J. Reynolds, *The Scientific Traveler: A Guide to the People, Places & Institutions of Europe*, John Wiley & Sons, Inc., New York, 1992.
3. Lavoisier in his laboratory conducting an experiment on the respiration of a man at work, Photogravure after M.A.P. Lavoisier, *ca.*, 1850, https://wellcomecollection.org/works/dfb7cw3n, accessed June 2019.

CHAPTER 9

Atoms Go South: The Italians Volta, Avogadro, and Cannizzaro (Italy)

9.1 A QUICK LOOK AT PLACES TO VISIT "TRAVELING WITH THE ATOM" IN ITALY

In Pisa: Ammannati House (Galileo birthplace), the Domus Galilaeana (Galileo and Fermi documents), the Campanile Pisa (leaning tower), Basilica di Santa Croce; in Florence: The Museo Galileo and the Basilica di Santa Croce (Galileo tomb) and the Cabinet of Physics at the Foundation for Science and Technology ; in Rome: the Giordano Bruno statue in Campo de' Fiori ("Field of Flowers") ; in Pavia: the Cabinet of Physics of Alessandro Volta in the Museum of the University of Pavia; in Como: Temple Voltiano (Volta Temple) , Life Electric , Statue in Piazza Alessandro Volta beside the Caffe Alessandro Volta , Plaque marking Volta birthplace on Via Alessandro Volta; Torre di Porta Nuovo or Torre Gattoni (tower where Volta carried out many of his experiments), Plaque beside Campora country-house, Volta Lighthouse ; in Camnago-Volta: Neoclassical Volta Tomb ; in Vercelli: Avogadro Bust ; in Quarenga. Avogadro mausoleum ; in Rome: The Museum

---

Traveling with the Atom: A Scientific Guide to Europe and Beyond
By Glen E. Rodgers
© Glen E. Rodgers 2020
Published by the Royal Society of Chemistry, www.rsc.org

of Chemistry "Primo Levi" ◉◉◉ *at the Sapienza University of Rome (Cannizzaro equipment and documents); in Palmero: Cannizzaro Tomb, Palermo Pantheon in the Church of San Doménico* ◉◉*; in Karlsruhe, Germany, The Baden Ständehaus* ◉*, the site of the 1st International Chemical Congress of 1860.*

> **In the Galilean Tradition: Voltaic Piles and the Role of Cannizzaro in Resolving the Dalton-Avogadro Controversy**
>
> - No chapter on Italian science could be complete without mentioning Galileo (1564–1642), whose works arguably mark the beginning of modern science. A quick reminder of Galileo's innovative, quantitative experiments precedes a note about his belief in atoms and heliocentrism, the resulting controversy with the church, and the fate of Giordano Bruno (1548–1600), who was burned at the stake for advocating similar views during Galileo's lifetime.
> - Alessandro Volta (1745–1827), who corresponded with Ben Franklin (1706–1790) and Joseph Priestley (1733–1804) about electricity, invented the first battery in 1799. The construction of his "voltaic piles" grew out of his skepticism regarding Luigi Galvani's "animal electricity". Volta's discovery changed, almost overnight, the approach to atoms, elements, and all of physical science. Napoleon invited him to Paris to demonstrate his battery.
> - During the first half of the nineteenth century, chemists and physicists were not willing to recognize that John Dalton (1766–1844) had stubbornly refused to acknowledge the contributions of Amedeo Avogadro (1776–1856). At the first International Chemical Congress, held in Karlsruhe, Germany, in 1860, Stanislao Cannizzaro (1826–1910) successfully promoted his countryman's hypothesis and thereby solved the problem of finally determining an accurate scale of atomic weights.

We cannot possibly start a chapter involving Italian science without mentioning Galileo, whose works, many say, mark the beginning of modern science. He was known not only for his innovative, quantitative experiments, but also for expressing his results in the language of mathematics. His study of swinging chandeliers, falling bodies, inclined planes, and the moon, Sun, and planets using his small telescope did not inform him about atoms, of course, but it's clear that, in fact, he did believe in them. (The idea of atoms had been around since the Greeks had described them some two millennia prior to Galileo.) Galileo maintained that ordinary solids and liquids were made up of infinitely small particles, that is, what we call atoms. On the right track, you say, and not particularly controversial. However, as it turned out, the church had a problem with atoms. The church believed the doctrine of transubstantiation, that is, that the communion bread and wine were literally the body and blood of Christ. How could this possibly be true if the bread and wine were composed of atoms? Wouldn't parishioners have to imagine that the atoms that originally made up Christ's body and blood were now miraculously transported to each communion table every time the Eucharist was served?

In the early seventeenth century, it was not a good idea to go against the teachings of the church. In 1600, Giordano Bruno (1548–1600) was burned at the stake for expressing his beliefs in heliocentrism, the possibility of life on other planets, atoms, and other heretical ideas. Galileo was 36 when Bruno was burned at the stake in Rome's Campo de' Fiori ("Field of Flowers") (41.895568°, 12.472049°). This outdoor market is a great place to visit and ponder the connections between science and religion. Eat at one of the many restaurants, have a craft beer or a glass of wine, buy some fresh fruits and vegetables, maybe some olive oil but, whatever you do, don't miss walking around the statue of the hooded Giordano Bruno to see the reliefs depicting his trial and execution and the inscription that translates to "And the flames rose up". Campo de' Fiori is also an important scientific/historical site because it was here that the 14-year-old Enrico Fermi, still recovering from the tragic death of his older brother, Giulio, bought a mathematical physics book and started his Nobel-prize-winning career. (See Chapter 17 for more about Enrico Fermi.)

As you might expect, there are a number of Galileo landmarks that scientific/historical travelers should consider visiting when they get the chance. In Pisa, these might include the Ammannati House, *Via* Giuseppe Giusti 24, (43.7165°, 10.4064°) where he was born in 1564, or perhaps the *Domus Galilaeana* at 26 *via* Santa Maria (43.718457°, 10.397207°). Founded in 1942 to mark the 300th anniversary of his death, the latter is a specialized history of science library, a center that houses Galilean documents and manuscripts and, as it turns out, some manuscripts by Enrico Fermi regarding the work he and his research group did in Rome. (Again, see Chapter 17.) These two places are not far from the famous leaning tower (Campanile Pisa), the legendary site of Galileo's experiments on falling bodies. (There is a Latin inscription over the entrance to this effect but, be careful, such inscriptions sometimes have to be taken with a mole of salt.) Finally, there is the Basilica di Santa Croce or Cathedral of Santa Maria where, according to tradition, Galileo's study of swaying lamps led to his conclusions about the isochronism of swaying pendulums (see Chart 9.1 to locate these sites).

The best place to see Galileo memorabilia is Florence. The Museo Galileo (43.76773°, 11.2559°), the former *Istituto e Museo di Storia della Scienza* (Institute and Museum of the History of Science), is located in Piazza dei Giudici, along the River Arno and close to the Uffizi Gallery. The museum houses a major collection of scientific instruments. On the first floor

Chart 9.1   Travel sites related to Galileo Galilei. These includes sites in Rome, Pisa, and Florence.

we find the Medici Collections (fifteenth to eighteenth century) that include an unparalleled collection of Galileo artifacts, including his two telescopes and the ebony- and ivory-framed objective lens from the telescope he used to discover the "Galilean moons" of Jupiter. Here, we also find the burning lens given to Grand Duke Cosimo III de' Medici in 1697. This is the lens used in 1814 by Humphry Davy to show that diamonds and graphite are both composed of carbon (See Chapter 5, p. 105). The second floor has the Lorraine Collections (eighteenth to nineteenth century), with items pertaining to electricity, electromagnetism, and chemistry, the Grand Duke Peter Leopold's chemistry cabinet, and the elaborate machines made to illustrate the fundamental laws of physics. Be sure to investigate the new Monumental Sundial, built in 2007, outside the museum.

Perhaps the most poignant Galileo site in Florence is the Basilica di Santa Croce or the Church of Santa Croce (43.768550°, 11.262280°). When he died in 1641, his will directed that he be buried here in a vault with other family members. However, the Holy Office in Rome forbade this public burial because he had "caused scandal to all Christendom by his false and damnable doctrine". (Presumably, this included his belief in atoms, as well as his advocacy of heliocentrism.) Accordingly, he was buried privately in the bell tower of the Novitiate of Santa Croce. It took nearly a century before, in 1737, a colorful and elaborate tomb was built for Galileo Galilei in the nave, directly across from the tomb of the great artist and architect, Michelangelo Buonarroti. Galileo's tomb is shown in Figure 9.1. Rather bizarrely, when Galileo's body was moved, various appendages, including three fingers, a tooth, and vertebrae were removed. One of these fingers, the middle finger of his right hand, is on display at the Galileo Museum. Presumably, he used his index finger to point to the Galilean moons of Jupiter when explaining his findings to church officials. When discussing the church's reception of his results, however, he more likely used the "Bras d'honneur" or "Iberian slap" instead of the lone finger on display. The Church of Santa Croce contains many paintings, frescoes, sculptures, sacred vestments, and stained-glass windows. There are many other graves here, including those of Rossini, Machiavelli and, for us travelers with the atom, a plaque honoring Enrico Fermi. There is magnificent statue of Dante Alighieri outside the entrance.

Atoms Go South 227

Figure 9.1 Galileo's tomb in the Basilica di Santa Croce, Florence. Reproduced from [https://commons.wikimedia.org/wiki/File:Galileo%27s_tomb.jpg] with permission from Benjamin Cousins, under the terms of a CC BY-SA 4.0 license [https://creativecommons.org/licenses/by-sa/4.0/deed.en].

Florence, as many readers will know, has many top-notch places to visit, including Michelango's David at the Accademia Gallery and the famous Uffizi Gallery. Scientific/historical travelers should include a visit to the Foundation for Science and Technology (Fondazione Scienza e Tecnica) at Via Giusti 29 (43.777280°, 11.263555°) in Florence. The cabinet of physics has over 3000 machines and appliances preserved in their original furnishings. It is the largest collection in Italy and one of the most complete on the European continent of scientific physics instruments from the second half of the nineteenth century.[1]

9.2 ALESSANDRO VOLTA (1745–1827)

> The Voltaic pile is "the most marvelous instrument which men have yet invented, the telescope and the steam engine not excepted", François Arago

How many scientists do you know that have had a magnificent neoclassical temple built in their honor? Not many, that's for sure. Alessandro Volta is the only one that we will encounter in our traveling-with-the-atom adventures. The Tempio Voltiano (Volta Temple) in the picturesque northern lake town of Como, Italy, is a magnificent place to visit. There are Volta landmarks (statue, lighthouse, square, cafe, street, plaque marking his birthplace, tomb) all around that scenic lakeside city and the nearby university town of Pavia. Let's investigate what Volta did to deserve these fantastic landmarks.

Alessandro Volta was one of nine children. His father, Filippo Volta, was of noble lineage and his mother came from a family of counts. He developed a bit slowly, not learning to talk until he was four. This slow development did not last long, however, and by the age of 7 he had become one of the brighter students in his class. He was fluent in many foreign languages including Latin, French, and English, and he had the ability to read German, Dutch, Spanish, Russian, and old Greek.

By the age of 14, he had decided he wanted to be a natural philosopher, or what we would call a physicist. Recall that Benjamin Franklin had encouraged Joseph Priestley, in 1771, to write his *History and Present State of Electricity* (See Chapter 3, p. 54). The publication of this book established Priestley's reputation and apparently spurred Volta's interest in electricity. By 1774, Volta was teaching in Como and writing to Franklin about his electrical experiments. In 1775, he invented what is called the "electrophorus", which is a simple electrostatic generator, not based on continuous friction and rotation. In 1779, Volta was appointed professor at the University of Pavia, where he was admired for his clear, lucid, lively, self-effacing, but imposing manner. As an aside, he also is credited with the discovery of methane gas in 1776 in Lake Maggiore. He isolated the gas in 1778. You might recall that John Dalton also investigated the occurrence of methane in a lake near him. This is celebrated in the Ford Madox Brown mural in Manchester town hall (see Chapter 4).

Volta was aware of the experiments of Luigi Galvani (1737–1798), a professor of medicine at the University of Bologna, who thought he had discovered "animal electricity" when, in the 1780s, he observed twitching in frogs' legs that he

Atoms Go South

had suspended from an iron railing using brass hooks. They twitched when subjected to an electrical current from an "electrical machine" or when they were put out in a thunderstorm. We probably don't want to know the details of how they came to be out in that thunderstorm. Galvani believed that animals generated electricity in their bodies and that a fluid within their nerves carried electricity to their muscles, causing movement. This was a fascinating idea, and it didn't take storytellers long to use it in their short stories and novels. For example, Mary Shelley, in 1817, wrote *Frankenstein,* in which a creature made of the body parts of dead people is brought to life by Doctor Frankenstein using electricity from a lightning storm.

Although Galvani's observations made front page news, Volta was skeptical. In 1794, he did the experiment without the frogs! Just putting the brass and iron together with a little moisture in between them generated a current. Galvani had enthusiastic followers and they defended the frog connection for years. Volta was galvanized (pun intended) into action, and soon made an ironclad argument for his point of view that did not involve "animal electricity". No brass knuckles were involved as far as we can tell.

Volta found that brass and iron were not the only metals that could produce an electrical current. Other metals he used included zinc, lead, tin, iron, copper, silver, gold, and even graphite. By looking at various combinations, he determined that the most effective pair of dissimilar metals to produce electricity was zinc and copper. One of his voltaic piles using alternating disks of these metals separated by cardboard or felt disks moistened with a salt solution is shone in Figure 9.2. When a wire is connected to the top and bottom of this assembly, an electrical current is continuously and steadily generated. In 1799, at the University of Pavia, Volta announced that he had invented a battery. Prior to this time, physicists had had to depend on static electricity to do their electrical studies. They generated the static electricity using frictional wheels of various types (sometimes called "electrical machines") and then stored it in Leyden jars. The competition between Galvani and Volta persisted. Ultimately, Galvani's point of view was not confirmed but the electrical current generated by a Voltaic pile was often called "galvanic electricity". We also still

Figure 9.2 A voltaic pile on display in the *Tempio Voltiano* (the Volta Temple) near Volta's home in Como. Reproduced from [https://commons.wikimedia.org/wiki/File:VoltaBattery.JPG] with permission from GuidoB, under the terms of a CC BY-SA 3.0 license [https://creativecommons.org/licenses/by-sa/3.0/deed.en].

use terms like galvanized steel (steel coated with a thin layer of zinc) and galvanometer, which is an instrument used for detecting an electrical current.

So voltaic piles were the first continuous source of current electricity, that is, they were the first batteries. But how, you ask, do they work? How is it that an electrical current is produced by these bimetallic systems separated by a salt (or an "electrolytic") solution. We can imagine that Volta had the same questions. A full explanation would have to wait for almost a century, when J. J. Thomson discovered the electron (See Chapter 5). Furthermore, we and Volta need to know about forces that move electrons, that is, "electromotive forces", which is a bit above the level of this travel book. Suffice it to say that the current is produced because the zinc atoms have a lesser attraction for electrons than the copper atoms. The zinc atoms lose their electrons,

Atoms Go South 231

which react with the ions in the electrolytic solution. The copper atoms, as it turns out, have no direct role in the chemical reaction that produces electricity. The greater the number of copper and zinc disks alternately stacked one upon another, the greater the potential (*i.e.*, the greater the "voltage") for electrons to flow and a current to be produced. Volta spoke of the "contact tension" that produced the electric current when the metals came in contact. He did not realize that a chemical reaction produced the electricity and that the salt solution played a significant role in that reaction. While he was a bit fuzzy as to how it all worked, he had, in fact, found a way to produce a steady electrical current and this knowledge changed the course of chemistry and physics almost immediately.

The young Napoleon Bonaparte (1769–1821), newly appointed as head of the "French Army of the South" and recently married, invaded northern Italy in 1796–1797. In 1805, Napoleon established a "Kingdom" of Italy that liberated it from the Austrians. Now, as it turned out Napoleon was a great fan of the sciences. Volta's discovery of an electric battery had quickly spread all over Europe and over to England. As we know, Humphry Davy (see Chapter 5) used a huge voltaic pile to isolate six elements from their oxides in just two years. It didn't take long for scientists to use a battery to separate water into its component elements, hydrogen and oxygen. Napoleon couldn't wait to see this great discovery for himself, so he invited Volta to Paris to demonstrate it to him and his Institute of France. As a result, one of the great paintings in the Temple Voltiano is the one by Giuseppe Bertini shown in Figure 9.3. Napoleon and Volta enjoyed a close relationship and the emperor conferred a number of honors on him including, in 1805, making him a Knight of the Legion of Honor. In 1810, Napoleon made him a Senator of the Kingdom of Italy and a count.

9.3 TRAVEL SITES RELATED TO LUIGI GALVANI

9.3.1 Galvani Statue

, Piazza Galvani, Bologna (44.491828°, 11.343043°). The statue, located behind the huge Gothic Basilica di San Petronio, shows Galvani contemplating one of his frogs. From the square, one

Figure 9.3 Alessandro Volta demonstrating his battery (called the "Voltaic Pile") to Napoleon, 1801.

finds the entrance to the panoramic terrace of the basilica, which affords a great view of the city. Inside the basilica are 22 art-filled side chapels.

9.4 TRAVEL SITES RELATED TO ALESSANDRO VOLTA

See Chart 9.2 for the location of these sites.

9.4.1 Pavia, Italy

9.4.1.1 The Cabinet of Physics of Alessandro Volta. ✹✹✹✹, The Museum of the University of Pavia, Corso Strada Nuova, 65 (45.186770°, 9.156107°). Volta was at the University of Pavia between 1778 and 1819. The collection includes various piles (including one hooked to a tube in which the electrolysis of water could be carried out), torpedo fish (a flattened fish capable of producing electrical discharges), eudiometers (graduated glass tubes in which mixtures of gases can be made to react by an electric spark, used to measure changes in volume of gases during chemical reactions), an electrophosphorus (a type of electrostatic generator), Leyden chairs of various shapes, his desk and chair, Volta "pistols" and a statue in a nearby courtyard.

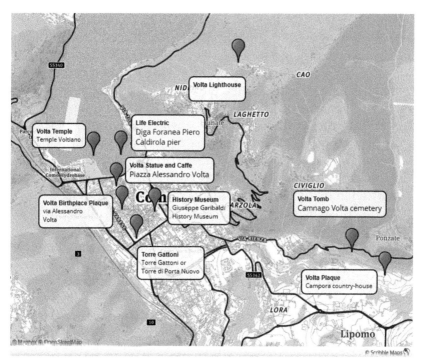

Chart 9.2 Travel sites in Como, Italy related to Alessandro Volta.

9.4.2 Como, Italy

As noted earlier, Como is a terrifically scenic town located on Lake Como in the Northern Lake District. A funicular connects the center of Como with Brunate, a small village (1800 inhabitants) on a mountain at 715 meters above sea level. The journey takes about 7 minutes and the view is worth the trip. The recreational and scenic lakeside and mountain-side attractions are all the more special due its celebration of the life of Alessandro Volta, one of its most famous men of science.

9.4.2.1 Tempio Voltiano (Volta Temple). (45.814866°, 9.075300°) Viale Guglielmo Marconi, 1. Originally, a grand celebration had been planned to celebrate the 100th anniversary of the invention of the battery in 1899. However, a huge fire destroyed the exhibition and severely damaged many of the Volta relics, which were subsequently stored, according to the temple guide book, in a "dismal room of the Civic Museum". The festivities were put off and planned to coincide with the 100th anniversary of Volta's death. For scientific travelers, this was a fortunate occurrence because the town decided to build a magnificent temple dedicated to Volta. The temple, shown in Figure 9.4, was built on a pedestal to protect it from floods. Inspired by the Pantheon in Paris, it was a cube with a cylindrical drum atop it and the whole thing was covered by a domed structure with a central light. The vestibule at the front has four Corinthian columns of *Aurisina* stone. Inside, a mosaic floor was constructed using marble from all over the world. Volta's relics, partly reconstructed, partly salvaged, are all grandly assembled and displayed in elaborate cabinets. Devices on display include eudiometer (a device in which to carry out reactions between gases, set off by a spark), phlogo-pneumatic pistols (inflammable air pistols that used the explosion to shoot bullets), electrostatic machines, Leyden jars, electroscopes and electrometers, devices to test Galvani's animal electricity, numerous examples of Voltaic piles and travel boxes. Four "stucco high-relief panels depicting episodes of Volta's life" are displayed in the balcony just under the dome. The temple was opened with a grand celebration in 1928. It was attended by a number of scientific celebrities of the day, including Enrico Fermi.

Figure 9.4 The Temple Voltiano on the shore of beautiful Lake Como surrounded by the Italian Alps. Photograph by Glen Rodgers.

Atoms Go South

9.4.2.2 Life Electric. (45.815387°, 9.080309°). This contemporary stainless-steel statue was built in 2015 and dedicated to Alessandro Volta. It is accessible to visitors at the end of the long Diga Foranea Piero Caldirola pier. The entrance can be reached from the Volta Temple by walking down the Lungolago Mafalda di Savoia. From here, walk south on the Viale Guglielmo Marconi and find the Piazza Alessandro Volta, with his tall statue.

9.4.2.3 Alessandro Volta Statue. (45.811788°, 9.079338°) beside the Caffe Alessandro Volta, Piazza Alessandro Volta. This is great place to have a coffee or tea or a stein of beer or glass of wine with lunch and discuss the life and accomplishments of the great Alessandro Volta, whose discovery of a battery changed, almost overnight, the approach to atoms, elements and all of physical science. From here, walk down to Via Alessandro Volta.

9.4.2.4 Volta Plaque. marking Volta's birthplace (45.807431°, 9.081356°), Via Alessandro Volta 62. Continue walking south to the corner of Viale Varese and Viale Carlo Cattaneo.

9.4.2.5 Torre di Porta Nuovo or Torre Gattoni. , Viale Varese, 11 (45.805471°, 9.082886°). This tower was part of the medieval walls that surrounded the city. It was used as a physics laboratory from 1783 to 1806. Giulio Cesare Gattoni was a mentor to Alessandro Volta. Volta carried out many of his experiments in this facility, but it can only be viewed from the outside. On the top of the tower there was a lightning rod, the first one ever installed in Italy.

9.4.2.6 Volta Plaque. beside the Campora country-house (Via Campora 1) (45.801145°, 9.126286°), now a private residence but a marker on the wall notes Volta's love of this place in summer and fall.

9.4.2.7 Volta Tomb. in Camnago Volta cemetery, Via Luigi Clerici, Camnago Volta (45.803669°, 9.120311°). *Via* a royal decree in 1863, the name Volta was appended to the name of the village, Camnago. Like the temple, this mausoleum is in the neoclassical style. It is adorned with statues and reliefs. For example, on either side of the entrance gate, two statues representing

Science and Religion guide the visitor in. Once inside, we see a relief behind the sarcophagus that shows Volta presenting his pile to Napoleon in Paris in 1801.

9.4.2.8 Volta Lighthouse. Volta lighthouse (Faro Voltiano), emits an alternating rotating green, white, and red light celebrating the Italian flag. This lighthouse is not on the lake shore as you might expect, but rather on top of a hill, it is said "to celebrate the grandeur of Alessandro Volta". Are there other scientists with lighthouses built in their honor? Perhaps, who would they be?

In preparation for visiting places associated with Stanislao Cannizzaro (see Section 9.7), travelers might want to visit the nearby Giuseppe Garibaldi History Museum at Piazza Medaglie d'Oro, 1, in Como (45.808500°, 9.086169°). This building was Garibaldi's headquarters during his campaigns for the Risorgimento, the campaign for the liberation and unification of Italy. Inside is a collection of dresses, weapons, documents, and furniture of the time. The museum is housed in a palace that dates back to the fifteenth century.

9.5　AMEDEO AVOGADRO (1776–1856)

> "My studies of the natural sciences have particularly involved that part of physics which looks at the atomic world …" Amedeo Avogadro

As if one Italian count was not enough, we now encounter Count Lorenzo Romano Amedeo Carlo Avogadro di Quaregna e di Cerreto! In Chapter 8, we discussed the proposals of Joseph Louis Gay-Lussac, Amedeo Avogadro and André-Marie Ampère, which would have greatly speeded up the progress of chemistry, specifically the problem of establishing accurate atomic weights, if only John Dalton had accepted them. Check Chapter 8 for all the details, but recall that Avogadro's Hypothesis would have led to the conclusion that water was H_2O and that hydrogen and oxygen naturally occur as the diatomic molecules, H_2 and O_2. Unfortunately, Avogadro published in an obscure journal

(*Journal de Physique*) and, unlike Volta, stayed put in Italy and did not have contact with other scientists. He certainly did not go to Paris to meet with Napoleon. International travel has definite advantages.

Amedeo Avogadro was born in Turin, Italy, on August 9th, 1776 (a good year from the author's perspective). His family background was aristocratic. His father, Filippo, was a magistrate and senator who had the title of Count. His mother was a noblewoman. Amedeo Avogadro inherited the title of Count from his father. Avogadro first published in 1803, only 3 years after that other count, Volta, first demonstrated his electric battery. By 1809, Avogadro had abandoned a successful legal practice (he had a doctorate in canon law) and was a teacher of natural philosophy at the Royal College of Vercelli. Here, he published his hypothesis in 1811. He published another paper in *Journal de Physique* in 1815 discussing atomic weights and the nature of compounds. In 1820 he became Professor of Mathematical Physics at the University of Turin. There was political turmoil, his professorship was interrupted for 10 years, but Avogadro persevered and published extensively in the late 1830s. It all fell on deaf ears. He died in 1856 (at age 80) without proper recognition.

9.6 TRAVEL SITES RELATED TO AMEDEO AVOGADRO

See Chart 9.3 for the location of these sites.

9.6.1 Vercelli, Italy

9.6.1.1 Avogadro Bust. (45.329133°, 8.417761°) is located between the beautiful Basilica Sant'Andrea and the railway station. His bust sits atop a modest plinth, where Via Galileo Ferraris meets the roundabout. Vercelli is where Avogadro spent "his most productive years as a professor at a local college". The statue was erected on the 100th anniversary of Avogadro's death. Vercelli is a city in the midst of a large rice-producing region. It is one of the oldest urban sites in northern Italy, founded around 600 BC, and is definitely off the beaten tourism path. Note that the statue is labeled Amedeo Avogadro di Quarenga.

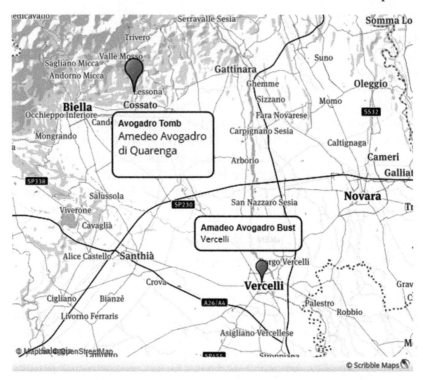

Chart 9.3 Travel sites related to Amedeo Avogadro. These includes sites in Vercelli and Quarenga.

9.6.2 Quarenga, Italy

9.6.2.1 Avogadro Family Tomb. ※ (45.582683°, 8.165058°) 13854 Quaregna Bl, Amedeo Avogadro di Quarenga and his wife are buried in this little cemetery in Quarenga, a few kilometers north of Vercelli. The tomb is past the entrance gate on the left. If you look through the grate you can see a terracotta bust of Avogadro and the tombstone he shares with his wife.

9.7 STANISLAO CANNIZZARO (1826–1910)

Cannizzaro was born in Palermo on the island of Sicily, well before the establishment of a united Italy (1861). He attended the University of Palermo, where he studied medicine but soon moved to Naples and then to the University of Pisa where he pursued chemistry. Returning home, he briefly fought in the Sicilian

Table 9.1 Two sets of Atomic Weights. Set 1 if Dalton correct; Set 2 if Avogadro correct.

Atom	If Dalton correct (1810)	If Avogadro correct
H	1	1
C	5.4	12
N	6	14
O	7	16
P	9	31
S	13	32
Fe	50	56

revolution, but when that faltered in 1849, he was forced to move from Sicily. For many years he was an itinerant chemist serving as a Professor of Chemistry at a variety of universities. He was professor of chemistry at the University of Genoa, when it became clear that something had to be done about establishing consistent and accurate atomic weights.

So, here's how things stood in about the mid-nineteenth-century. The so-called "Dalton–Avogadro debate" was in full swing. Dalton had died a famous but stubborn man in 1844. The revered and highly respected Swedish chemist Jöns Jacob Berzelius (See Chapter 19) who, like Dalton, had rejected Avogadro's idea, had died in 1848. Avogadro himself had died in 1856, his work unrecognized. Cannizzaro came across Avogadro's Hypothesis in 1858 and wrote a paper entitled *"Sunto de un corso di filosofia chimica"* ("Sketch of a course in chemical philosophy"). In his "Sunto" Cannizzaro demonstrated how Avogadro's Hypothesis would allow chemists to measure atomic weights unambiguously. The question of accurate atomic weights was certainly unresolved. Table 9.1 shows two sets of values, one if Dalton was correct, water was HO and hydrogen and oxygen existed as just atoms. The second set is if Avogadro was correct, water was H_2O, and hydrogen and oxygen existed as diatomic molecules, H_2 and O_2. These are distinctly different values. Clearly, some decisions had to be made.

The 1st International Chemical Congress was held in 1860 in Karlsruhe, at that time placed in the small kingdom of Baden, just across the Rhine from France. August Kekulé (he who discovered that benzene was a 6-membered ring of carbon atoms) and several others convened the by-invitation-only Congress for the express purpose of resolving the roles of atoms and molecules and determining, once and for all, a system of atomic weights. Among the 127 invitees, we travelers-with-the-atom would recognize the names Lothar Meyer, Dmitri Mendeleev, Robert Bunsen, and Emil Erlenmeyer. No invitations were sent to the United States. So ... what happened in Karlsruhe? At first, it seemed that there would be no resolution of these crucial questions. Cannizzaro gave an animated speech about the validity of Avogadro's conclusions about atoms and molecules but left the conference probably feeling that he had failed to persuade anyone of Avogadro's solution to the problem. However, copies of his "Sunto" were reprinted as a pamphlet and widely distributed to the conferees. As these leading chemists of the day returned to their homes, it soon became obvious to many that Cannizzaro and Avogadro were, in fact, correct. For example, Lothar Meyer, who with Dmitri Mendeleev would shortly go on to formulate the periodic table, said, "I too got a copy, which I put in my pocket to read on my long journey home [to Breslau]. I read it again and again and was astonished at the light which the writing threw on the most important points at issue. The scales fell from my eyes, doubt vanished, and a sense of the calmest certainty took its place." So, in the end, Avogadro's ideas were accepted. Elements could exist in the form of molecules, rather than individual atoms. The hydrogen atom, H, is the smallest portion of an element which enters into an H_2 molecule, which makes up the H_2 compound. Formulas were liberated from Dalton's principle of simplicity and the second set of atomic weights in Table 9.1 were established.

Cannizzaro had to leave the Congress earlier than he intended so he could return to Sicily to rejoin the effort to establish a united Italy. In 1861, he became a professor of inorganic and organic chemistry at the University of Palermo, which soon became a well-known center of chemical education. In 1871, he moved to the University of Rome, where he was appointed the chair of the new chemistry department and became a member of the Senate in the new Kingdom of Italy. At the university he transformed the old Convent of San Lorenzo (in Panisperna) into the first Italian Institute of Chemistry. Stanislao Cannizzaro was a hero of revolutionary movements – one that put the discipline of chemistry in a solid footing and one that resulted in the unification of Italy. He died in Rome in 1910, but his body was brought back to his native city of Palermo in 1926, the centenary of his birth. He was buried with due honor in the Palermo Pantheon (the church of San Doménico).

9.8 TRAVEL SITES RELATED TO STANISLAO CANNIZZARO AND KARLSRUHE

9.8.1 The Museum of Chemistry "Primo Levi"

The Sapienza University of Rome (41.902421°, 12.513858°) Piazzale Aldo Moro 5, Rome preserves and exhibits historical scientific equipment. Specifically, it houses a collection of chemicals and documents that belonged to Stanislao Cannizzaro and his group. Primo Levi wrote a variety of intriguing books. A book that every scientific/historical traveler should read is *Uncle Tungsten: Memories of a Chemical Boyhood*.[2]

9.8.2 Cannizzaro Tomb, Palermo Pantheon in the Church of San Doménico

Via S. Lorenzo, 1 – Villa Castelnuovo, Palermo. (38.116667°, 13.366667°). This is located in Saint Dominic Square in the neighborhood of La Loggia in the historic center of the city. During the Sicilian revolution of 1848, the Sicilian Parliament met here. In 1853, the church became the pantheon of illustrious Sicilians. The baroque-era church was recently renovated to restore its original gold and white colors.

9.8.3 The Baden Ständehaus

Ständehausstr. 2 (49.008849°, 8.399931°) in Karlsruhe, Germany, was the site of the 1860 Chemical Congress. The Congress was held in the meeting room of the second Chamber of State in the old Ständehaus. It was destroyed in the World War II, but the new Parliament building was built on the same site and has a similar central tower. Situated in the rotunda on the ground floor of the Ständehaus memorial, the Ständehaus Information System gives information about the state parliament. At this writing the new Ständehaus contains photos, displays, and other records of the original. The 49th parallel runs through the city center of Karlsruhe. Its course is marked by a stone and painted line in the *Stadtgarten* (municipal park).

9.9   SUMMARY

In this chapter we briefly discuss the influence of Galileo Galilei and some of the best sites in Pisa, Florence, and Rome to visit that commemorate his scientific contributions. Galileo was an atomist (and a heliocentrist, of course) but neither of these views was acceptable to the church. In 1799, Alessandro Volta, trying to disprove the "animal electricity" experiments of Luigi Galvani, constructed a frog-less "voltaic pile", which was the first example of a battery. His revolutionary discovery quickly changed the approach to atoms and the physical sciences. The city of Como in the Italian Lake District celebrates the life of Alessandro Volta in myriad ways – from a temple to a lighthouse and everything in between! Starting in 1811, Amedeo Avogadro had proposed ideas that would resolve the problem of determining unambiguous atomic weights. Unfortunately, his work had been dismissed by the famous but stubborn John Dalton, who maintained to his dying day that water was HO and diatomic molecules could not exist. Dalton's opposition delayed the determination of an accurate set of atomic weights by some 50 years (1808 to 1860). We visit Avogadro sites in Vercelli and Quarenga. It was another Italian, Stanislao Cannizzaro, a Sicilian patriot and itinerant professor of chemistry, who resurrected Avogadro's ideas and convinced his fellow chemists of their value. At the first Chemical Congress held in Karlsruhe, Germany in 1860, the charismatic Sicilian presented his ideas, but left the Congress feeling that he may have not succeeded. As the community of invited participants to the Congress returned to their homes, they realized that Avogadro and Cannizzaro were right. Water is H_2O, hydrogen and oxygen are diatomic molecules in their natural states, and that they all could agree on a uniform set of atomic weights. We visit sites in Rome, Palermo, and Karlsruhe, Germany where Cannizzaro's accomplishments are celebrated.

ADDITIONAL READING

- A. Longatti, *The Volta Temple in Como, A Guide*, Rotary Club, Como Baradello, 2006.
- J. Marshall, *Rediscovering Atoms: An Atomic Travelogue: A Selection of photos from sites important in the history of atoms*, ACS Symposium Series, American Chemical Society, Washington, DC, 2010, vol. 1044, DOI: 10.1021/bk-2010-1044.ch008.
- J. L. Marshall and V. R. Marshal, The Road to Karlsruhe, The Hexagon, 2007, https://digital.library.unt.edu/ark:/67531/metadc111216/, accessed June 2019.
- C. J. Giunta, *Atoms in Chemistry: From Dalton's Predecessors to Complex Atoms and Beyond*, American Chemical Society, Washington, DC, 2010, vol. 1004, DOI: 10.1021/bk-2010-1044.

REFERENCES

1. Benvenuti alla Fondazione Scienza e Tecnica Firenze, http://www.fstfirenze.it/, accessed June 2019.
2. O. Sacks, *Uncle Tungsten: Memories of a Chemical Boyhood*, Alfred A. Knopf, a Division of Random House, New York, 2001.

CHAPTER 10

Questioning the Reality of Atoms on the Ground: Loschmidt, Mach, Boltzmann, and Ostwald (Germany and Austria)

10.1 A QUICK LOOK AT PLACES TO VISIT "TRAVELING WITH THE ATOM" PRINCIPALLY IN VIENNA, GRAZ, AND GROSSBOTHEN

In Vienna: the Arcades of the University of Vienna honoring Josef Loschmidt, Ludwig Boltzmann, Christian Doppler, Erwin Schrödinger, and others; Zentralfriedhof (Central Cemetery) with the graves of Loschmidt and Boltzmann (and the "honorary" graves of many famous musicians); bust of Ernest Mach in the Rathauspark; Landtman Pub where atomists and anti-atomists alike gathered; In Graz, Physics Institute, University of Graz (the UniGraz@Museum Collection); In Munich, The Hofbräuhaus; In Duino, Trieste, Italy, Duino Castle where Boltzmann died by suicide; In Grossbothen, Germany, Ostwald's Energiehaus in the Wilhelm Ostwald Park and Museum; In Riga, Latvia, the Wilhelm Ostwald Monument.

Traveling with the Atom: A Scientific Guide to Europe and Beyond
By Glen E. Rodgers
© Glen E. Rodgers 2020
Published by the Royal Society of Chemistry, www.rsc.org

How Real are Atoms?

- The physical existence of atoms remained in question until the end of the nineteenth century. After the Chemical Congress of 1860, an accurate set of atomic weights had been determined. However, this did not mean that everyone was convinced that atoms were real.
- Some scientists were "chemical atomists" who believed that the concept of atoms was essentially a calculational device, or a fictional construct that only served the purpose of calculating mass relationships in compounds and writing chemical equations that accurately represented how compounds are formed.
- "Physical atomists", on the other hand, did believe that atoms actually existed. Not only did they serve the functions attributed to "chemical atoms", they were hard, massy, unsplittable, impenetrable spheres that combined in definite geometrical arrangements.
- Josef Loschmidt, in Vienna, a physical atomist, contributed to the kinetic-molecular theory (KMT) of gases by calculating the number of gaseous atoms or molecules in one cubic centimeter of a gas and the approximate size of those molecules.
- Ernest Mach and Ludwig Boltzmann, in the last decade of the nineteenth century, vociferously debated whether atoms existed at all. Mach ridiculed those who believed in atoms.
- Boltzmann had built his entire professional life around their existence. The Maxwell–Boltzmann distribution of energies in the KMT depended upon the physical reality of atoms, as did his explanation of entropy embodied in his famous equation: $S = k \log W$.
- Boltzmann also debated the adamant anti-atomist, Wilhelm Ostwald, who believed that energy, not atoms, was the ultimate component of reality.
- Suffering from various debilitating physical conditions and fearing he was losing the atomic debate, Boltzmann died by suicide in 1906.
- In the same year, Wilhelm Ostwald retired to his Energiehaus in Grossbothen and took up the study of color theory. After converting to atomism, he was awarded the Nobel Prize in Chemistry in 1909 for his work in catalysis and investigations into principles governing chemical equilibria and rates of reaction.

10.2 PHYSICAL *VERSUS* CHEMICAL ATOMS (CHEMICAL ATOMISTS, PHYSICAL ATOMISTS, ANTI-ATOMISTS)

After the 1st Chemical Congress of 1860, atoms were now generally (but not universally) accepted and an unambiguous set of atomic weights had been established. A few examples, rounded off to whole numbers, are shown in Table 10.1.

Using these values, we could account for the combining weight ratios of various substances. For example, water, now finally accepted (at last!) to be of formula H_2O, was pictured as a molecule containing two hydrogen atoms and one oxygen atom. Using Table 10.1, the two hydrogen atoms would contribute two units of mass and the one oxygen atom would contribute 16 units. The oxygen to hydrogen mass ratio would be 16 to 2 or 8. Indeed, this result conforms to the fact that if we combine two grams of hydrogen with 16 g of oxygen, we get 18 g of water.

We could now write, with greater authority, chemical equations to represent how compounds form. Let's look again at the equation for the formation of water that we discussed in some detail in Chapter 8.

$$2H_2 + O_2 \rightarrow 2H_2O$$

This equation can be interpreted on several levels. For example, it tells us that two diatomic molecules of hydrogen (H_2) combine with one diatomic molecule of oxygen (O_2) to form two molecules of water (H_2O). Or, expanding the equation a little by adding the state of matter for each reactant and product (gases for the reactants, liquid for the product),

$$2H_2 \,(g) + O_2 \,(g) \rightarrow 2H_2O \tag{10.1}$$

Table 10.1 Some accepted atomic weights.

Element	Atomic symbol	Atomic weight
Hydrogen	H	1
Carbon	C	12
Nitrogen	N	14
Oxygen	O	16
Phosphorus	P	31
Sulfur	S	32
Iron	Fe	56

we could say, for example, that 4 (2 × 2) grams of hydrogen gas combine with 32 (2 × 16) grams of oxygen gas to produce 36 grams of liquid water. So, in this way, the concept of atoms and molecules, combined with known atomic weights, could account for many chemical compounds and the reactions that produce them. Atoms now seemed to be fully established as a useful concept that helped to explain chemical phenomena. But were they real? Did they physically exist? Let's explore that issue in just a bit more detail.

Scientists were of two minds when it came to atoms. "Physical atomists" maintained that, though we could not actually see atoms, they do, indeed, *physically* exist. Alan Rocke in his book, *Chemical Atomism in the Nineteenth Century: Dalton to Cannizzaro*,[1] says they believed in "physical atoms". Dalton was the original physical atomist but Jon Jacob Berzelius, the man who, in 1813, devised our present-day symbols for elements (and their atoms), was also an early advocate. (See Chapter 19 for the complete details about this famous Swedish chemist and atomist and the places we can visit that commemorate him.)

"Chemical atomists", on the other hand, maintained that atoms were merely a useful construct that could explain chemical behavior, but weren't real. These "chemical atoms" were essentially calculational devices or fictional constructs that served the purpose of calculating mass relationships in molecules and predicting chemical behavior that could be represented by chemical equations. Such calculations come under the heading of stoichiometry, a topic that many readers may have studied in high school and/or college with varying degrees of interest and proficiency. Do you recall such calculations? Humphry Davy was certainly an early chemical atomist. Davy seemed to agree with Dalton's basic principle but the details about the nature of atoms – hard, indivisible, spherical – he tended to regard as unsubstantiated speculation.

These two camps vigorously, even vociferously, disagreed with each other. In the late nineteenth century, the three or four decades after the 1860 Congress, chemical atomists seemed to have the upper hand. Physical atoms were described as hard, massy, unsplittable, impenetrable spheres, that combined in definite geometrical arrangements. Chemical atomists referred to such descriptions as unnecessary, indeed as "atomic fictions" or "hyperhypothetical trivialities". (Now that's a fairly sophisticated insult!)

10.3 JOSEF LOSCHMIDT (1821–1895)

> "His work forms a mighty cornerstone that will be visible as long as science exists ... Loschmidt's excessive modesty prevented his being appreciated as much as he could and should have been." Eulogy for Loschmidt delivered by Ludwig Boltzmann.

As these debates continued, more evidence kept coming to light. Some of this related to the kinetic molecular theory of gases. Recall from Chapter 5 that James Clerk Maxwell, in 1860, was working out the mathematics surrounding this theory that pictured the atoms or molecules of a gas moving at great speeds, slamming into each other and the walls of its container billions of times per second, thereby exerting a pressure. Clerk Maxwell had proposed that there was a distribution of atomic/molecular velocities in a gas. Loschmidt took this work a little further and, in 1865, calculated the number of atoms or molecules that would be present in a cubic centimeter (cc) of a container of a gas at what are called standard temperatures and pressures (STP). Loschmidt's estimate was 1×10^{18} molecules per cc and that is startlingly close to modern value of 2.69×10^{19} molecules per cc. (This value is still called the "Loschmidt Number".) He also estimated that the typical diameter of a gas particle would be about 1×10^{-9} m; this also is impressively close to the modern value. Suddenly, atoms and molecules in a gas, albeit incredibly numerous and small, were coming to life. Physical atomists were encouraged.

10.4 TRAVEL SITES RELATED TO JOSEF LOSCHMIDT

10.4.1 Vienna, Austria

10.4.1.1 The Arcades of the University of Vienna. (48.213144°, 16.360049°). Loschmidt became professor of physical chemistry at the University of Vienna in 1868. He and many other atomists are honored here at the Arcades of the University of Vienna. The physics department at the University of Vienna was founded by Christian Doppler (1803–1853), he of the famous Doppler Effect. He is also honored at the Arcades. Doppler is buried in Venice at the Isola di San Michele (45.448968°, 12.347046°) in the Venetian Lagoon. It is an amazing island to visit. Give it a try when you visit Venice.

10.4.1.2 The Gravesite of Josef Loschmidt. Zentralfriedhof (Central Cemetery), 11 Simmeringer Hauptstrasse 234 (48.153274°, 16.440856°). We will have occasion to visit the Arcades and this cemetery in connection with other atomists. One excellent way to get to the cemetery is the No. 71 bus that leaves you off at the main entrance (Tor 2). A custodian at the gate can direct you to Loschmidt's grave. Loschmidt was a good friend of Ludwig van Beethoven, who is buried in Section 32A, Grave 29, which is among the "honor graves" (Ehrengräben) that include not only Beethoven but also Franz Schubert, Johannes Brahms, Johann Strauss II, and Arnold Schoenberg. A memorial to Wolfgang Amadeus Mozart is also in this section, although he is buried in the nearby St. Marx Cemetery.

10.5 ERNST MACH (1838–1916) *VERSUS* LUDWIG BOLTZMANN (1844–1906)

The debate about the physical existence of atoms would rage on for the next three decades of the nineteenth century. It involved strong personalities on both sides, the most prominent among these being Mach and Boltzmann.

Ernest Mach was born in Czechoslovakia, but his family soon moved to Vienna. He earned his PhD at the University of Vienna in 1860 and taught at the University of Graz before moving to Prague in 1867. As a professor of experimental physics there, he studied the flow of air over moving objects. We all know the name Mach from the "Mach Number" used to indicate the speed of an aircraft. "Mach 1" is the speed of sound. Mach always impressed his colleagues as a true genius, but they also reported that he was a charming man with an unassuming demeanor. Mach became a Professor of History and Theory of the Inductive Sciences (Whew! Quite a title!) at the University of Vienna in 1895.

Remarkably, at least from our modern perspective, he was a severe, indeed an adamant, anti-atomist. He didn't believe in *any* atoms, chemical, physical, or otherwise. He believed only the facts, that is, only the direct observations of science. In his opinion, the use of unseen objects (like atoms and molecules) to explain any physical or chemical phenomenon was strictly prohibited, that is, *verboten*. Using tiny billiard-ball type atoms to explain disciplines like thermodynamics, in his not-so-humble opinion, just introduced an unneeded mystical element.

Ludwig Boltzmann was born in Vienna on Shrove Tuesday, the day before the Christian season of Lent begins on Ash Wednesday. He suffered from what we term "bipolar disorder", and liked to tell the story that his sudden changes in spirit back and forth from happiness to depression was because he was born during the dying hours of a Mardi Gras ball. He too took his PhD from the University of Vienna. He recalled that, as he started his work, his advisor, Franz Stefan, immediately handed him a collection of John Clerk Maxwell's papers, essentially hot off the Scottish press, all written in English and all exceptionally difficult to fathom. Fascinatingly, Stefan knew that Boltzmann did not understand one word of English and so included a copy of an English grammar book for his reference! You have to start somewhere!

Boltzmann made significant progress in independently elaborating on Maxwell's work on the kinetic-molecular theory (KMT) of gases and graduated in 1866. At the age of 25, he became full professor of mathematical physics at the University of Graz and stayed there for a total of 14 years. Here, in 1873, he met his future wife, Henriette, a young woman 10 years his junior who had long blonde hair and blue eyes. Boltzmann himself was a short, stout man with an impressive black mustache and beard. In 1876, Boltzmann proposed marriage to Henriette in a letter in which he noted that "permanent love cannot exist if [a wife] has no understanding and enthusiasm for her husband's efforts and is just his maid and not the companion who struggles alongside him". Taking this cue, she enrolled in a mathematics course, but the university soon passed a regulation to exclude women students. An exception was made for her but ultimately, after much back and forth with the authorities, she had to replace mathematics with a course in cookery. Such were the attitudes at that time. Nevertheless, it was a fascinating relationship. One other story about the Boltzmanns: even then, professors were underpaid. Henriette once saw a sewing machine in a store-front display. She said to Ludwig, "you are a genius, why don't you build me one of those?" He did. It was a most successful marriage, producing two sons (Ludwig Hugo and Arthur) and three daughters (Henrietta and Ida and, born in Munich after they left Graz, Elsa). The Boltzmanns entertained frequently and the guests included his students. He was an excellent pianist and he often played at these gatherings.

At Graz, he continued to work on improving the KMT and had a strong hand in developing what is known as statistical mechanics. This discipline builds on Loschmidt's calculation of the truly mind-boggling number of atoms and/or molecules in a gas. Boltzmann concluded that these molecules have a distribution of energies that can be roughly represented on a bell-curve (or a Gaussian distribution) that you may have encountered in a statistics class, or perhaps when professors grade on a curve. Most of the molecular energies will be near the middle or the mean with some on either side of that. All of this is at a given temperature. If the temperature of the gas increases, the most probable or mean energy increases, the molecules move faster and slam into each other and the walls of the container with a greater force, thereby increasing the pressure of the gas. This is now known as the Maxwell–Boltzmann distribution of energies and the mathematics supporting it is very challenging. However, scientific travelers should note that the whole basis of this argument is the physical reality of the atoms and molecules of a gas.

Boltzmann's main claim to fame was his work in accounting for entropy. The second law of thermodynamics states that spontaneous processes, that is, those that occur on their own without prodding, are accompanied by an increase in entropy. Originally formulated around 1877, the Boltzmann equation, $S = k \log W$, tells us that entropy (S) increases as the number of ways (W) the atoms or molecules in a "thermodynamic system" can be arranged and still have the same energy. The system might be defined as a closed container at some constant temperature. For example, in a solid, the atoms or molecules are in contact with each other and are quite orderly and therefore W is relatively small. If the temperature of the system is increased and the substance is converted to a liquid and then a gas, the atoms and molecules, now separated from each other, begin to randomly fly around the container, and can assume a much larger number of different arrangements that have the same energy. W is now much larger and therefore the entropy of a gas is much larger than the solid. k is the "Boltzmann constant" and is sometimes called a scaling factor between the molecular level (the W side) and the entropy (S) we measure in the laboratory. Again, the whole basis of this argument is also the reality of the atoms and molecules that make up the compound.

Boltzmann never stayed long in one academic position and held appointments at seven universities during his career, including three separate stints at Vienna. He taught a wide variety of subjects and was, at various times, a professor of mathematics, mathematical physics, theoretical physics, and experimental physics. He was an excellent teacher. His lectures were delivered without notes, accompanied with excellent board work, and were "crystal clear". He was lively, fascinating, witty, and humorous. He had an excellent rapport with his students, including Lise Meitner (See Chapter 16), who took his mechanics course in 1902. She was very impressed. He was kind-hearted, completely open to questions, and even criticism.

When his doctoral advisor, Franz Stefan, died in 1894, Boltzmann took his position at Vienna University. This turned out to be quite a risk. He and Henriette and the family had left Graz four years earlier, just as the university had just completed a new institute of physics that provided an ideal culture for high-quality research. They had settled in at Munich and now they were headed back to Vienna. The very next year, 1895, Ernest Mach also joined the Vienna faculty and for the next few years a vociferous debate raged between these two men. The polemics often occurred during heated Thursday night debates. Iona James, in his fine collection of biographical essays about "remarkable physicists", says that "for Boltzmann it was unbearable to have a 'malevolent' colleague who openly fought the very theory to which he had devoted his entire professional life". Mach infamously often remarked when discussing atoms, "But who has ever seen one!" He basically ridiculed the idea of atoms and regarded Boltzmann as "the last pillar of that bold edifice of thought", that is, atomic theory. There was certainly no love lost between these two highly intelligent men who had unalterably opposing views about the reality of atoms. During these days, Boltzmann's health further declined, and he became more and more depressed as he feared that atomism was about to collapse and all that he had worked for his entire life might well be for naught.

Boltzmann lasted in Vienna until 1900 when, at the invitation of Wilhelm Ostwald, he accepted a position at Leipzig, where this early founder of physical chemistry had built a leading research center. This was a definite improvement for Boltzmann, because he and Ostwald were on significantly better terms than he and Mach. However, it was still a difficult decision because Ostwald, while a personal friend, was also an adamant anti-atomist. Boltzmann could not seem to escape the "perfect storm" building against him and atomism. Ostwald was an "energetist". He did not believe that atoms existed but rather that energy was the essential component of reality. We will return to the details of Ostwald's "energism" in the next section.

Ernest Mach retired after suffering a stroke in 1901. Boltzmann immediately returned to Vienna and was warmly received. His first lecture on "natural philosophy" in 1903 drew enormous attention, was held in the largest lecture hall and, even then, it was "standing room only". He had an audience with Emperor Franz Joseph, who was impressed to hear how crowded his lectures were. Occasionally, he even dined with the emperor. One might think this would be a great culinary experience but, unfortunately for Boltzmann, the emperor ate very little. According to court etiquette, when the monarch finishes, you are finished, and all the plates are promptly removed. Boltzmann, a slow eater, often did not get to enjoy his meal. Hopefully, he enjoyed the conversations.

Boltzmann continued to suffer from fits of depression, some of them due to the cumulative stress of the attacks from Mach and Ostwald. He also suffered from asthma, heavy headaches, and failing eyesight, to the point that he had to have someone read scientific articles to him. Henriette took to writing his manuscripts. It was just too much to bear. In 1906 he, Henriette and his youngest daughter, Elsa, went to Duino, a village near Trieste, famous for its castle perched on a rocky promontory on the Adriatic coast. Here, at the age of 62, Boltzmann died by suicide on September 5, 1906. The following day he had been scheduled to return to Vienna to start his classes.

On September 8 there was a "splendid memorial ceremony" celebrating his life, a life that he spent in the unwavering support for his beloved atoms. It is not an exaggeration to call him a martyr for atomism. He is sometimes called "the man who trusted the atom". In the end, both Mach and Ostwald saw the light and converted to atomism, but they did this after Boltzmann's death.

10.6 TRAVEL SITES RELATED TO ERNST MACH AND LUDWIG BOLTZMANN

See Chart 10.1 for the location of these sites.

10.6.1 Bust of Ernst Mach in Vienna Rathauspark

, Vienna (48.211992°, 16.360060°). This statue in the northeast corner of the park was unveiled in 1926 to mark the 10th anniversary of Mach's death. Note that he "stands" alone here and is not honored in the Arcades of the University of Vienna, as are physicists like Loschmidt and Boltzmann. This is what happens when you lose one of the biggest arguments in the history of science!

10.6.2 Physics Institute, University of Graz

, Universitätsplatz 5, Graz, Austria (47.078384°, 15.448603°). Boltzmann was professor here from 1876 to 1890. This is commemorated by a plaque on the staircase of the Physics Institute. In the 1990s, a multitude of instruments dating from Boltzmann's time was discovered and are termed the UniGraz@ Museum Collection . Some of these instruments were built by Boltzmann himself. They've been beautifully restored and form an outstanding collection, reputed to be one of the best of its kind in the world.

10.6.3 Duino Castle

, Duino, Trieste, Italy. Boltzmann died by suicide near here in 1906. In 2006, a plaque honoring him was placed at the former Hotel Ples (45.773739°, 13.598960°). It simply notes that in this building, the great physicist Ludwig Boltzmann passed away and includes his formula: $S = k \log W$. When we scientific travelers see this simple plaque, we will know why this was such a tragic event. There are two Duino castles here on the Adriatic

Questioning the Reality of Atoms on the Ground 255

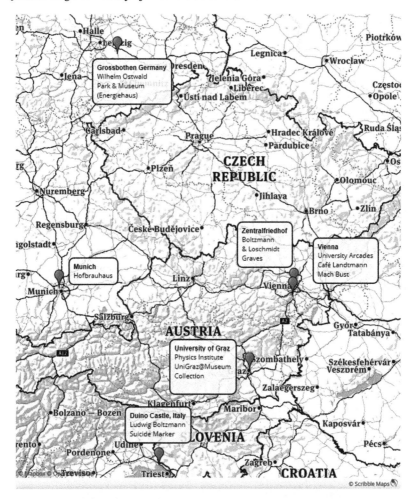

Chart 10.1 Travel sites in Austria and Italy related to Ernst Mach, Josef Loschmidt, and Ludwig Boltzmann. These include places in Vienna and Graz in Austria, Munich and Grossbothen in Germany, and Duino, near Trieste, in Italy.

Coast. The older castle, now in ruins, was built in the 11th century. Shown in Figure 10.1, the "new" castle dates back to 1389 and is open to visitors. Below the ruins of the ancient castle is a white rock that projects into the sea. It is said to resemble a veiled woman, and has inspired a number of legendary tales. This castle is owned by the Thurn and Taxis family and there is a fee for taking a tour. If you haven't visited the Thurn and Taxis castle (49.014465°, 12.092061°) in Regensburg, this is definitely worth putting on your Germany visiting list. It is the site of an extraordinarily beautiful Christmas Market in December.

10.6.4 Boltzmann's Grave

, Zentralfriedhof (Central Cemetery), Vienna. "$S = k \log W$" is permanently etched on Boltzmann's grave in Section 14C, No.1 (48.15172°, 16.43884°). Stand in front of his gravestone, look at the face of this man, and think about all he suffered and all he accomplished. He was a man who stood up for the atom the best he could, but felt he was losing the battle. Note the names of his wife Henriette and his son, Ludwig, who are also buried here. If you are with someone today, be sure they know about Boltzmann's amazing life. (Tell them about the proposal he wrote to Henriette and the "sewing machine story".) The Zentralfriedhof also has the gravesite of Josef Loschmidt and many important musicians such as Brahms, Beethoven, Schubert, Mozart, and the Strausses.

10.6.5 Café Landtmann

, Universitätsring 4, Vienna (48.211561°, 16.361516°). This famous café was established in 1873 and most likely hosted

Figure 10.1 The Thurn and Taxis Duino Castle, Trieste, Italy. Ludwig Boltzmann took his own life near here. Reproduced from ref. 2 [https://commons.wikimedia.org/wiki/File:Castello_di_Duino_0904.jpg] under the terms of a CC BY-SA 3.0 license [https://creativecommons.org/licenses/by-sa/3.0/deed.en].

Questioning the Reality of Atoms on the Ground 257

chemical atomists, physical atomists, and anti-atomists. It was also a favorite of Sigmund Freud, whose house, now the Sigmund Freud Museum at Berggasse 19 (48.218610°, 16.363060°) is fairly close by. The Café Landtmann is definitely a place to take students, friends, and companions to discuss arguments about the physical reality of atoms.

10.6.6 The Hofbräuhaus

Munich (48.137570°, 11.580290°). Boltzmann would meet his students and colleagues here once a week to discuss academics and have a beer. Boltzmann, a rather itinerant scholar, as we have seen, was a professor in Munich for four years, from 1890 to 1894. He left here to go to Vienna, where Mach joined him on the faculty a year later. There are common tables here, where you can drink a beer with folks from all over the world. If the opportunity presents itself, be sure to tell them about Ludwig Boltzmann and Ernest Mach and their atomic debates. The nearby Beer and Oktoberfest Museum (48.135315°, 11.580117°) is definitely worth a visit! Figure 10.2 shows the author and his wife lifting a toast to Ludwig Boltzmann at the Hofbräuhaus.

Figure 10.2 Glen and Kitty Rodgers lifting a toast to Ludwig Boltzmann in the Hofbräuhaus in Munich.

10.7 WILHELM FRIEDRICH OSTWALD (1853–1932)

"What we call matter is only a complex of energies which we find together in the same place", Wilhelm Ostwald

As noted above, Ostwald was one of the founders of the sub-discipline we call physical chemistry. He also coined the term "mole", the molecular weight of a substance in grams, that so many chemistry students know so well from their study of stoichiometry. Some of Ostwald's early work involved experimentally verifying the dissociation theory of Svante Arrhenius. [Arrhenius was the first to propose that a salt like sodium chloride, NaCl, breaks up into sodium and chloride ions (Na^+ and Cl^-) when it dissolves in water. See Chapter 19 for an account of how this early support of Arrhenius eventually played a hand in Ostwald finally receiving the Nobel Prize in Chemistry in 1909.] Ostwald invented (and patented in 1902) what is still known as the "Ostwald process", which is used to manufacture nitric acid from ammonia. In this case, a platinum catalyst is needed to speed along the reaction. Investigating the effects of temperature, pressure, and catalysts on the rates of reactions, the area of "chemical kinetics" is one of the principal areas of investigation for physical chemists. With the advent of the "Haber–Bosch process" (completed about 1911) for preparing ammonia, the two processes meant that Germany would be able to make its own munitions (based on nitrates derived from nitric acid) in World War I and thereby extend the war, it is estimated, by two years. Nitrates are also one of the bases of fertilizers, so the Ostwald and Haber–Bosch processes have also had a huge positive effect in increasing crop yields all over the world. Fritz Haber was one of the most fascinating characters in the history of chemistry. We will discuss the man, his work and places to visit in Dahlem outside Berlin in Chapter 17. Carl Bosch and the museum in Heidelberg named after him will be discussed in Chapter 12.

As noted above, Ostwald was committed to picturing energy, not atoms, as the chief component of the universe. Moreover, not unexpectedly, he also saw energy as the key to personal happiness. He even had a "Happiness Formula" that he claimed yielded a measure of the "degree of happiness", G.

$$G = k(A + W)(A - W),$$

where A denotes the energy flow associated with welcome activities that produce a pleasant or happy feeling and W the energy spent on unwelcome activities that produce displeasure or unhappiness! The constant k (definitely not the Boltzmann constant by the way!) expresses the conversion of the energetic terms into the psychological feeling that we perceive as happiness. It's difficult to imagine, of course, how the magnitudes of A, W and k would actually be measured but it speaks to Ostwald's character that he desired to express his state of mind in mathematical terms.

The very same year (1906) that Boltzmann took his own life after despairing about his poor health and perhaps a fear of losing the atomic debates, Ostwald also decided he had had enough of these stressful dialogues. He had spent one term as the first "Exchange Professor" at Harvard University in 1904–1905, came back and taught one semester at Leipzig and then retired at the age of 53. He settled in his country estate in Grossbothen, Germany, about 25 miles from Leipzig. Not surprisingly, he called his principal residence the "Energiehaus". Once retired, Ostwald started another whole set of scientific investigations in color theory. He built his own equipment, devised instruments for measuring colors, produced a vast set of paint specimens, founded a factory that produced paint boxes, and wrote several textbooks on the theory of color and its history. Moreover, he was an accomplished amateur painter.

Of course, during his entire professional life, he was an antiatomist. (He even tried unsuccessfully to derive the laws of definite and multiple proportions without using the atomic hypothesis.) Ultimately, however, he had to abandon his anti-atomism. The evidence continued to mount and became overwhelming (or was it over-Wilhelming?). J. J. Thomson had discovered the electron in 1897 (See Chapter 5). Ernest Rutherford ("the crocodile") and his "boys" had thoroughly investigated nuclear transformations in 1902 (See Chapter 6). Albert Einstein, in one of his seminal papers of 1905, had explained Brownian motion in terms of atoms and molecules colliding with macroscopic particles (See Chapter 15). Moreover, it was becoming apparent that Ostwald was a strong candidate for one of the newly instituted Nobel Prizes. The conferees at the Swedish Royal Academy were not inclined to give one of these prizes to an anti-atomist (See Chapter 19.) He relented and, accordingly, received the Nobel Prize in Chemistry in 1909 for his work in catalysis and investigations into principles governing chemical equilibria and rates of reaction. Somewhat ironically, Ostwald went on to serve ten years on the International Committee on Atomic Weights from 1906 to 1916. Other Nobel laureates that served on the commission include Frederick Soddy, Niels Bohr, Otto Hahn, and Marie Curie.

10.7.1 Wilhelm Ostwald Park and Museum

Grimmaer Str. 25, Grossbothen, Germany (51.1922°, 12.7428°). The centerpiece of the park is the "Energiehaus", shown in Figure 10.3. The museum is in the Energiehaus and here travelers can see Ostwald's many paintings, his extensive library, and his truly unique homemade laboratory equipment including lamps, spatulas, cutters, weighing and dispensing boats, *etc.*, many of which were fashioned from card stock. He devised an elaborate color system and infused the resulting paints into threads, water colors and oils. His color cubes, wheels, and portable patches are on display, as is his portable paint box with its ink pen and reservoir. He also had extensive correspondence with a variety of other scientists and philosophers. This correspondence is fully catalogued and, particularly if you make arrangements in advance, you may be able to request to see his correspondence with another scientist of your choosing. (Boltzmann would be a good choice.) In addition to the museum in the Energiehaus, other houses called the Glückauf, Werk, and Waldhaus provide modern facilities for meetings, guest accommodations and dining. (The Werkhaus was originally the factory where Ostwald's various colored paints were manufactured.) By the way, in the Energiehaus, look for the plaque that acknowledges Allegheny College's support of this facility. Allegheny decided to offer this support after the author's group of students visited here in 2004. Ostwald lived a long life after his retirement and died

Figure 10.3 The Energiehaus in the Wilhelm Ostwald Park and Museum, Grossbothen, Germany. Photograph by Glen Rodgers.

in 1934. His ashes and those of his wife and various family members are interred in an open quarry on the grounds of the Wilhelm Ostwald Park.

10.7.2 Wilhelm Ostwald Monument

near the Art Noveau Building, the Art Noveau Centre, Riga, Latvia (56.959249°, 24.108349°). Ostwald was born in Riga, Latvia (then part of the Russian Empire) and was a professor at what now is called Riga Technical University from 1881 to 1887. Outside the fence along Krišjāṇa Barona Street there is a bas-relief memorial to Ostwald, who won the Nobel Prize in Chemistry in 1909.

10.8 SUMMARY

The 1st Chemical Congress of 1860 finally established a consistent set of atomic weights. Using these, chemists could account for the combining weight ratios of various substances and write equations for their production. Atoms had been fully established as a useful concept to explain chemical phenomena, but scientists were still divided about whether they actually existed or not. After all, who had ever seen one! "Physical atomists" believe they did exist but "chemical atomists" thought they were merely convenient fictions. Josef Loschmidt, a physical atomist, extended James Clerk Maxwell's kinetic-molecular theory (KMT), calculated the number of atoms or molecules in one cubic centimeter of a gas and estimated the diameter of a typical gaseous atom or molecule. Ernest Mach, he of the "Mach Number", was a prominent and articulate anti-atomist who believed that thinking in terms of atoms introduced an unneeded mystical element. Ludwig Boltzmann, a champion of physical atomism all his life, working in Graz and Vienna, continued the extension of the KMT and worked out the mathematics of what became known as the Maxwell–Boltzmann distribution of molecular energies in a gas. He also was able to explain the concept of entropy, the subject of the second law of thermodynamics, in terms of the randomness of a collection of atoms and molecules. Boltzmann and Mach, at the University of Vienna from 1895 to 1900, engaged in vitriolic debates that raised emotions on all sides. Boltzmann, who suffered from undiagnosed bipolar disorder, as well as asthma and failing eyesight, moved to Leipzig to escape this confrontational atmosphere, but soon found himself embroiled in an extension of the atomic debates with Wilhelm Ostwald, an energetist who believed that the universe should be explained in terms of energy and certainly not in terms of atoms. When Mach retired in 1902, Boltzmann returned in relative triumph to Vienna. However, the cumulative stress of the long-standing debates and his ill health soon led him to take his own life while on holiday in 1906 near the Duino Castle in Italy. He was literally a martyr for atomism. Ostwald retired to his country estate in Grossbothen in 1906, converted to atomism and took up a whole new scientific career in color theory. The Arcades of the University of Vienna, the Zentralfriedhof in Vienna, the Physics Institute in Graz, the Duino Castle in Trieste, Italy, and the Wilhelm Ostwald Park and Museum (with Ostwald's "Energiehaus") are among the sites that travelers with the atom are encouraged to visit.

ADDITIONAL READING

- I. James, *Remarkable Physicists: From Galileo to Yukawa*, Cambridge University Press, Cambridge, 2004, Ludwig Boltzmann, pp. 168–176.
- C. Cercignani, *Ludwig Boltzmann, The Man Who Trusted Atoms*, Oxford University Press, Oxford, 1998.
- W. L. Reiter, Vienna: A Random Walk in Science, in *The Physical Tourist: A Science Guide for the Traveler*, ed. J. S. Rigden and R. H. Stuewer, Birkhäuser Verlag AG, Basel, Boston, Berlin, 2009.

REFERENCES

1. A. J. Rocke, *Chemical Atomism in the Nineteenth Century: Dalton to Cannizzaro*, Ohio State University Press, Columbus, 1984.
2. Castello di Duino 0904, https://commons.wikimedia.org/wiki/File:Castello_di_Duino_0904.jpg, accessed June 2019.

CHAPTER 11

Lighting the Dark Path to Atomism: Spectroscopy Shows the Way: Fraunhofer, Bunsen, and Kirchhoff (Germany I)

11.1 A QUICK LOOK AT PLACES TO VISIT "TRAVELING WITH THE ATOM" IN MUNICH AND HEIDELBERG

In Benediktbeuren, Fraunhofer Glassworks or Historische Fraunhofer-Glashütte ; In Munich: Fraunhofer Glassmaker Relief , The Municipal Museum or Münchner Stadtmuseum , Statue of Fraunhofer , Alter Südfriedhof and the Deutsches Museum ; In Göttingen: Plaque at Bunsen Residence ; In Bad Durkheim Salinen/Gradierbau, ; In Heidelberg: Bunsen Statue and Bunsen and Kirchhoff Plaque on Hauptstrasse, Bunsen's Laboratory and Residence and Bunsen Plaque on Plöck; Stadthalle (Town hall) on Neckarstaden (Portraits of Robert Bunsen, Hermann von Helmholtz, and Gustav Kirchhoff); Robert Bunsen Grave Bergfriedhof, Bunsen Exhibits at the Hörsaal of the Chemistry Department , Kirchhoff Exhibits at the Kirchhoff-Institut für Physik , Philosopher's Way (Philosophenweg) , Mendeleev

---

Traveling with the Atom: A Scientific Guide to Europe and Beyond
By Glen E. Rodgers
© Glen E. Rodgers 2020
Published by the Royal Society of Chemistry, www.rsc.org

Residence, August von Kekulé plaque, Pharmacy Museum (Apothekenmuseum) in Heidelberg Castle.

> **Spectroscopy Reveals the Fingerprints of Atoms**
> - Joseph von Fraunhofer, one of the best glassmakers of the early nineteenth century, passed a narrow beam of sunlight through his excellent prism and discovered the black lines or missing wavelengths in the solar spectrum that still bear the name "Fraunhofer lines".
> - Robert Bunsen and Gustav Kirchhoff were familiar with the "flame tests" that characterized various salts and metals. Bunsen invented a laboratory burner that produced a clean, almost colorless flame that would not interfere with these flame tests. They passed a narrow beam from these flames through a prism and found that each element had a characteristic "bright-line" spectrum, which functioned as a fingerprint of the atoms making up these elements.
> - They used this chemical spectroscopy to identify two new elements, cesium and rubidium, whose names were derived from the prominent lines in their spectra.
> - Kirchhoff found that the bright-line spectra of many elements exactly matched the dark-line spectra of Fraunhofer lines and therefore he could identify what elements existed in the Sun.
> - Using spectroscopy, Anders Ångstrom identified hydrogen in the Sun. Helium was discovered in, and named after, the Sun by Pierre Janssen.
> - Other elements including thallium, indium, and gallium were discovered spectroscopically.
> - Johann Balmer constructed his empirical equation that used integers to generate the wavelengths of hydrogen line spectra. No one could explain why this empirical equation worked.

11.2 JOSEPH VON FRAUNHOFER (1787–1826)

"Look here, I have succeeded at last in fetching some gold from the sun" Joseph von Fraunhofer

"Hence the path is opened for the determination of the chemical composition of the Sun and the fixed stars." Robert Bunsen.

Sometimes, what seems to be an early-life tragedy turns into extraordinarily good fortune. Joseph von Fraunhofer was born in Bavaria, the eleventh child of a glazier, who worked to cut and install glass products. He was orphaned at age 11 and apprenticed to a glassmaker. In 1801, when he was 14, the building where he was working collapsed and he was buried alive for four hours. His rescue was led by the Prince Elector, who would become King Maximilian Joseph I of Bavaria. After his rescue, the future king provided Joseph with books and persuaded his rather exacting employer to allow the young man time to study them. With this help, he was able to learn to read and write and begin to study some science. In 1806, he went to the Institute at Benediktbeuren, a secularized Benedictine monastery that specialized in glassmaking. He ultimately became a partner and then director of the glassmaking firm.

In 1814, he passed a beam of sunlight through a narrow slit before allowing it to be refracted by a prism. By "refraction", we mean that each wavelength is bent or diverted from its pathway by a different angle. The most common place to see the result of refraction is a rainbow, which is formed when light rays passing through water droplets are bent at different angles depending on their color. Fraunhofer's prisms were so well made that it was possible to note that some wavelengths of the solar spectrum were missing and resulted in black lines. These became known as "Fraunhofer lines". He identified nearly 700 of them. He did not know it, but he had discovered the fingerprints of atoms found on the Sun. As shown in Figure 11.1 in a stamp issued by the German Post Office in 1987, he used letters to indicate the position of the more prominent lines. These letters are still used today. Fraunhofer was at a loss to explain the origin of these lines.

He also was one of the first to use a grating as a refracting device. He used a diamond point and manually fabricated an optical diffraction grating with 310 lines per inch. Such a grating would have generated spectra of resolution far better than that available using a glass prism. Fraunhofer had essentially produced the components of the first spectroscope, that is, a device for examining a spectrum. He quickly became a celebrity, with many scientists coming to Benediktbeuren to talk with him and order the high-quality instruments that he produced there. He moved to Munich in 1819 and, at the order of the now King Joseph, was made a Knight of the Order of Civil Service of the

Lighting the Dark Path to Atomism: Spectroscopy Shows the Way

Figure 11.1 Solar spectrum with Fraunhofer lines. Drawn and colored by Joseph von Fraunhofer, with the dark lines named after him. (1987 Deutsche Bundespost stamp issued on the 200th anniversary of Fraunhofer's birth). Image courtesy of Mitrofanov Alexander/Shutterstock.com.

Bavarian Crown in 1824. He died two years later at the early age of 39 due, it is said, to his frail health and the toxic fumes involved in his glassmaking activities.

11.3 TRAVEL SITES RELATED TO JOSEPH VON FRAUNHOFER

See Chart 11.1 for the location of these sites.

11.3.1 Benediktbeuren and Munich

11.3.1.1 Fraunhofer Glassworks or Historische Fraunhofer-Glashütte. ▒▒ (47.706407°, 11.401127°) Fraunhoferstrasse 2, Benediktbeuren, right behind the former Benedictine monastery and near an eighteenth-century Rococo chapel. This is a small museum containing two huge furnaces used by Fraunhofer for glassmaking, a few small telescopes, pictures and other memorials to the town's most famous citizen. The monastery has a beautiful basilica and cloister. There is a nearby vegetable and herb garden, a guest house, biergarten, and cemetery.

11.3.1.2 Fraunhofer Glassmaker Relief. ▒▒ (48.137941°, 11.574175°) Thiereckstrasse 3, Munich. A relief on the side of a house at this address shows how the young Fraunhofer was rescued after the collapse of the glassmaking factory. Not far from here, at Frauenplatz 9, is the famous Nürnberger Bratwurst Glöcki am Dom restaurant.

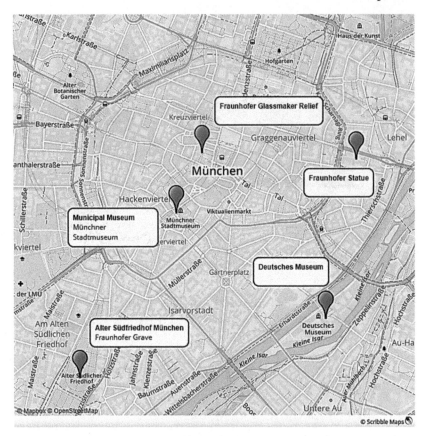

Chart 11.1 Travel sites in Munich related to Joseph von Fraunhofer.

11.3.1.3 The Municipal Museum or Münchner Stadtmuseum. (48.135030°, 11.572310°) St.-Jakobs-Platz 1, Munich. The Munich municipal museum has a display of Fraunhofer's workshop, some documents and a death mask.

11.3.1.4 Statue of Fraunhofer. (48.137640°, 11.585740°) Maximilianstrasse 42, Munich, in front of the Five Continents Museum (formerly the Bavarian State Museum of Ethnology), directly across the street from the statue of Count Rumford (the American physicist, Benjamin Thompson). Fraunhofer holds a prism and behind his feet there is a telescope. The Maximilianstrasse is one of the city's four royal avenues. A statue of Maximilian I Joseph (48.139827°, 11.578087°), who had led the rescue of the young Fraunhofer when the building where he was apprenticing collapsed on top of him, is found, appropriately, at the Max-Joseph-Platz. This is at the western end of the Maximilianstrasse and is less than a ten-minute walk from the statues of Fraunhofer and Rumford. The Hofbrauhaus where Ludwig Boltzmann held forth with his colleagues and students is close by as well (See Chapter 10 for more information on Boltzmann.) An additional great place for a beer is the Andechser am Dom, which serves brews from Kloster Abbey. It would be appropriate to find one of these places and raise a toast to Joseph von Fraunhofer and Max Joseph, the King of Bavaria.

11.3.1.5 Alter Südfriedhof München. Thalkirchner Str. 17, Munich. Fraunhofer is buried here (48.127027°, 11.564053°). A telescope and the Sun are carved on his memorial stone. Nearby is a stone bowl that is engraved *Approximavit sidera* ("He approached the stars"). Stand here and recall that this man saw the fingerprints of atoms in the light that came from the Sun. Georg Simon Ohm (of Ohm's Law fame) and Justus von Liebig are also buried here. (See the end of Chapter 14 for information for visiting Liebig's famous educational laboratory in Geissen.)

11.3.1.6 Deutsches Museum. Munich, Museumsinsel 1 (48.129899°, 11.583463°). This is the largest museum of technology and natural sciences in the world and includes over 13 acres of exhibitions. There is a display of Fraunhofer's optical instruments including his 9″ refracting telescope. The visitor can also see sun spots and Fraunhofer lines here. Their solar telescope is mounted in a stairwell between the third and sixth floor of the museum. The lens of the solar telescope produces an image of the Sun that is projected onto a tabletop on the 3rd floor. A visitor can observe a "live spectrum" of the Sun with its dark Fraunhofer absorption lines. The electricity and magnetism section of the museum is being renovated and will reopen in 2020. The Hall of Fame includes busts and portraits of important inventors, engineers, industrialists, and scientists including Fraunhofer, Bunsen, and Kirchhoff, and others like Wilhelm Conrad Röntgen, Max Planck, Otto Hahn, Werner Heisenberg, and Albert Einstein, who we cover in other chapters of this book.

11.4 ROBERT BUNSEN (1811–1899) AND GUSTAV KIRCHHOFF (1824–1887)

Robert Bunsen was a chemist, whereas the younger Gustav Kirchhoff was a physicist. Together they made one of the most significant contributions to atomic science, chemical spectroscopy, that is, spectroscopy used to identify chemical elements and compounds. Bunsen had started off as an organic chemist but lost an eye in an explosion and twice nearly died due to the organic-arsenic compounds, the extremely foul-smelling "cacodyl" compounds, that he was investigating. This was evidently enough to convert him to an inorganic chemist. Like John Dalton, Bunsen never married and, like Dalton, cited the same reason: he could never find the time. Nothing converted him from this point of view. He was known for having a total disregard for pain. He could pick up the lid atop a glowing crucible with his bare hand. When he was blowing red-hot glass, sometimes one could smell his burning flesh. Given his history of accidents, perhaps he should have been just a bit more careful!

Bunsen was an accomplished chemist before he turned to spectroscopy. He analyzed the gases produced in blast furnaces, invented a carbon-zinc battery, and discovered an antidote for arsenic poisoning. After the explosion of the organic-arsenic compound described above, Bunsen used his antidote to save his own life. He produced large amounts of magnesium and showed that it could be burned to produce an extremely bright light source that is extremely useful in photography. How many of us recall magnesium flash bulbs like the one shown in Figure 11.2? Ask your traveling companions if they do. Bunsen invented this type of flash photography.

Bunsen was recruited to the University of Heidelberg in 1852 as a professor of chemistry and recommended Gustav Kirchhoff for an opening in the physics department in 1854. He had known Kirchhoff from their time together at the University of Breslau (now Wroclaw, Poland). Bunsen knew that heating various salts and metals produces intense colors. (For example, sodium salts characteristically produce yellow flames, lithium compounds burn with a rose-red flame, whereas potassium salts produce lavender flames.) "Flame tests" were used to qualitatively identify some metals. Some of these colors were difficult to observe due to the fundamental color of the flame before inserting the

Lighting the Dark Path to Atomism: Spectroscopy Shows the Way

Figure 11.2 A magnesium flash cube on a Kodak Instamatic Camera 104 (sixty million of these cameras were sold Kodak in the 1960s and 1970s).

sample. To solve this problem, Bunsen, in 1857, invented his celebrated burner. A Bunsen burner mixes the proper amount of air and gas before ignition and therefore produces a clean, almost colorless flame. *Voilà* (or the German equivalent), his flame tests were dramatically more reliable using his burner. Bunsen and his burner are shown in Figure 11.3.

It was Kirchhoff who suggested that the light from the flames might be passed through a slit and then a prism to see what information might be obtained. They purified various elements and systematically heated them to incandescence and mapped the bright emission lines given off. The result, as shown in Figure 11.4, was that characteristic line spectra or "emission spectra" were produced for each element. They had invented chemical spectroscopy. This discovery created a sensation among scientists all over the world.

Starting in 1860, Bunsen and Kirchhoff had the occasion to analyze the mineral waters from the springs of the nearby spa town of Bad Dürkheim, which literally means the bath of Dürkheim. Rather amazingly, the emission line spectra from this water revealed the fingerprints of two previously unknown atoms. (Imagine becoming aware that you most likely had uncovered the existence of two "brand new" elements.) A previously unobserved sky-blue emission line was due to a new element that Bunsen called cesium. (He named the element based on the Latin word *caesius*, meaning "bluish gray".) Two dark-red lines

Figure 11.3 "Burner", drawing by William B. Jensen (courtesy of William B. Jensen).

Figure 11.4 Characteristic line spectra for the elements (top to bottom) hydrogen, helium, neon, sodium and mercury. Reproduced from ref. 1 [https://courses.lumenlearning.com/astronomy/chapter/spectroscopy-in-astronomy/] under the terms of a CC BY 4.0 license [https://creativecommons.org/licenses/by/4.0/].

were due to the element that he called rubidium (from *rubidus*, meaning "deepest red"). Bunsen went on to concentrate 40 tons of this mineral water from which he obtained weighable amounts of crude salts containing these metals.

In another stunning development, Bunsen and Kirchhoff were able to correlate some of the still unexplained Fraunhofer lines with elements found here on earth. For example, the bright yellow line of sodium appeared at exactly the same wavelength as a prominent dark line in the solar spectrum. Bunsen and Kirchhoff had found that there was sodium in the sun. Other Fraunhofer lines were eventually identified as being due to calcium, iron, and magnesium. The Bunsen-Kirchhoff spectroscope went on to contribute to the discovery of other new metals including thallium, indium, gallium, germanium, and at least seven lesser known ones. Discoveries due to spectroscopy produced a significant increase in the number of known elements. In addition to bright-line emission and dark-line absorption spectra, Kirchhoff identified what he called "black-body radiation". It was in trying to explain black-body radiation that Max Planck was forced, against his will, to propose the quantization of energy (see Chapter 15).

11.4.1 Plaque at Bunsen Residence

Untere-Masch-Strasse 30, Göttingen. (51.535381°, 9.929478°).

11.4.2 Salinen/Gradierbau, Bad Dürkheim

Kurbrunnenstrasse 32 (49.465537°, 8.174552°) Bunsen and Kirchhoff found cesium and rubidium in the waters of this mineral spa. Here you can visit a museum and the historic sanatorium. The town is still known for its spa facilities, but also for its sausages and Riesling wines.

11.5 BUNSEN AND KIRCHHOFF TRAVEL SITES IN ALTSTADT HEIDELBERG

See Chart 11.2 for the location of some of the following sites.

11.5.1 Bunsen Statue

Anatomiegarten, Hauptstrasse 49 (49.410874°, 8.698000°), Heidelberg. Bunsen's physical characteristics are apparent in this statue. He was a large man for his time, fully six feet tall and, some have said, "built like Hercules". In the United States we might say he was built like a linebacker. The plaque in front of the statue says in German: "Robert Bunsen: professor in Marburg, Breslau, and Heidelberg; director of the (university's)

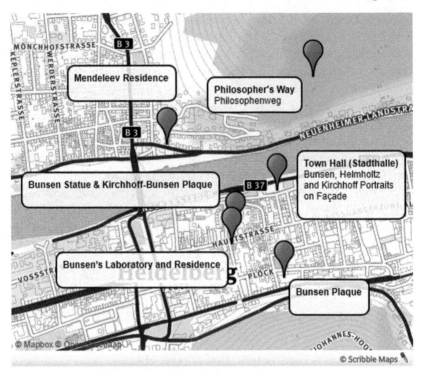

Chart 11.2 Some of the travel sites in Heidelberg related to Bunsen and Kirchhoff (the Bunsen grave at Bergfriedhof and the Bunsen exhibits at the University of Heidelberg are omitted due to scale).

chemical laboratory; founder of chemical analysis; developed the chromic-acid battery and fused-salt electrolysis for the production of magnesium; created spectral analysis technique with Gustav Kirchhoff; discovered the chemical elements of cesium and rubidium (1860)." On either side of Bunsen's statue are two figures representing the Rivers Rhine and Danube. The beautiful cream-colored building behind the statue is called the Friedrichsbau or the Friedrichs Building. It was initially a monastery, but in 1804 was purchased by the Grand Duke Karl Friedrich the First. This is where Bunsen worked when he first came to Heidelberg in 1852. There was no gas or water available here. They used alcohol lamps and charcoal fires and water from an outside pump. Kirchhoff lived in this building during his time at Heidelberg. In 1864, the building was renovated and became the home to the departments of mathematics, physics, and physiology of the University. Behind the Friedrichsbau is a building that housed the Department of Anatomy and Zoology, thus the name Anatomy Garden. Rather amazingly, this is the site of one of the Heidelberg Christmas Markets during the holidays.

11.5.2 Bunsen and Kirchhoff Plaque

Haus zum Riesen, Haupstrasse 52 (49.410409°, 8.698159°). High on the wall, the plaque in German translates to "Within this building in 1859, Kirchhoff and Bunsen determined a spectral analysis of the sun and nearby stars, opening the study of the chemical composition of the universe". Bunsen had been promised a new laboratory (and an adjoining house) when he was recruited to Heidelberg in 1852. This was completed in 1855. Heidelberg had installed gas pipelines which had been used for lighting the streets since 1852. Bunsen decided to use this gas in his new laboratory. It was then that he started his quest to design a burner that could use this new gas line. As noted earlier, he invented his famous Bunsen burner because he needed a flame that had very little color of its own. The building was originally a Baroque palace built in 1707. From 1797 to 1819 it was an inn, a beer brewery and a schnapps distillery. (The name of the inn "Zum Riesen" means "At the Ogre's".) It was here that Bunsen and Kirchhoff did their research that established chemical spectroscopy.

11.5.3 Bunsen's Laboratory and Residence

Plöck 55 (49.4103°, 8.6978°). "Behind the Haus zum Riesen, a new building was completed that provided working space for 50 students and functional laboratories for Bunsen's own research. At the southern end of the building, along the Plöck and opposite the Friedrich-Ebert-Platz (FEP) Bunsen's private house was built."[2] This building is one block south of the Hauptstrasse 52. It was built in 1855 and extends the entire block north (along Akademiestrasse) reaching the "Zum Riesen" building. The exterior of the Bunsen home and laboratory look very much today as they did in Bunsen's time in the 1800s. The home now houses an institute for German as a Second Language. In 2011, on the 200th anniversary of Bunsen's birth, a new beautiful blue plaque (in German) was installed on the wall of the house.

11.5.4 Bunsen Plaque

Plöck 41 (49.409384°, 8.700722°). This also says that Robert Bunsen worked here. Bunsen was succeeded as the chair of the chemistry department by Viktor Meyer and then Theodor Curtius, who also lived here. (It may be that this is where Bunsen lived until moving into his new residence at Plöck 55.)

11.5.5 Stadthalle (Town Hall)

Neckarstaden 24, (49.412191°, 8.700340°). The portraits of Robert Bunsen, Hermann von Helmholtz, and Gustav Kirchhoff are incorporated left to right into the façade of the "Stadthalle", which was built at the turn of the nineteenth century. Their stone-carved faces look down onto the terrace over the main entrance on the western side of the building. Hermann von Helmholtz was a German physiologist and physicist. He conducted pioneering studies of both the eye and the ear, applied principles of science to the art of music, and is often given the majority of the credit for the Law of Conservation of Energy.

11.5.6 Robert Bunsen Grave

Bergfriedhof, Rohrbacher Strasse, Heidelberg (49.39722°, 8.69028°). Incidentally, Gustav Kirchhoff is buried in Alter Sankt-Matthäus-Kirchhof (Old St. Matthew's Churchyard) in Berlin. St. Matthew's is also the burial site of the Brothers Grimm – Jacob and Wilhelm Grimm – who were responsible for such fairy tales as "Cinderella", "The Frog Prince", "Hansel and Gretel", "Rapunzel", and "Snow White".

11.5.7 Bunsen Exhibits at the Hörsaal of the Chemistry Department

. In the hallway of the Hörsaal Zentrum Chemie (Lecture Hall), Chemistry; Im Neuenheimer Feld 252 (49.418633°, 8.673453°). Here we see display cases containing an original spectroscope, Bunsen burner, preparations of rubidium and cesium, a chromic acid battery, other instruments, historic letters, and a Bunsen death mask. At the end of the hall is a life-size enlargement of the famous photograph of Kirchhoff, Bunsen, and Roscoe. Henry Enfield Roscoe helped Bunsen design his

Lighting the Dark Path to Atomism: Spectroscopy Shows the Way 277

Figure 11.5 Helen Beatrix Potter's watercolor entitled "A Dream of Toasted Cheese", showing mice toasting a piece of cheese in a Bunsen burner. The original appeared on the Fall, 2008 cover of *The Hexagon* (Volume 99, Number 3). Appreciation is extended to Jim and Jenny Marshall, who produced this enhanced copy.

new burner. Roscoe's niece, Helen Beatrix Potter, did a water color painting of a mouse and a Bunsen burner, which she presented to Roscoe upon the publication of his book "Inorganic Chemistry for Beginners", published in 1893. Her painting, entitled "A Dream of Toasted Cheese", is shown in Figure 11.5. Looking closely, we can see some mice using a platinum wire to suspend a piece of cheese in the flame of the burner.

11.5.8 Kirchhoff Exhibits at the Kirchhoff-Institut für Physik

, Im Neuenheimer Feld 227, Heidelberg (49.416266°, 8.672059°), 1st Floor. Here we find a permanent exhibition about the life and work of Kirchhoff.

11.6 OTHER SCIENTIFIC/HISTORICAL SITES IN HEIDELBERG

11.6.1 Philosopher's Way (Philosophenweg)

This famous walking pathway is located on the north side of the River Neckar. There are several ways to get up on to it. The walk up from the Alte Brucke Carl Theodor Old Bridge is scenic, but some say the best way is *via* Albert Überle Strasse, which starts at the corner of Neuenheimer Landstrasse (49.413837°, 8.694197°). It's a bit steep and cobblestoned but well worth the effort. The view of the Old Castle across the river is spectacular. You can be assured that Robert Bunsen, Gustav Kirchhoff, Hermann v. Helmholtz and chemists like Emil Erlenmeyer and his student Dmitri Mendeleev spent time walking this amazingly scenic pathway.

11.6.2   Mendeleev Residence

Schulgasse 2 (49.411094°, 8.707355°). Dmitri Mendeleev lived in this beautiful baroque house with a Madonna statue to the right of the main entrance of Jesuitenkirche from 1859 to 1860 while he was a student of Emil Erlenmeyer. (You remember using Erlenmeyer flasks in your chemistry classes, right?) In 1860, Mendeleev attended the 1st Chemical Congress in Karlsruhe (See Chapter 10) and then returned to St. Petersburg. Later, he became famous for his periodic table of the elements. (See Chapter 18 for extensive background on Mendeleev and the intriguing places in St. Petersburg that scientific/historical travelers will want to visit if at all possible.) The Jesuit Church here in Heidelberg has an elegant, bright white interior and is worth a quick visit to rest and get one's bearings.

11.6.3   August von Kekulé Plaque

Hauptstrasse 4 (opposite Kaufhalle) (49.409914°, 8.694455°). This commemorates his private laboratory. It was here that Kekulé proposed that the carbon atom had a valence of four, that is, that it could be bound to as many as four other atoms.

11.6.4   Pharmacy Museum (Apothekenmuseum)

(49.410918°, 8.715678°). The museum, located in the Heidelberg Old Castle (Heidelberger Schloss), has a number items of interest, particularly to pharmacists and chemists. This baroque-era chemistry laboratory has an array of glassware, especially a variety of distillation equipment. One fun way to get up to the castle is the steep funicular railway (bergbahnen) from the Old Town. The railway was opened in 1890. Two split-level cars carry 50 persons each.

11.6.5 University Library (Universitätsbibliothek)

(49.409722°, 8.705950°). Busts of famous scientists line the staircase landings.

11.6.6 Carl Bosch Museum Heidelberg

Schloss-Wolfsbrunnenweg 46 (49.415467°, 8.730023°). Not directly related to the history of the atom, this museum is dedicated to the life and career of Carl Bosch, Nobel laureate in chemistry in 1931. His residence, Villa Bosch, is located next door to the museum. The Haber-Bosch process for producing industrial quantities of ammonia gas from nitrogen and hydrogen gases was part of a series of processes in which nitrates for fertilizers, explosives, and munitions could be readily produced in Germany in the beginning of the twentieth century. These processes include the Ostwald Process discussed in Chapter 10.

11.7 SPECTROSCOPY'S ROLE IN DISCOVERING MORE ELEMENTS AND THEIR ATOMS

Following in Bunsen and Kirchhoff's footsteps, the Swedish physicist Anders Ångström (1814–1874), in 1862, identified hydrogen in the Sun by spectroscopic means. He also examined the spectrum of the aurora borealis and detected in it a bright line due to oxygen. He reported his data in units of 10^{-10} meters, a unit that is now known as an Ångstrom. In 1868, the French astronomer Pierre Janssen (1824–1887) observed a new spectral line in the chromosphere of the Sun. After conferring with fellow astronomers, they proposed that it was due to a new element, one which had not yet been identified here on Earth. This proposal was greeted with a great deal of skepticism, but they had discovered helium, named after the Latin *hēlios* for Sun. Helium was the Sun's new element come to light on Earth. In the late 1880s this same element was found to be liberated from the mineral uraninite. Recall from Chapter 6 (pp. 153–154) that Rutherford and Geiger, in the first decade of the twentieth century, proved that alpha particles given off by uranium were just helium atoms stripped of their two electrons.

The element thallium was discovered by spectroscopy in 1861 by Sir William Crookes, whom you may associate with Crookes' tubes and the initial work that led J. J. Thomson to discover the electron (See Chapter 5, pp. 132–133). Crookes, while "scooped" on the electron, analyzed selenium ores to find a beautiful green line previously unidentified. The element takes its name from *thallos*, the Greek word for "green twig". Indium was also discovered by spectroscopy in 1863 by Ferdinand Reich and his student Hieronymus Theodor Richter, who identified an indigo emission line in a zinc ore. It is interesting to note that although Richter identified his new element by color, he was in fact color-blind. Gallium, Mendeleev's eka-aluminum, was discovered by Paul Émile Lecoq de Boisbaudran in 1875, 5 years after the Russian chemist left a blank in this table for it and predicted its properties (See Chapter 18). De Boisbaudran identified the element in a sample of zinc ore by using spectroscopy.

After Kirchhoff had analyzed the Fraunhofer lines of the solar spectrum, it became apparent that the lines assigned to hydrogen were spaced in a pattern that begged to be analyzed: as shown in Figure 11.6, they became closer and closer together as the wavelength increased. In 1885, at the age of 60, Johann Balmer (1825–1898), a Swiss mathematician and physicist, using the wavelengths of the solar hydrogen lines determined by Ångström, proposed a formula, known as the Balmer equation, that demonstrated a relationship among the known wavelengths (λ) of hydrogen that used a series of integers starting with 2.

$1/\lambda = R\,(1/2^2 - 1/n^2)$, where R = Rydberg constant, n = integers greater than 2.

Rydberg (1854–1919) worked out a more general relationship and was able to demonstrate that the Balmer equation was a specific case.

It is important to note that the Balmer relationship is purely *empirical*, *i.e.*, it is based solely on experiment without any theoretical basis whatsoever. Balmer predicted other values of the series of lines in the visible region of the electromagnetic spectrum. These new lines were found in the spectra of white stars

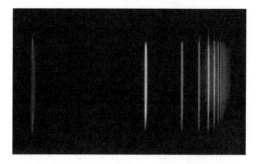

Figure 11.6 The Balmer series of hydrogen. Photograph by Theodore W. Hansch and used with his permission.

and presented a striking corroboration of his equation. It was soon found that there were other series in the spectrum of hydrogen. These could be calculated where the 2 of the Balmer equation is replaced with 1, 3, 4, 5,... These series were a bit more difficult to find and measure as they are found in the ultraviolet and infrared regions of the spectrum.

The Lyman series, named after its discoverer, Theodore Lyman (1874–1954), an American spectroscopist and physicist, was found in the ultraviolet region. The challenge, of course, was to explain what was going on here. What model can be conceived to explain or account for these phenomena and these rather strange and esoteric equations? All of this will be discussed in the next chapter.

11.8 SUMMARY

Joseph von Fraunhofer, an orphan apprenticed to a demanding glassmaker in Munich, was rescued from the building that collapsed around him by the future King Maximilian Joseph I of Bavaria. In Benediktbeuren, he learned how to make excellent prisms and soon used them to discover the black lines of what are still known today as Fraunhofer lines of the Sun. We can visit multiple Fraunhofer sites in Munich and nearby Benediktbeuren, where his laboratory was located for many years.

In Heidelberg, Robert Bunsen and his younger colleague Gustav Kirchhoff invented the technique of spectroscopy. They passed the light from "flame tests" through their prisms to produce characteristic "bright-line" spectra that proved to be the fingerprints of atoms. As part of this technique, Bunsen invented his famous burner that emitted very little light of its own. Using spectroscopy, Bunsen and Kirchhoff discovered the elements cesium and rubidium in the waters of Bad Dürkheim, a little spa town 25 miles west of Heidelberg that we can visit. They named these elements in honor of the prominent lines in their spectra. Kirchhoff found that the bright-line spectra of many elements exactly matched the dark-line spectra of Fraunhofer lines and therefore he could identify what elements existed on the Sun. Heidelberg honors Bunsen and Kirchhoff with impressive statues, informative plaques, original residences, intriguing portraits, graves, and exhibits of original equipment and multiple documents.

Many other scientists lived and worked in Heidelberg. These include Dmitri Mendeleev, August von Kekulé and Carl Bosch. Travelers are invited to go up to the Philosopher's Way and walk in the footsteps of some of the greatest scientists of all time. From this walking path, one is afforded a magnificent view of the old Heidelberg Castle across the river. Once back in the city, it is fun to use the funicular railway and visit this castle that houses an elaborate apothecary museum.

Using spectroscopy, Anders Ångstrom identified hydrogen in the Sun. The previously unknown element we now call helium was discovered in the Sun by Pierre Janssen. Other elements including thallium, indium, and gallium were discovered spectroscopically. Johann Balmer constructed his empirical equation that used integers to generate the wavelengths of hydrogen line spectra. No one could explain why this empirical equation worked. The Balmer equation will be of great importance when we discuss the work of Niels Bohr in the next chapter.

ADDITIONAL READING

- J. Teichmann, M. Eckart and S. Wolff, Physicists and Physics in Munich, *Phys. Perspect.*, 2002, **4**, 333.
- P. J. T. Morris, Modern Conveniences, Robert Bunsen and Heidelberg, 1850s, in *The Matter Factory: A History of the Chemical Laboratory*, Reaktion Books Ltd., London, 2016, ch. 5.

- J. L. Marshall and V. R. Marshall, *Rediscovery of the Elements: Mineral Waters and Spectroscopy*, Indianapolis, Indiana, 2008, https://digital.library.unt.edu/ark:/67531/metadc111227/, accessed June 2019.

REFERENCES

1. Spectroscopy in Astronomy, https://courses.lumenlearning.com/astronomy/chapter/spectroscopy-in-astronomy/, accessed June 2019.
2. Sightseeing in Heidelberg, A stroll through the history of chemistry along the Hauptstrasse, http://www.uni-heidelberg.de/institute/fak12/texte/histor2.engl.html, accessed June 2019.

CHAPTER 12

The Danes Jump in: Ørsted and Bohr (Denmark)

12.1 A QUICK LOOK AT PLACES TO VISIT "TRAVELING WITH THE ATOM" IN DENMARK

In Rudkøbing, on the island of Langeland, the Hans Christian Ørsted Statue 🌼🌼 in the square near the house where he was born and a memorial to Ørsted and his brother 🌼🌼; in Copenhagen: Ørsted statue 🌼🌼🌼 guarded by the three goddesses of fate in Ørstedparken, Ørsted Statue 🌼 at Sølvgade 83, Ørsted Plaque 🌼 at Norregade 21 where he lived from 1819 to 1824 and discovered electromagnetism, Ørsted Plaque 🌼 at Studiestraede 6 where he lived from 1824 to 1851 and discovered aluminum, Assistens Kirkegård Cemetery 🌼🌼 where Ørsted, Bohr, and Hans Christian Andersen are buried. Plaque at Niels Bohr Birthplace 🌼, Plaque at Niels Bohr Childhood Home 🌼, Bohr Statue, University of Copenhagen 🌼🌼, Niels Bohr's Nobel Prize Gold Medal 🌼🌼, Niels Bohr Institute 🌼🌼🌼🌼, The Carlsberg Honorary Residence 🌼🌼; in Hillerød, Niels Bohr Bronze Bust and Nobel Prize Gold Medal 🌼🌼 at the Danish Historical Museum of Fredriksborg.

Traveling with the Atom: A Scientific Guide to Europe and Beyond
By Glen E. Rodgers
© Glen E. Rodgers 2020
Published by the Royal Society of Chemistry, www.rsc.org

Hans Christian Ørsted Discovers Electromagnetism (1820) and Niels Bohr Brings Quantization to the Atom (1913)

- Hans Christian Ørsted discovers electromagnetism in 1820.
- In 1825, Ørsted produces the element aluminium for the first time.
- In 1907, Niels Bohr receives a gold medal from the Danish Academy of Sciences for his work in experimentally determining the surface tension of water.
- In 1911, Bohr, with his mother's help as his "pen", completes his PhD thesis on the role of electrons in metals. Bohr goes to Cambridge to work with J. J. Thomson but, failing to find a fit there, moves to Manchester to work with Ernest Rutherford.
- In 1913, Bohr applies the new quantum ideas of Planck and Einstein to the energy of the electron in a hydrogen atom and publishes his "trilogy" describing the Bohr–Rutherford atom.
- In 1916, Niels Bohr appointed as University of Copenhagen's first professor of theoretical physics.
- In 1921, the Institute for Theoretical Physics is opened in Copenhagen; in 1965, it is renamed the Niels Bohr Institute.
- In 1922, Bohr is awarded the Nobel Prize in Physics.
- In 1932, Niels and Margrethe move to the Carlsberg Honorary Residence.

12.2 HANS CHRISTIAN ØRSTED (1777–1851)

"Our physics would thus be no longer a collection of fragments on motion, on heat, on air, on light, on electricity, on magnetism, and who knows what else, but we would include the whole universe in one system." Hans Christian Ørsted

Ørsted was born in Rudkøbing on the island of Langeland, located between the Great Belt and the Bay of Kiel. There were no schools in the town when Hans was born so he was educated through home-schooling and private tutors. Fortunately, these lessons included several foreign languages. He was the son of an apothecary and was drawn to the sciences, working as his father's assistant for a time before going to the University of Copenhagen. At that time, neither chemistry nor physics were offered as separate subjects at the University, so Ørsted obtained a pharmaceutical degree. He graduated in 1796 at the age of 19, continued on to study philosophy and, in 1799, actually graduated with a PhD, which, as many of us know, stands for Doctor of Philosophy. Nowadays, a PhD is earned in a particular discipline like chemistry, physics, English, history, and, yes, philosophy. However, Ørsted actually earned a PhD in philosophy studying the work of Immanuel Kant!

Studying and then advocating the philosophy of Immanuel Kant (1724–1804) almost led Ørsted astray. For one thing, Kant was an early anti-atomist. He believed that matter *could* be divided infinitely, that is, that one never reached the point where particles were a-tomic or indivisible. Instead, Kant believed that matter was constructed of only two fundamental, opposing forces, called the principle of acidity and the principle of alkalinity. In the end, Ørsted did not pursue this line of reasoning partly because, like so many others, he was soon intrigued by the work of Alessandro Volta, who discovered his battery (or voltaic pile) in 1800 (see Chapter 9). He soon was assembling his own pile and setting to work. Given his fluency in several languages, the Danish government sponsored him to study in Germany, France, and Holland for several years. His studies led him to believe that natural philosophers (not yet called "scientists") should look for ways to unify their understanding of the forces of nature. In particular, he wondered if electricity and magnetism might be closely related. When he returned home in 1806, he was appointed professor of physics and chemistry at University of Copenhagen.

Ørsted was known for his enthusiastic public lectures. In 1819, he was lecturing on electricity and magnetism and had arranged to do a demonstration that he thought might show a connection between the two. Professor Ørsted liked to think about his experiments. He was one of the first to speak of "thought experiments" or *gedankenexperiments* that were made so famous by Albert Einstein many years later (see Chapter 15). As it turned out, he was not able to try the demonstration before his lecture started and gave some thought to postponing it. However, as the lecture progressed, he became more confident that the demonstration just might work. He went ahead and placed a platinum wire carrying an electric current (generated by his Voltaic pile) over a compass. Figure 12.1 illustrates how he might have done

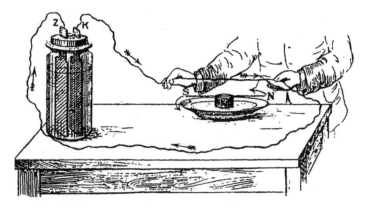

Figure 12.1 An illustration from a Danish schoolbook showing how Ørsted discovered electromagnetism. He stretches a wire carrying a current generated by his battery over a compass needle. Reproduced from ref. 1 with permission from Springer Nature, Copyright 2009.

this. The needle of the compass was deflected feebly in a direction perpendicular to the flow of electricity. When he reversed the flow of electricity, the compass was diverted in the opposite direction. The demonstration did not make much of an impression on his audience. (Occasionally, as both teachers and students know, demonstrations can be a little underwhelming.) A few months later he did the experiment again, this time with an audience of his friends and colleagues. He used stronger electrical forces and found that, in fact, he had discovered that the electric current did produce a magnetic force and deflected the compass. Electricity and magnetism were, in fact, connected and he had discovered what we now call electromagnetism. When he published his result in 1820, it set off a flurry of activity. Starting eleven years later, around 1831, Michael Faraday developed his set of electromagnetic principals, including the concepts of "lines of force" and magnetic and electric fields (see Chapter 5). From there, starting in 1855 or so, James Clerk Maxwell extended electromagnetism to great heights. Clerk Maxwell even made the amazing discovery that light was just electromagnetic radiation! (See Chapter 5).

Ørsted did not pursue electromagnetism as much as we might expect, preferring to stick more closely to philosophy and chemistry. In 1825, he made a significant contribution to chemistry by producing the element aluminium for the first time. While an aluminium-iron alloy had previously been developed by Humphry Davy, Ørsted was the first to isolate the element by treating aluminium chloride with an amalgam of potassium and mercury. Recall that Davy was the first to isolate potassium in 1807 when he subjected potassium hydroxide ("potash") with his Voltaic pile. Potassium itself is very reactive in the air and water and could be stabilized by incorporating it in a mercury amalgam, that is, an alloy of mercury and potassium.

You might wonder if there was a connection between Hans Christian Andersen (1805–1875) and Hans Christian Ørsted (1777–1851). Although H. C. Andersen was 28 years younger, the two H. C. s were great friends. Hans Christian Andersen called Ørsted the "Store Hans Christian" (Great Hans Christian) and had a standing invitation to dine once a week at the Ørsted home. Ørsted helped get Andersen's fairy tales published in 1835. Ørsted's daughter, Sophie, was one of Andersen's female lovers. Andersen's fairy tales included "The Ugly Duckling", "The Princess and the Pea", and "The Emperor's New Clothes". He also wrote plays, travelogues, novels, and poems. A statue of Hans Christian Andersen can be found at the southern corner of Rådhuspladsen (City Hall Square) in central Copenhagen (55.675330°, 12.569117°). Busts of Andersen and Niels Bohr are found in the grand hall of the City Hall.

12.3 TRAVEL SITES RELATED TO HANS CHRISTIAN ØRSTED

12.3.1 Rudkøbing, Island of Langeland, Denmark

12.3.1.1 Ørsted Statue. Gaasetorvet 5900 (54.936954°, 10.708435°). This impressive statue shows Ørsted standing tall in a little square opposite the Old Apothecary Shop, the house where he was born. The statue was erected in 1920 to mark the centennial of Ørsted's discovery of electromagnetism. There is a small museum in the shop. A plaque on the outside wall indicates that Hans and his brother Anders were born here.

12.3.1.2 Ørsted Brothers Memorial. (54.937810°, 10.710622°). From the statue, walk down Ramsherred ("Street of the Poor") to a small city park where this memorial is located. Anders Ørsted, one year younger than Hans, became a lawyer and ultimately, for a short time, the prime minister of Denmark.

12.3.2 Copenhagen, Denmark

See Chart 12.2 for the location of these Copenhagen sites.

12.3.2.1 Ørstedparken, Ørsted Statue. (55.680841°, 12.567379°) surrounded by three goddesses of fate! This bronze statue, as shown in Figure 12.2, depicts Ørsted demonstrating the effect of an electric current on a magnetic needle. At the foot of the statue we find the three goddesses of destiny, Urðr (the past), who is noting Ørsted's name on a tablet, Verðandi (the present), who is spinning the thread of fate, and Skuld (the future), who is silently awaiting the fullness of time with a rune stick in her hand. Runes are the characters of the alphabet used by Vikings from about the second to the fifteenth centuries. A rune stick is used to permanently inscribe these characters onto rune

Figure 12.2 Left: statue of Hans Christian Ørsted in Ørsted Park. At the base are the three goddesses of fate, from left to right, Urðr (the past), Skuld (the future), and Verðandi (the present). Right: at the top, Ørsted holds his platinum wire, which he will suspend over the needle of a compass. Left reproduced from ref. 2 [https://commons.wikimedia.org/wiki/File:Statue_of_Hans_Christian_%C3%98ersted,_%C3%98erstedsparken,_Copenhagen.jpg] under the terms of a CC BY-SA 3.0 license [https://creativecommons.org/licenses/by-sa/3.0/deed.en]. Right side: Photograph by G. E. Rodgers.

Figure 12.3 Niels Bohr and his mother, Ellen, taken in 1902. From the Niels Bohr Archive, courtesy AIP Emilio Segrè Visual Archives.

stones, large monuments honoring the names of past important Norsemen. Among the donors who contributed to a fund to erect this statue was Hans Christian Andersen.

12.3.2.2 Hans Christian Ørsted Statue. (55.688729°, 12.574668°) Sølvgade 83. In the courtyard of the former Polyteknisk Läreanstalt, Sølvtorvet, which was founded by Ørsted in 1829. It is now called the Technical University of Denmark and has moved to a site north of Copenhagen. The buildings here in Sølvgade have become the home of the Institute of Microbiology of the University of Copenhagen.

12.3.2.3 Ørsted Plaque. (55.680471°, 12.570998°) Norregade 21. Ørsted lived here from 1819 to 1824 and in 1820 discovered electromagnetism in the auditorium here. (The auditorium no longer exists.) There is an inscription on the *Telefonhuset* (The Telephone House) that translates to "In a house on this spot, the physicist Hans Christian Ørsted discovered electromagnetism in the year 1820".

12.3.2.4 Ørsted Plaque. (55.679118°, 12.570585°) Studiestraede 6. The plaque is located above the gate on the building. Ørsted lived here from 1824 to 1851. Working in the chemistry laboratory of the university in 1825, he discovered aluminium.

12.3.2.5 Assistens Kirkegård (Cemetery). ●●● (55.689426°, 12.552981°) Kapelvej 2. Here, in this beautiful cemetery, we find the final resting places of the physicists Hans Christian Ørsted (Section E) and Niels Bohr (Section Q), writer Hans Christian Andersen (Section P), and philosopher Søren Kierkegaard (Section A). There is an information office at the main entrance where charts of the cemetery are available.

12.4   NIELS BOHR (1885–1962)

> When asked "Why does an electron jump?" Bohr would respond "You don't want to know. You're too young to know."

Niels Bohr was born in Copenhagen to a prosperous, academic family. His father, Christian Bohr, was a famous professor of physiology who was thrice nominated for a Nobel Prize. His mother, Ellen, came from a wealthy Jewish banking family. Niels was the second of three children. The family highly valued education and all three of the Bohr children graduated from the University of Copenhagen. His older sister, Jenny, was a teacher of history and Danish. His younger brother, Harald, became a prominent mathematician. Niels was a good athlete and enjoyed skiing, biking, hiking, and sailing. His colleagues remember the youthful Niels as a bear of a man who always moved fast. Both Harald and Niels were accomplished soccer players. Niels was a goal-keeper and Harald captained the Danish team that won an Olympic silver medal in 1908. During his school years, Niels studied other scientific disciplines in addition to physics. In inorganic chemistry he earned a reputation for producing explosions in the laboratory and reportedly broke a record amount of glassware. He also carried out an experimental project in which he determined the surface tension of water by measuring the waves produced by oscillating fluid jets. In 1907, he was awarded a gold medal from the Danish Academy of Sciences for this project.

Niels was rather different in his thinking and comprehension. Particularly as a young man, he suffered from a sometimes-paralyzing self-doubt and an inability to follow a straight line of inquiry from beginning to end. This made writing nearly impossible and, starting with his mother and his brother and then his wife and a life-long series of younger colleagues, he always had to have someone available to put his words to paper. He constantly reviewed and reworked his writings to the point that making the decision to send them out to a publisher was a tormenting process. Compounding the problem, he was not especially good with languages in general and even found it difficult to write in his own native Danish. With these difficulties in mind, we can understand that the dictation and constant reworking of his doctoral thesis promised to be an arduous process. His mother Ellen had assured Niels that she would "certainly with the greatest pleasure, be your pen, my dear boy". Her pleasure as her son's amanuensis most assuredly had to be balanced with extraordinary patience as he reportedly went through 14 drafts of the thesis before finally determining that it would have to do as written. (Even then, when the thesis was bound, Niels insisted that blank pages for eventual corrections and additions should be inserted!) As if these writing trials were not enough, Niels' father died suddenly in February 1911. When the doctoral thesis was accepted in April 1911, it included a dedication to his father, who had been his academic role model. It could well have also been dedicated to his mother. In any case, he defended the thesis in May and formally received his degree a few months later. It's a significant understatement to say that the publication of his PhD thesis on the role of electrons in metals, dedicated to his beloved father, transcribed by his dedicated mother, written in his native Danish as required by the University of Copenhagen, was a major accomplishment.

Fortunately, Bohr's father had pre-arranged for Niels to receive a post-graduate fellowship from the Carlsberg Foundation, which ran (and still runs) the great brewery in Copenhagen. The foundation had been established in 1876 when the brewery was transferred to the Royal Danish Academy of Sciences and Letters. Under that aegis, it supported basic research in the natural sciences, social sciences, and humanities. The fellowship enabled Niels to study abroad. During the summer, he spent some time sailing and hiking with his fiancée, Margrethe Nørlund, and then set off for the University of Cambridge to study with J. J. Thomson at the Cavendish Laboratory of Physics. Recall from Chapter 5 that Thomson had discovered the electron in 1897, formulated the "plum-pudding" model of the atom in 1904, received the

Nobel Prize in Physics in 1906, and by 1907 had turned his attention to "canal rays", which he hoped Bohr would continue to investigate during his time at the Cavendish. Bohr loved following in the footsteps of Newton and Clerk Maxwell at Cambridge, but working with Thomson proved difficult. In the first place, the young Dane reportedly started his initial face-to-face meeting with Thomson by opening J. J.'s book *Conduction of Electricity Through Gases*, pointing to a certain formula and politely saying: "This is wrong." This evidently did not sit particularly well with his new mentor.

Bohr's thesis, originally written in Danish and therefore not readily accessible to the international physics community, had been imperfectly translated to English by a friend who had little knowledge of physics. Niels felt a strong need for an improved translation. He hoped that Thomson would help him in this endeavor. J. J. evidently did not and, much to Niels disappointment, was most reluctant to read the thesis, even though it was about the behavior of electrons in metals. Adding to the young Dane's frustration, things were not going well in the laboratory, which he found poorly organized. Others had reported that it was a place of "molecular disorder". The high entropy and lack of results were disappointing. But then something quite exciting happened: In December, Ernest Rutherford came to Cambridge to give the after-dinner speech at the festive Cavendish Research Students' Annual Dinner (see Chapter 5, pp. 129–130). He was introduced as the once-young Cavendish physicist who "could swear at his apparatus most forcefully!"

Recall that Rutherford (a bit to his chagrin) had received the 1908 Nobel Prize in Chemistry for his work in Montreal on nuclear transformations and half-lives. He had moved to Manchester in 1907 and in March 1911, after the 1909 alpha-particle/gold-foil experiments by Geiger and Marsden (see Chapter 6, p. 155) had announced his nuclear model of the atom. Rutherford's dynamic personality, his good humor, and his reputation for relating well to his younger co-workers, made a strong impression on Bohr. Rutherford remarked early on (most likely in his characteristic blustery and loud voice) that "Bohr's different", "He's a football player!" They immediately bonded in a near father–son relationship that lasted the next 25 years. Bohr was quick to write requesting that he leave Cambridge and come work for Rutherford in

Manchester. By the Ides of March 1912, after an introduction to the laboratory by Ernest Marsden, Bohr was at work studying how alpha rays were absorbed by atoms and would soon start his seminal work accounting for the stability of the Rutherford atom.

First, however, he returned to Denmark to marry Margrethe. It was a 2-minute civil ceremony held on the 1st of August 1912, in the town hall of Slagelse, Margrethe's home town, about 60 miles west of Copenhagen, close to the Great Belt. Originally, the couple had planned to honeymoon in Norway, but Niels could not stay away from completing his article about the absorption of alpha-rays, and now he and Margrethe could work on it together. So, after the short ceremony and an abbreviated – and most likely not very romantic wedding dinner in an industrial association assembly hall – they boarded the ferry that headed across the Great Belt (see Chart 12.1). The first stop was Cambridge, where they stayed together in the apartment where Niels had lived when he first arrived there. Here the newlyweds, shown in Figure 12.4, with Margrethe now acting as Niels' "pen", completed the alpha-particle paper.

Then they were off to Manchester where they were warmly received by Ernest and May Rutherford, and Niels delivered his paper to his new mentor. Finally, they went off to Loch Lomond for a few uninterrupted days before returning to Copenhagen.

In those days, the nuclear atom with its electrons orbiting around the nucleus was not taken seriously by anybody who knew their physics. Thomson certainly didn't believe it; even Rutherford had trouble justifying it. How could such a model be true? It was totally inconsistent with the famous and by then well-accepted Maxwell's equations (see Chapter 5, pp. 115–116) that predicted that the electron orbiting the proton of the hydrogen atom, as shown in Figure 12.5, would emit energy and collapse in 10^{-8} seconds with a flash of light. What could prevent the electron from instantaneously spiraling into the nucleus?

Bohr's answer involved applying the new quantum ideas of Planck (1900) and Einstein (1905) to the planetary atom. (See Chapter 15 for the details of these revolutionary discoveries and the places we can visit that celebrate them.) Bohr postulated that the classical ideas (represented here by Maxwell's laws) just could not be valid inside the incredibly small atom. As we will discuss

The Danes Jump in: Ørsted and Bohr (Denmark)

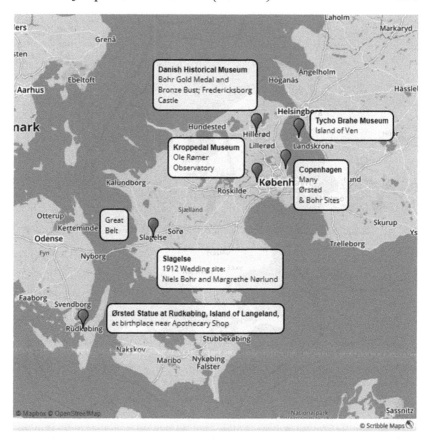

Chart 12.1 Travel sites in Denmark and Sweden related to Hans Christian Ørsted, Niels Bohr, Ole Rømer, and Tycho Brahe.

in greater detail in Chapter 15, Planck and Einstein had concluded that the energy of light is quantized, that is, that it comes in only certain allowed "particles" or "chunks" or "bursts" given by Planck's equation, $E = nh\nu$. In this formula, E is the energy of the light, ν is frequency of the light when it is characterized as a wave with wavelength and frequency, h is Planck's constant and n is an integer or what became known as a "quantum number". Planck called these bursts "quanta of action" or just "quanta", a word that comes from the Latin *quantus* (plural *quanta*), for "how much". Bohr decided that if light energy was quantized, why not extend that assumption to the energy of electrons circling around the nucleus.

Figure 12.4 Margrethe and Niels, Easter 1912, used courtesy of the Niels Bohr Archive.

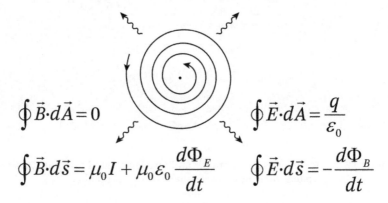

Figure 12.5 Maxwell Equations predict that the electron of a hydrogen atom would spiral into the nucleus in 10^{-8} seconds and emit a flash of light.

Now that's quite a paragraph you just read! Many readers will be a more than a little dismayed that they really don't understand these quantum ideas or how they might be applied to the structure of the atom! To put this into perspective, the reader should realize that you are in great company if you don't understand. Bohr's contemporaries didn't understand it either! When it comes to quantum ideas, I used to tell my students that "if you think you understand, then it's clear you haven't been listening!" For the moment, suspend your dismay and keep reading!

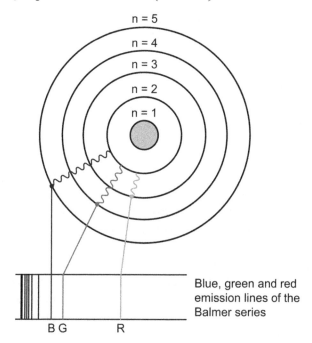

Figure 12.6 The Bohr atom and the explanation of the Balmer series of the Hydrogen spectrum.

With little justification, Bohr stated essentially by fiat that the electron of a hydrogen atom can only orbit at certain allowed radii with corresponding energies. That is, he asserted that the orbits and energies of the electron in the hydrogen atom are quantized. Figure 12.6 shows the Bohr model of the atom. Rather amazingly, he used the masses and charges of the proton and electron, the well-known knowledge of the forces (centripetal and Coulombic) that would keep the electron in orbit, asserted his quantum condition mathematically, and then theoretically derived the Balmer equation that generates the exact values of the Balmer emission lines. (The Balmer equation was discussed at the end of Chapter 11 and Balmer series of hydrogen emission lines are shown in Figure 11.6.) Recall that the Balmer Equation, a purely empirical relationship with no theoretical basis whatsoever, generated the known wavelengths (λ) of hydrogen using a constant and a series of integers starting with 2. Here it is again for us to consider.

$1/\lambda = R\,(1/2^2 - 1/n^2)$, where R = Rydberg constant, n = integers greater than 2

Bohr was able to calculate a correct value for the Rydberg constant and, more importantly, could "account", at least in part, for the origin of the integers in the equation. The integers were the result of the quantum condition (initially proposed by Planck and Einstein), now asserted by Niels Bohr to apply to the structure of atoms. By mid-1913, Niels and Margrethe were hard at work putting his thoughts on paper. Drafts of these papers in Margrethe's clear and legible hand are available.

So, according to Bohr, how were these beautiful visible wavelengths (the "Balmer series") of the hydrogen emission spectra produced? To see how he explained it, we will need the diagram of the Bohr model of the atom that is shown in Figure 12.6. Here, we see the lowest five allowed orbits (or what Bohr called "stationary states") for the electron. Each orbit corresponds to a given allowed amount of energy and to a quantum number, n = 1, 2, 3, 4, and 5. Bohr postulated that electrons can only gain and lose energy by instantaneously moving (he did not call this "jumping" in his early papers) from one allowed stationary state to another, absorbing or emitting electromagnetic radiation, *i.e.*, light, with a frequency v determined by the energy difference (ΔE) of the levels according to the Planck relation, $\Delta E = hv$. So, for example, as shown in Figure 12.6, when the electron "jumps" from the fifth level to the second level, it emits the blue emission line of the Balmer series; when it "jumps" from the fourth level to the second, it emits the green line, and when it "jumps" from the third level to the second, it emits an energy corresponding to the red line. Restated a different way, each line of the Balmer series corresponds to the *difference* in energy as an electron moves from one allowed "stationary state" to another.

But what about the stability of the electron in these "stationary states"? How, you might logically ask, did he explain that there are only certain allowed orbits and *why* an electron occupying one of them, particularly the lowest one (with n = 1, what Bohr called the "ground state") doesn't spiral into the nucleus as described by Maxwell's equations. With what we still view as overwhelming audacity, he offered no proof whatsoever for these ideas but just said, essentially, "I say there are only certain allowed radii and stationary states and the ground state is, in fact, stable". You cannot argue with these assumptions, he boldly proclaimed, because he could theoretically derive Balmer's Equation including incredibly accurate values for the Rydberg constant and the wavelengths of the emission lines. Amazingly, particularly for hydrogen, he was right. The Bohr theory met with tremendous initial success. In the end, for any atom heavier than hydrogen, it had deficiencies and limitations that, despite many efforts by Bohr and others, could not be overcome. The interactions between multiple numbers of electrons in heavier atoms introduced complexities that were difficult for the Bohr model (sometimes called the Rutherford–Bohr model) to explain. However, Bohr had successfully introduced the quantum condition into the structure of the atom, and the rest is history.

Bohr presented his ideas in three articles (often called his "Trilogy") published in the latter half of 1913. The first one addressed hydrogen only, while the second and third discussed multi-electron atoms in which electrons might be found in a series of shells that had definite and discrete energy levels of their own. Bohr discussed the possibility that the occupancy of these shells might somehow be correlated with an atom's place in the periodic table. These papers created an uproar amongst physicists. Some found the assumptions just too bold and fantastic, but were nevertheless mightily impressed with his calculation of the Rydberg constant, a feat which was hailed as a great accomplishment. Skeptics abounded, that's for sure. For years, J. J. Thomson ignored these quantum ideas and would not include them in his lectures on atomic theory. Others said, "If this is the way to reach the goal [of determining the role of electrons in atoms], I must give up doing physics." Even Rutherford had big-time questions. One of the biggest was, "how does the electron decide which orbit to 'jump' to?" Bohr believed the answer was beyond human imagination and most always responded "You don't want to know. You're too young to know." Rutherford remained cautious but realized that old-time (now called "classical") physics was just not up to the task of explaining the internal structure of atoms.

One thing we do know for sure. These articles were the product of the new collaboration between the newlyweds, Niels and Margrethe. This partnership lasted throughout their lives together. Even though Margrethe did not have a background in physics or the sciences, she was often his trusted sounding board. Particularly in their younger years, she took over as Niels' "pen", putting his thoughts and words on paper. Even more importantly, she always played a crucial role in helping Niels crystallize and clarify his ideas both scientifically and personally.

Bohr became famous fast. He was invited to present seminars all over Europe, including in Manchester, Munich, Vienna, Berlin, and Göttingen. In Berlin, Bohr was credited with a "stroke of genius" for proving that Planck's constant, h, was a key for understanding the atom. In Göttingen, the denizen of physics there, Max Born, was quoted as saying "All of this is absolutely queer and incredible, but this Danish physicist looks so like an original genius that I cannot deny that there must be something to it ...." A rather unusual reaction by a scientist, to say the least. Other reactions followed. One of the most intriguing was a poem entitled "Hail to Niels Bohr", written in the early 1920s by a Russian theoretical physicist, Vladimir A. Fock, and presented as a free English translation in George Gamow's *Biography of Physics*.[4]

With a healthy touch of irony and considerable wit and good-natured humor, the first stanza gives us an inkling of the stature that Bohr had achieved in the physics community.

> Fill up the tankards and brighten the fire!
> Drink to him – drink to him – toast him once more!
> Plucking the strings of our latter-day lyre
> We sing of our hero and idol – Niels Bohr!
>   Prosperous days to you.
>   Honor and praise to you,
> Bohr, the Colussus we fear and adore!

Fock goes on to say that Bohr's theory is "madly obscure" but, because it works so well, "we dassent defy it". The poet then addresses the idiosyncratic role of the electron in an atom and says that "each tiny electron" in orbit "examines its plight ... flies in a flutter ... [and leaps] from his mother orb ... to another orb ... seeking, if vainly, a refuge in flight". It's clear from the final stanza that no one knows (they're too young to know, after all!) why electrons should behave this way. Nevertheless, the poem ends with a toast to the mighty Dane.

> Hail to Niels Bohr from the worshipful nations!
> You are the Master by whom we are led.
> Awed by your cryptic and proud affirmations,
> Each of us, driven half out of his head,
>   Yet remains true to you,
>   Wouldn't say boo to you,
> Swallows your theories from alpha to zed,
>   Even if – (Drink to him!
>   Tankards must clink to him!)
> None of us fathoms a word you have said![4]

The Bohrs spent some time in Copenhagen but then, although travel was hampered by the onset of World War I, returned to Manchester for a few years. Rutherford was soon working on anti-submarine warfare but Bohr, a citizen of a neutral country, could not participate in such activities. In 1916, only four short years after Niels had started his work with Rutherford, word came that he was to be appointed as the University of Copenhagen's first professor of theoretical physics. They headed for home. Niels and Margrethe were expecting their first child. Ultimately, they were the parents of six sons! Margrethe's duties setting her husband's words to paper were soon severely curtailed.

After presenting himself formally to King Christian X, as all new professors were required to do, Niels found himself bicycling between his home in Hellerup, a Copenhagen suburb, and his exceedingly small office. Never shy about advocating for himself and physics, early in 1917 he was already proposing the establishment of an Institute for Theoretical Physics. His proposal was supported by the government and the Carlsberg Foundation. Niels was known in scientific circles all around the world, so it is only moderately surprising that, also backed by private outside investment, property for the institute was purchased in 1918. Despite the end of the war and labor strikes during a time of unrest in Denmark reminiscent of the Russian revolution, the first building of the institute was ready in January 1921. The first letter Bohr wrote in his new office was, of course, to Sir Ernest Rutherford! Soon, young physicists from all over the world were travelling to Copenhagen to work with Niels Bohr. In 1922, only ten years after he left Copenhagen to study with J. J. at Cambridge, Niels Bohr was awarded the Nobel Prize in Physics "for his services in the investigation of the structure of atoms and of the radiation emanating from them".

There is a great story about two Nobel Prize gold medals that were under Niels Bohr's care when the Nazis took over Copenhagen in 1940. In those days, these medals were made of 23-karat gold and weighed 200 g. The Nazis had declared that no gold should leave their jurisdiction, which now included Denmark. Two German physicists, the outspoken Hitler critic, Max von Laue (who won the 1914 Nobel Prize "for his discovery of the diffraction of X-rays by crystals") and the Jewish James Franck (who won the 1925 Nobel Prize with Gustav Hertz "for their discovery of the laws governing the impact of an electron upon an atom") had illicitly smuggled their medals to Copenhagen from Germany and had entrusted Bohr to keep them safe. (Niels had previously auctioned off his 1922 gold medal to support the unsuccessful Finnish war effort against the Nazis.) But what about the von Laue and Franck medals? Their names were on the back of their medals and, if they were found, von Laue and Franck could face capital punishment from the Nazis. Georgy de Hevesy, a Hungarian radiochemist (a future Nobel Laureate in chemistry in 1943 for "his key role in the development of radioactive tracers"), after consulting with Bohr, decided that these medals should be dissolved! (Leave it to a chemist! Chemists always have solutions!) As many know, gold is a "noble" (not Nobel) metal, that is, it is reluctant to react with anything. De Hevesy knew it would be difficult to dissolve these huge gold medals, so he turned to one of most powerful acids known, "aqua regia" (literally, "royal water"), a mixture of hydrochloric and nitric acids. He was dissolving the medals in a beaker (see Figure 12.7) as Nazi soldiers were marching in the streets of Copenhagen. It was an agonizingly slow process, but by the time the Nazis arrived at the Institute, the resulting bright orange solution was sitting in a flask on a high laboratory shelf with many other chemicals. In 1943, both Bohr and de Hevesy were forced to leave Denmark. Once the Nazis had been defeated, the physicists returned to the Institute to find the beaker undisturbed. The metallic gold was recovered, and sent back to the Swedish Academy who recast the medals and re-presented them to von Laue and Franck in a ceremony in 1952.

Niels Bohr went on to make many more contributions to physics and the broader world that would soon have to cope with another major world war and the advent of atomic warfare. We will describe many of these contributions in context in later chapters. These will include his "Complementarity Principle" (1927) regarding wave-particle duality (Chapter 16) and his liquid-drop model of the nucleus (1934) that helped explain nuclear fission (Chapter 17). We also will have the occasion to describe his role in helping Jewish scientists find new positions as they escaped Nazi Germany (starting in 1934), his own dangerous escape in the bomb bay of

Figure 12.7 Georgy de Hevesy dissolves two Nobel Prize gold medals in aqua regia. Reproduced from ref. 3 with permission from Benjamin Arthur.

a Mosquito (a British combat aircraft) to Britain (1943) and then to the United States, his role in the Manhatten Project, and his strong advocacy for peaceful uses of atomic energy.

12.5 TRAVEL SITES RELATED TO NIELS BOHR

12.5.1 Copenhagen, Denmark

See Chart 12.2 for the location of these Copenhagen sites.

12.5.1.1 Plaque at Niels Bohr Birthplace. (55.677486°, 12.581502°) Ved Strauden 14. On the façade of the building, the plaque reads: "I DETTE HUS FØDTES/ATOMFYSIKEREN/NIELS BOHR, 7.10.1885". It was placed here on Bohr's 75th birthday in 1960. This was the childhood home of Niels' beloved mother, Ellen, who served as one of Niels' first "pens" (see Figure 12.3.) This is great place to pause and think about Ellen's contribution to her son's efforts to achieve a PhD in physics.

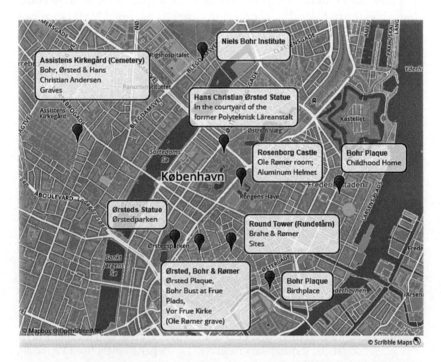

Chart 12.2 Travel sites in Copenhagen related to Hans Christian Ørsted, Niels Bohr, Ole Rømer, and Tycho Brahe.

The Danes Jump in: Ørsted and Bohr (Denmark) 305

12.5.1.2 Plaque at Niels Bohr Childhood Home. (55.685983°, 12.591798°) Bredgade 62. The family moved here in 1886. Niels lived here until he received his PhD in 1911. His father's physiology laboratory was located directly behind this building. It was in this laboratory that Niels investigated the surface tension of liquids, work that won him a gold medal from the Royal Academy of Sciences and Letters in 1909.

12.5.1.3 Bohr Bust, University of Copenhagen. University of Copenhagen Main Building (55.679685°, 12.572630°) Frue Plads. In front of the building there are six busts, portraits of the philologist Vilhelm Thomsen, the philologist J. N. Madvig, the theologian H. N. Clausen, the botanist J. F. Schouw, the zoologist Japetus Steenstrup, and the physicist Niels Bohr (from left to right).

12.5.1.4 Niels Bohr Bronze Bust and Nobel Prize Gold Medal. the Danish Historical Museum of Fredriksborg, Hillerød (north of Copenhagen) (55.935011°, 12.301261°). When Bohr's gold medal was auctioned off to support the Finnish Relief fund in 1940, the anonymous bidder gave it to the museum which holds it to this day. It is not always on display in the permanent exhibitions but it's possible to see it by appointment. The museum also has a fine bronze bust of Bohr on display in the section on Danish history in the twentieth and twenty-first centuries. The museum is located in the Frederiksborg Castle (*Frederiksborg Slot*), once the royal residence of Danish kings starting with King Christian IV of Denmark-Norway in the early seventeenth century. Visitors can view the castle's state rooms, including the Valdemar Room, the Great Hall, and the Chapel and Audience Chamber. There is also a large formal Baroque-era garden. What a fun spot for scientific/historical travelers and their companions.

12.5.1.5 Niels Bohr Institute. (55.696822°, 12.571554°) Blegdamsvej 17. In 1921, the Institute for Theoretical Physics of the University of Copenhagen opened its doors. Its opening was scheduled for 1920 and that is the year we find on the front of the building. It opened five years after Bohr was appointed as the first professor of theoretical physics. It was renamed the Niels Bohr Institute in 1965. In the 1920s, young theoretical physicists from around the world studied quantum mechanics here. In the

1930s, the center of attraction was nuclear physics. By 1930, over 60 physicists from 16 countries had spent time in Copenhagen. When Bohr died in 1962, more than 400 visitors had spent at least a month in Copenhagen. It is still an international center for physics where young scientists investigate problems in a variety of disciplines related to physics. Visitors are encouraged to contact the Niels Bohr archive at www.nbarchive.dk.

A special attraction is Niels Bohr's Office which is housed in a building constructed as living quarters for the Bohr family in 1926. He used the office until the year of his death in 1962 and it has been preserved pretty much as he left it. Portraits above his desk include Galileo Galilei, Michael Faraday, Max Planck, J. J. Thomson, and the Bohr's eldest son, Christian. On the mantel above the stove is a carving of Ernest Rutherford originally commissioned by Peter Kapitza, who also commissioned the alligator carving in honor of Rutherford at the Old Cavendish Laboratory (see Chapter 6, pp. 165–166). Some people have criticized it for being too modern, but Bohr reacted so positively to it that Kapitza asked the artist to make a second relief to send to Bohr. Visitors can also visit the original Auditorium A, where so many prominent physicists have lectured over the years. It is also preserved in its original configuration with multiple sliding blackboards.

12.5.1.6 The Carlsberg Honorary Residence. (55.664215°, 12.530625°) Gamle Carlsbergvej 11, Valby. The Bohrs lived here from 1932 to the end of Niels's life. They entertained colleagues and visitors from all over the world. This villa was the original home of Carl Jacobsen, for whom the brewery was named. After he died, it was used as an Honorary residence by "a man or a woman deserving of esteem from the community by reason of services to science, literature, or art, or other reasons". The Bohrs lived here for 30 years! The Carlsberg Visitor's Center is a combined museum and exhibition center. The tour of the old brewery provides a glimpse of the residence. Be sure retire up to the beautiful pub with its long and curved, spectacular bar, order a Carlsberg Pale Lager or the richer Elephant Beer, and raise a toast to Niels and Margrethe Bohr.

Speaking of elephants, during your stay in this area be sure to walk by the Elephant Gate that is the landmark entrance to the brewhouse and now the Carlsberg Museum. The Elephant Gate, also referred to as the Elephant Tower, is the most famous landmark of the Carlsberg area. Built in 1901, at the same time as the present brewhouse, it consists of four granite elephants each bearing the initial of one of Carl Jacobsen's children, Theodora, Paula, Helge, and Vagn. The elephants support a small tower built of red, ornamental brick and topped off with a copper-clad onion dome. The swastikas on the elephants are from the pre-Nazi era when they symbolized good fortune. Niels Bohr was awarded the Knight of the Order of the Elephant in 1947. Knights of the order are granted a place in the 1st Class of the Danish order of precedence. It is Denmark's highest-ranked honor.

12.5.1.7 Assistens Kirkegård (Cemetery). (55.689426°, 12.552981°) Kapelvej 2. As noted above in the Ørsted travel sites, this historical cemetery is the final resting place of the physicist Ørsted (Section E), writer Hans Christian Andersen (Section P), and philosopher Søren Kierkegaard (Section A). There is an information office at the main entrance where charts of the cemetery are available. The Bohr section (Section Q) was founded when Christian Bohr, Niels' father died in 1911. A monument with an owl on a pillar sits above the graves of Niels (ashes), his wife Margrethe, his parents Christian and Ellen, his brother Harald, and his eldest son, Christian. This is an excellent place to sit and think about the roles of these people in Niels' life.

12.6 OTHER SCIENCE-RELATED SITES IN THE COPENHAGEN AREA

Two astronomers of note lived and worked in or near Copenhagen. Tycho Brahe (1546–1601) was the greatest of the "naked eye astronomers" (working without telescopes) and one of the last great astronomers to believe in the Aristotelian or geocentric system. Tycho observed a "new star", what we now call a supernova, in 1572 and won the support of Frederick II of Denmark who, fearing Tycho would not stay in Denmark, subsidized the building of a new, richly equipped observatory called the "Uraniborg" (heavenly castle) for Tycho on the island of Ven (55.908206°, 12.695550°), between Denmark and Sweden. Tycho was known for accurate and wide-ranging observations. He was also an astrologer and alchemist. In the end, after a dispute with Christian IV, Tycho was exiled from Denmark in 1597 and became the imperial astronomer to Rudolph II, Bohemian

King and Holy Roman Emperor, in Prague. Johannes Kepler inherited Tycho's astronomical data and used it to formulate his three laws of planetary motion and confirm the Copernican heliocentric system.

Ven (or Hven) is a beautiful island for scientific/historical travelers and their companions to visit. From Copenhagen, one can take the 10-mile-long Örsendbro bridge over to Malmö, Sweden and proceed up to the town of Landskrona. A ferry takes you over to the island. Alternatively, one can take a boat from Copenhagen over to the island. The Tycho Brahe Museum is housed in the All Saints Church and is small but well done. There are some great reconstructions of Tycho's equipment and the information is posted in both English and Swedish. There's a magnificent statue of Brahe observing the heavens, and also a Renaissance garden. The reconstructed underground observatory, the Stjärneborg ("Star Castle"), features an amazing multimedia show of astronomic equipment. Enjoy some refreshments at the Café Tycho Brahe.

Ole Rømer (1644–1710) studied with Jean Picard (not Jean-*Luc* Picard!) at the Royal Observatory in Paris where, among other things, he made careful observations of Jupiter's moons and also, incidentally, designed the fountains of Versailles for Louis XIV. Inaccuracies in his predictions of the eclipses of these moons led him, in 1676, to describe what he called the "hesitation" or what we now call the speed of light. In 1681 he returned to Copenhagen where he became the director of the observatory at the Round Tower (Rundetårn) (55.681371°, 12.575741°). He also had an observatory in Kroppedal to the west of Copenhagen. The small Kroppedal Museum (55.684925°, 12.311643°) now includes information on Ole Rømer and his observatory. In Copenhagen, one can visit the Vor Frue Kirke (the "Church of Our Lady"), near the university (55.679351°, 12.572348°) where Rømer was given a state funeral and buried. Rømer's grave, a little difficult to find along the right-side aisle, is marked with "he measured the speed of light". The Ole Rømer room in the basement of Rosenborg Castle (55.685843°, 12.577259°) has his planetarium, eclipsarium, and the national standard prototypes of the weights and measures system he introduced. Also, in the castle, there is Frederick VII's general's helmet made of aluminum, made back in the day when aluminum was a precious metal. In the Round Tower you can walk up the 209 m long spiral ramp leading to the astronomer's study. The walk was designed so horse-drawn wagons could be used to bring heavy equipment up to the observatory. At the end of the ramp is a planet plotter designed by Rømer. There is a fine bust of Tycho Brahe at the base of the tower. See Charts 12.1 and 12.2 for the locations of many of these sites.

12.7   SUMMARY

Hans Christian Ørsted discovered electromagnetism in 1820 and isolated aluminum in 1825. In Copenhagen, Ørsted was experimenting with his "voltaic piles" when he noticed that a compass needle was deflected when it was brought near a wire carrying an electrical current. He had discovered electromagnetism and his announcement set off a flurry of activity. Michael Faraday immediately started researching the topic and that was followed by the seminal theoretical work of James Clerk Maxwell. Danes celebrate Ørsted big time and we visit sites on the island of Langeland, his birthplace, and in Copenhagen where plaques and statues celebrate his life and discovery. Niels Bohr is one of the most famous names in the development of the atomic concept. We discuss his early difficulties with setting his words on paper and the role of his mother, wife, and colleagues over the years as his amanuensis. After earning his PhD in 1911, he moved to Cambridge to serve a postdoctoral appointment with J. J. Thomson. Not finding a good fit there, he moved to Manchester to work with Ernest Rutherford. Every physicist knew that Clerk Maxwell's equations predicted that Rutherford's nuclear atom could not be stable. In Manchester, Bohr worked feverishly to devise a new model that would explain the stability of the atom. In the Bohr–Rutherford atom, he boldly asserted that classical laws are not valid inside the atom and needed to be replaced with a quantum condition. We spend some time explaining his theory and how it was received by his peers. We then discuss the many Bohr sites in Copenhagen including his birthplace, childhood home, university plaza, residences, and the Bohr family graves in the cemetery where Ørsted, Hans Christian Andersen, and Soren Kirkegard are also buried. When we finally visit the Niels Bohr Institute ●●●●●, one of the best travels-with-the-atom sites in this book, we all can more fully appreciate his extraordinary influence on the physics of his time. At the end of the chapter, we briefly talk about two famous Danish astronomers, Tycho Brahe and Ole Rømer, and describe some relevant sites to visit including the extraordinary island of Hven where Brahe set up his naked-eye observatory called Uranienborg or the "Castle of the Heavens".

ADDITIONAL READING

- F. Aaserud and F. Pors, A Guide to historical sites of physics in Copenhagen, http://www.nbi.ku.dk/hhh/historiske_steder/fysikvidenskaben/fysiktur_box/fysiktur_folder.pdf, accessed September 2019.
- F. Aaserud and J. L. Heilborn, *Love, Literature, and the Quantum Atom, Niels Bohr's 1913 Trilogy Revisted*, Oxford University Press, Oxford, 2013.
- Niels Bohr, Famous Scientists, http://www.famousscientists.org/niels-bohr/, accessed September 2019.
- Richard Rhodes, *The Making of the Atomic Bomb*, Simon & Schuster, Inc., New York, 1988.
- A. Pais, *Niels Bohr's Times, in Physics, Philosophy and Polity*, Clarendon Press, Oxford, 1991.

REFERENCES

1. F. Pors and F. Aaserud, Historical Sites of Physical Science in Copenhagen, in *The Physical Tourist*, ed. J. R. Rigden and R. H. Stuewer, Birkhäuser Basel, 2009.
2. Statue of Hans Christian Øersted, Øerstedsparken, Copenhagen, https://commons.wikimedia.org/wiki/File:Statue_of_Hans_Christian_%C3%98ersted,_%C3%98erstedsparken,_Copenhagen.jpg, accessed June 2019.
3. R. Krulwich, Dissolve My Nobel Prize! Fast! (A True Story), https://www.npr.org/sections/krulwich/2011/10/03/140815154/dissolve-my-nobel-prize-fast-a-true-story?t=1561632967993, accessed June 2019.
4. G. Gamow, *Biography of Physics*, Harper & Row Publishers, New York, 1961.

CHAPTER 13

Röntgen Rays Revolutionize Physics and Lead to the Inner Atom: (Germany II)

13.1 A QUICK LOOK AT PLACES TO VISIT "TRAVELING WITH THE ATOM" IN CENTRAL GERMANY (LENNUP, GIEßEN, AND WÜRZBURG)

Deutsches Röntgen Birthplace and Museum in Ramscheid-Lennup ●●●●, *Röntgen-Memorial in Würzburg* ●●●●, *Röntgen Grave, Alter Friedhoff cemetery in Giessen* ●● *and the Justus-Liebig-Museum in Giessen.*

> **Röntgen's Discovery of X-rays Begins a New Age in the Study of Atoms.**
>
> - In November 1895, Wilhelm Röntgen discovers X-rays, also known as Röntgen rays, in Würzburg.
> - On December 22, 1895, Röntgen made his first X-ray photograph of his wife Bertha's hand.
> - On New Year's Day, 1896, Röntgen announced his discovery.
> - On January 13, 1896, The Emperor of Prussia, Wilhelm II, presented Röntgen with the highest possible medal: the Prussian Order of the Crown, Second Class.
> - On Jan 23, 1896, Röntgen presented his results at the only public lecture he ever presented on X-rays.
> - Röntgen received the 1901 Nobel Prize in Physics, the first Nobel prize ever presented.

13.2 WILHELM RÖNTGEN (1845–1923)

"If the hand be held between the discharge-tube and the screen, the darker shadow of the bones is seen within the slightly dark shadow-image of the hand itself.... For brevity's sake I shall use the expression 'rays'; and to distinguish them from others of this name I shall call them X-rays." Wilhelm Röntgen

13.2.1 Some Background and Perspective

A quick look at the "Travelers Guide Timeline" in Appendix A reveals that 1895 was a watershed year in the history of the atomic concept. In the first decade of the nineteenth century, Dalton's Atomic Theory (Chapter 4) had organized the laws of conservation of mass, definite proportion, and multiple proportions to establish the utility of the atomic concept, including a scale of relative atomic weights. Unfortunately, his obstinate opposition to Gay-Lussac's Law of Combining Volumes (1808), Avogadro's Hypothesis (1811), and the existence of diatomic molecules left the values of atomic weights in limbo for more than 50 years. This difference of opinion, as we discussed in Chapter 9, p. 239, is sometimes called the Dalton–Avogadro debate.

In the 1860s, some solid steps had been taken that certainly seemed to establish the physical existence of atoms. James Clerk Maxwell (Chapter 5), Josef Loschmidt and Ludwig Boltzmann (Chapter 10) had maintained that atoms and molecules colliding with the walls of their containers were responsible for the internal pressure of gases. Also, about that time, Robert Bunsen and Gustav Kirchhoff (Chapter 11) had shown that spectroscopy seemed to provide fingerprints of atoms in the form of emission spectra. After Stanislao Cannizzaro's presentation at the Chemical Congress in Karlsruhe (Chapter 9), the Dalton–Avogadro debate was resolved, and an accurate scale of atomic weights was firmly established. In the mid-1870s, Boltzmann also used an atomic argument in his explanation of entropy, summarized by his equation, $S = k \log W$ (See Chapter 10, pp. 251–252). Nevertheless, rather astonishingly, there were still those, like Ernest Mach and Wilhelm Ostwald, who doubted the physical existence of atoms and even would play a role in driving the avowed atomist Boltzmann to commit suicide (discussed in Chapter 9). What would it take to carry the day and firmly establish the physical existence of atoms? Events in 1895 began a cascade of scientific results that would resolve the problem.

With perfect 20/20 hindsight, we now realize that Röntgen's discovery heralded the dawn of a new age in the story of the atomic concept. It was this discovery, and those that closely followed, that ultimately led to our present understanding of the internal structure of atoms. Within a few years, the physical reality of atoms would no longer be questioned. In Chapter 1, we noted that the pathway to the modern atom is one characterized by ingenious experiments and clear-headed observations as well as serendipitous accidents. The discovery of X-rays was not the result of an ingenious experiment but rather an example of a worker who observed something (a "serendipitous accident" essentially) really quite bewildering and then doggedly studied it until he felt he had done everything he could to characterize it before finally communicating with his colleagues. Many times, scientific discoveries are not a "Eureka" experience, but rather a "that seems strange" experience. Let's investigate how, where and by whom this discovery was made and the special places we can visit today that celebrate it.

Wilhelm Röntgen was born in Lennep im Bergischen (Remscheid-Lennep), the only child of a textile merchant in this once-Prussian, cloth-manufacturing town. He had a scholarly German father and a loving Dutch mother who, together, moved the family to Apeldoorn, Holland, when Wilhelm was only three years old. He was educated in Apeldoorn and later in Switzerland. It was not always easy for him to find the type of education he desired. He was expelled from one technical school in Ultrecht because he would not tell on another boy who had made a caricature of an unpopular teacher and then blamed him. He failed to qualify at another school and finally decided to seek admission to the Polytechnic at Zurich, the Eidgenössige Technische Hochschule (the ETH), in Switzerland. This required him to take a special examination, but on the day of the exam he came down with an eye infection. Luckily for him and us, he was admitted anyway. In 1868, he graduated from the ETH with an undergraduate degree in mechanical engineering, but went on to secure his PhD in physics. (ETH, you may recognize, is also the *alma mater* of Albert Einstein. He graduated in 1900.)

Wilhelm was a little luckier in love, although there were some mild setbacks along the way. He met Anna Bertha Ludwig in her father's café in Zurich. She was charming, tall and slender, but six years older than Wilhelm. Unfortunately, she became ill and was hospitalized in a sanatorium. In a story right out of a romance novel, Ioan James in his book *Remarkable Physicists*,[1] says that on her 30th birthday in 1869, Röntgen hired "a carriage drawn by four perfectly matched horses and driven by a top-hatted coachman of splendid livery, and drove up to the sanatorium to present [Bertha] with a huge bouquet of red roses. It was agreed they would marry once his future was more secure." Once he had his PhD, however, Wilhelm's parents, Constance and Frederich, who were well into their 60s by then, "inspected" Bertha and decided she was not quite ready to marry their son! She was required to go to Apeldoorn to learn cooking and other house-keeping skills. After Wilhelm published his first scientific paper in 1872, he and Bertha were finally permitted to marry. They were unable to have children but, in 1887, adopted their niece, 6 year-old Josephine Bertha Ludwig, the daughter of Bertha's brother Hans.

A bit of an itinerant physicist, Wilhelm took several short-term appointments, but then spent ten years at the University of Giessen, where he firmly established his reputation by verifying an electromagnetic prediction made by James Clerk Maxwell predicated on work done by Michael Faraday (See Chapter 5). As a result, Röntgen, in 1888, had the opportunity to take a position at the Royal University of Würzburg as a professor of physics and director of the new institute there. As part of the deal, the director was given a nine-room apartment immediately above the laboratory. Wilhelm, Bertha, and Josephine moved in and settled down for what seemed like an excellent position.

Wilhelm was an able physicist, but certainly not a genius. He has been described as amiable and courteous, modest and dignified, shy, intellectually honest, thorough, retiring, and unassuming. He had no patience with selfishness in others, disliked lecturing, and spoke in low voice that was difficult for his students to understand. He loved photography, hiking, and other outdoor occupations. He was reluctant to have an assistant and preferred to work alone. He built much of his own equipment and put it to work with great ingenuity and experimental skill.

In 1894, he started a series of experiments with cathode rays, repeating some of the experiments by William Crookes and J. J. Thomson, among others. This was a fertile area of research, with many scientists working in the area. Recall from our discussion in Chapter 5 that Thomson was able to divert the path of the cathode rays with a strong electric field and thereby went on, in 1897, to discover what became known as the "electron". Röntgen used a slightly different tube for his experiments. Figure 13.1 shows a simple cathode tube (CRT). Röntgen extracted the air from his tube with a vacuum pump for two days before a sufficient vacuum was achieved. When a current was generated by connecting an "induction coil" (a static machine) between the cathode and anode, cathode rays came directly off the cathode at right angles and produced a blue fluorescent glow when they slammed into the other side of the tube. (The glow is green for tubes made of English lime glass and blue for the lead glass of the Germans.) On November 8, 1895, when he carried out this experiment, he happened to notice a flash of light coming from a poster on the other side of the room. The poster was painted with a compound called barium platinocyanide, $Ba[Pt(CN)_4]$, which was known to fluoresce when hit by cathode rays. However, it was known that cathode rays can only travel two or three inches in ordinary air, so something else was causing this poster to light up. But what was it? He turned off his CRT and the poster stopped glowing! He enclosed the CRT with a piece of conically shaped black paper to make sure that the cathode rays were, in fact, blocked. He turned the tube back on and the poster lit up once again. What on earth was going on here? He quickly went upstairs and told Bertha that he had discovered a type of *radiation no one else in the world knew about*.

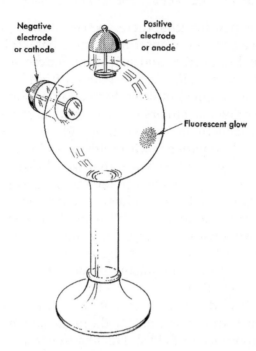

Figure 13.1 A simple cathode ray tube (CRT). Reprinted by the permission of the Estate of Alfred Romer, Physics Department, St. Lawrence University.

He told her that "if he spoke of it without the most convincing evidence, people would say, 'Röntgen has gone mad'". Asked some years later what he thought when he made his discovery he said, "I didn't think; I experimented".

He worked for 6 weeks and ate and slept in his lab during that time. Bertha, now well-schooled in home-economics (!), brought him food from their apartment above the laboratory. Most likely, Josephine, now a teenager, came down once in a while as well. Since the nature of this radiation was unknown, he called it X-rays, X being the usual symbol for the unknown. Somewhat ironically, when you think about it, they are still called X-rays. He tried everything he could think of. He tried to block the rays with various materials, but not much stopped them except lead. He was unable to divert them with magnetic and electric fields. He set up the X-ray tube in front of a door leading to an adjacent room. He put a glass photographic plate behind the door and observed many stripes and wondered what these were. It turned out the door was painted with lead paint

Röntgen Rays Revolutionize Physics and Lead to the Inner Atom

and these were stripes from the curved portion of the door moldings which had an extra thick covering of paint. The rays had penetrated through the door! They could also pass through books and papers on his desk. He was particularly struck that when he held up his hand, the screen showed his bones, encased in darker shadows. Can you imagine what this was like – to see the bones of his own hand? It would be akin to seeing a ghost. As was his accepted experimental practice, he did this experiment over and over again.

On December 22, just days before Christmas, he made his first X-ray photograph of Bertha's hand (see Figure 13.2). This photograph took 12–15 minutes to take – she had to hold her hand perfectly still under the X-ray tube and over a photographic plate. It showed the bones of her hand and her wedding ring! This was the first "röntgenogram" ever taken. There was, of course, no protection from those rays. When she saw the picture, she said "I have seen my death".

By the end of 1895, Röntgen was ready to report his discovery in an aptly worded article entitled "On a new kind of ray:

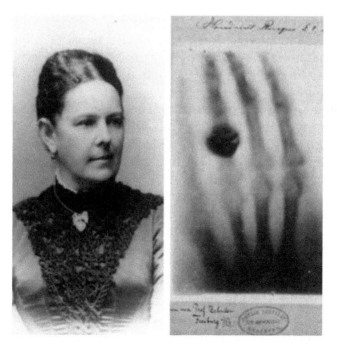

Figure 13.2 X-ray photograph of Bertha Röntgen's hand. Reproduced with permission of Jeremy Norman's HistoryofScience.com.

A preliminary communication". He sent copies of the article and a few photographs to leading scientists in the field (physicists and medical men in Europe, England, and America) on New Year's Day 1896. This article reported all the fundamental properties of X-rays. A newspaper carried the story of the discovery on January 5. The Emperor of Prussia, Wilhelm II, read the account of the discovery in the paper and telegraphed to Würzburg to ask if it was true and then invited Röntgen to Berlin where, on January 13, 1896, he decorated him with the highest possible medal: the Prussian Order of the Crown, Second Class. Röntgen demonstrated the X-rays to the emperor as part of the ceremony. Presumably, Bertha and Josephine were in the audience. This decoration carried with it a mark of personal nobility. He accepted the decoration, but turned down the nobility. He declined to speak before the Reichstag.

On January 23, 1896, he presented his results at the only public lecture he ever presented on X-rays. When he finished his talk, he called for a volunteer, and Rudolf Albert von Kölliker, almost 80 years old at the time, stepped up. The result was a famous photograph showing the bones, according to Isaac Asimov, "in beautiful shape for an octogenarian". "There was wild applause, and interest in X-rays swept over Europe and America." Kölliker was a Swiss anatomist and physiologist, a professor at the University of Würzburg for half a century, and one of the founders of modern embryology.

Röntgen presented only 3 papers on the subject. One of three stipulations in his will was that all of his scientific papers were to be burned. His friends accommodated this wish. Only 3 papers were left. (It is not clear if these are the 3 papers on X-rays.) Although Röntgen called them X-rays, others (over Röntgen's great objections) referred to them as Röntgen rays. They are still referred to as such in many languages including German, Danish, Polish, Swedish, Finnish, Estonian, Russian, Japanese, Dutch, and Norwegian. It is now known that X-rays are high-frequency electromagnetic radiation, that is, a form of light.

People all over the world repeated these experiments. Nikola Tesla (he for whom the modern electrical cars are named) was one of those. In an article, "On Roentgen Rays", published in the *Electrical Review* in 1896, he noted that "an outline of the skull is easily obtained with an exposure of 20 to 40 minutes. In one instance an exposure of 40 minutes gave clearly not only the outline, but the cavity of the eye, the chin and cheek and nasal bones, the lower jaw and connections to the upper one, the vertebral column and connections to the skull, the flesh and even the hair. By exposing the head to a powerful radiation strange effects have been noted. For instance, I find that there is a tendency to sleep and the time seems to pass away quickly. There is a general soothing effect, and I have felt a sensation of warmth in the upper part of the head." Don't try this at home!

Much to Röntgen's dismay, the responses to his discovery were immediate. X-ray facilities were quickly set up all over the world to detect fractures, find foreign bodies (like safety pins), gallstones, dental caries, *etc.* Only four days after news of the discovery reached America, X-rays were used to locate a bullet in a patient's leg. By the end of 1896, more than 1000 scientific articles on X-rays had appeared. Röntgen and his family moved to Munich in 1900. He received the first Nobel Prize in Physics in 1901 "for the discovery of the remarkable rays subsequently named after him". On receiving the prize, he did not give the expected talk on his discovery but only said "Thank You". He worked on X-rays for only 10 months and never patented his work.

Inevitably, there were a series of frivolous responses to the discovery. Asimov writes that "panicky members of the New Jersey legislature tried to push through a law preventing the use of X-rays in opera glasses to protect maidenly modesty – about par, perhaps, for legislative intelligence." In his *Uncle Tungsten: Memories of a Chemical Boyhood*,[2] Oliver Sacks says that lead-lined underclothes were put on sale to shield people's private parts from the all-seeing rays. He quotes a ditty that appeared in the journal *Photography* ending,

> I hear they'll gaze
> Through cloak and gown – and even stays
> Those naughty, naughty, Roentgen rays.

Sacks also notes in a footnote that "shoe shops everywhere in my boyhood were equipped with X-ray machines, fluoroscopes, so that one could see how the bones of one's feet were fitting in new shoes. I loved these machines, for one could wiggle one's toes and see the many separate bones in the foot moving in unison, in their almost transparent envelope of flesh".

In summary, Röntgen's was a seminal discovery, one that set physics and medicine on new and revolutionary paths. It is important, however, to note that he did not set out to revolutionize physics or find a technique that would change medicine forever. He was merely doing some basic research using Crookes tubes, vacuum pumps, and induction coils to see if he could replicate some work on cathode rays.

Briefly, what happened after Röntgen's discovery in 1895? Antoine Becquerel, and Marie and Pierre Curie, curious about Röntgen's results, discovered radioactivity in 1896 (discussed in Chapter 14), J. J. Thomson discovered the electron in 1897 (discussed in Chapter 5, pp. 132–133), Ernest Rutherford characterized alpha and beta radiation in 1898 (discussed in Chapter 6, p. 148), Rutherford and Frederick Soddy characterized nuclear transformations and half-lives between 1898 and 1907 (discussed in Chapter 6, p. 150), Rutherford, using the result of the alpha-particle/gold-foil experiment of Hans Geiger and Ernest Marsden of 1909, proposed the nuclear model of the atom in 1911 (also discussed in Chapter 6), Max Planck in 1900 and Albert Einstein in 1905 proposed the first forms of quantum mechanics (discussed in Chapter 15), Niels Bohr applied quantum mechanics to the structure of the atom in 1913 (discussed in Chapter 12), and Erwin Schrödinger proposed the modern wave theory of the atom in 1926 (discussed in Chapters 2 and 15). With Röntgen's discovery, we were off to the races.

13.3 TRAVEL SITES RELATED TO WILHELM RÖNTGEN

See Chart 13.1 for the location of these sites.

13.3.1 Deutsches Röntgen Birthplace and Museum

Gänsemarkt 1, Remscheid Lennup (51.193539°, 7.256866°). This is about 40 km east of Dusseldorf. The house where Röntgen was born and spent his first three years has recently been renovated. The museum is a short walk away at Schwelmer Strasse 41, Remscheid (51.193780°, 7.259576°). Opened in 1932 and extensively renovated between 2006 and 2015, this is a modern, hands-on museum suitable for all ages and backgrounds. It is housed in a timber-framed house with green shutters and white trim (a "Bergisch-style mansion", typical of this region).

Röntgen Rays Revolutionize Physics and Lead to the Inner Atom

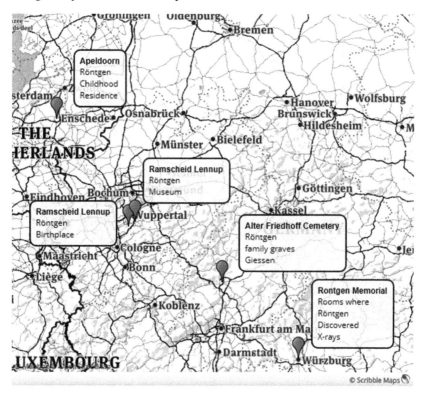

Chart 13.1 Travel sites in Apeldoorn, Lennup, Würzburg, and Giessen related to career of Wilhelm Röntgen.

Röntgen's equipment and personal items include his writing desk and bookcase, the clock which hung in his Würzburg laboratory, his hunting and travelling equipment, spectacles and a record of the award of the 1901 Nobel Prize for Physics. Nearby is a room with X-ray equipment as might have been used by "a practitioner of the day" and a laboratory setting where visitors can carry out their own experiments. Hand-held audio guides are available in German and English. The museum can only display a quarter of its collection. At this writing, arrangements are being made to rehouse the archive. Inquiries should be sent to https://roentgenmuseum.de/en/home-en/. Items in the archive include 600 iconic X-ray tubes (1896–1939), the first dental X-ray equipment (1897), a shoe fluoroscope collection (1930–1970s), a collection of X-ray generators and control panels (1905–1980s), and a collection of mobile X-ray equipment (1897–1980s).

13.3.2 Röntgen-memorial

8 Röntgenring, Würzburg, (49.799978°, 9.931319°). The inscription on the side of the house translates to: "In this house in the year 1895 W. C. Röntgen discovered the rays named after him." The road was renamed Röntgenring in 1909 in Röntgen's honor. The memorial is situated in the actual rooms where the discovery was made and that is what makes it so special. Here we see his desk and his laboratory all set up and ready to go. The equipment on display would have been typical of a physics lab in the late 1890s. There are cathode ray tubes, conically shaped pieces of black paper, induction coils (including the one Röntgen built himself), and a vacuum pump that employs mercury. The pump had to be operated for two days to produce a sufficient vacuum in the cathode ray tube. There is a replica of his Nobel Prize medal and also a nicely done video on Röntgen and X-Strahlen (X-rays). On Jan 23, 1896, he presented his results in a lecture hall then located about where the movie/audio is shown today. When the building was renovated the desk and chairs were moved to what is now known as "Wilhelm-Conrad-Röntgen-Hörsaal". Like the author and his wife, you can sit in the lecture hall and think about Wilhelm and Bertha as they contemplated Wilhelm's discovery.

13.3.3 Röntgen Grave, Alter Friedhoff Cemetery

Giessen (50.582136°, 8.687322°). Wilhelm Röntgen's grave is well marked in the directory that includes a number of famous graves of professional people, members of the military, doctors, and other notables. As you walk around the cemetery, there are several signs pointing visitors directly to Röntgen's grave. ("Gegrabnisstatte Prof. Wilh. Konrad Roentgen 1879–1888 Prof in Giessen"). His parents Constance and Frederich Röntgen and his wife Bertha are also buried here. His parents died while he was a professor at Giessen. This is a good place for the traveler with the atom to sit back and think about these people and the adventures they had together. It is university-connected and maintained. In season, many of the graves, including Röntgen's, are adorned with well-cared-for flowers.

13.4 AN ADDED ATTRACTION IN GIESSEN

Occasionally in this book, we mention places that are not directly related to the history of the atom *per se*, but are among those sites a scientific-historical traveler should not miss. These have included The Pasteur Institute in Paris, were Louis Pasteur lived, worked, died, and is buried (See Chapter 8, p. 218), the Down House in Downe, Kent, England, homestead of Charles Darwin (see Chapter 7, pp. 191–192), the Royal Observatory, Greenwich, England (also in Chapter 7), and the Tycho Brahe Museum and reconstructed underground observatory on the island of Ven in Sweden (See Chapter 12, p. 308). Here, we add Justus-Liebig-Museum, which is only a 10 minute drive from the Alter Friedhoff Cemetery.

13.4.1 Justus-Liebig-Museum

Liebigstrasse 12, Giessen (50.580918°, 8.666397°). Justus von Liebig (1803–1873) held the title of professor *extraordinarius* at the University of Giessen from 1824 to 1852. One of his chief claims to fame was his modern laboratory-oriented teaching method. He is regarded as one the greatest chemistry teachers of all time. His laboratory became a model institution for the teaching of practical or applied chemistry. The laboratory is virtually unchanged from Liebig's days. It ranks as among the 10 best most important museums in the history of chemistry. In some ways, this laboratory is the birthplace of modern organic chemistry.

Liebig has been described as the "father of the fertilizer industry" developing artificial chemical fertilizers and for his emphasis on nitrogen and certain minerals as essential plant nutrients. He was also very active in applying chemistry to agriculture, plant and animal metabolism, and food production. There is one atomist connection. In 1831, he helped perfect a method of analyzing organic compounds for carbon and hydrogen (and by difference, oxygen). In this regard, he invented the Kaliapparat, an ingenious array of five glass bulbs, used to trap oxidation products. An organic compound consisting of carbon, hydrogen, and oxygen was first burned (combusted) to convert the carbon to carbon dioxide and the hydrogen to water. The water was absorbed

Figure 13.3 Liebig's Kaliapparat (left) and the seal of the American Chemical Society that incorporates its design. Left used with permission from Justus Liebig-Gesellschaft zu Gießen. Right used with permission.

in a bulb of hygroscopic calcium chloride, which was weighed to measure hydrogen. Carbon dioxide was absorbed in a potassium hydroxide solution in the three lower bulbs and used to measure carbon. Oxygen was calculated from the difference. As shown in Figure 13.3, the American Chemical Society, founded in 1876, incorporated the design of the apparatus into its seal.

Once the percentage of carbon, hydrogen, and oxygen in a compound had been determined, a table of atomic weights could be used to determine the relative numbers of carbon, hydrogen and oxygen atoms in a compound and therefore come up with a formula and perhaps a structure for the compound. Tables of atomic weights were varied and not particularly reliable in the years before 1860 (See Chapter 9) so this was certainly an inexact art.

Liebig also popularized an earlier invention for condensing vapors in a distillation apparatus. This is still known as a Liebig condenser. He proposed a process for silvering that eventually became the basis of modern mirror-making, and created a meat-extracting company that produced beef bouillon for the first time. As a measure of his ability to train young chemists, 44 of the first 60 Nobel Prizes in Chemistry went through the Liebig system in one way or another.

The laboratory was first established in a guard room of a disused army barracks. The present museum preserves the various rooms in almost mint condition. These include the "old laboratory", the weighing room, Liebig's private laboratory and writing room with his desk and chair, the pharmaceutical laboratory where pharmacists were trained, the library containing all the works of Liebig and others, and the auditorium or lecture hall that includes a hood for doing demonstrations and a traditional lecture bench. The student benches were constructed in a series of "U"s around the lecturer.

13.5 SUMMARY

Wilhelm Röntgen's discovery of X-rays ("Röntgen Rays") in 1895 heralded the dawn of a new age in physics and chemistry. This discovery quickly led to the basic tenets of our present understanding of the internal structure of the atom. It's a prototypical example of a totally unexpected and truly bewildering, serendipitous experimental result that was followed by six weeks of furious experimentation. Röntgen realized that he very well might be the only person in the world who knew about these rays – rays that could penetrate both common objects and human flesh. His announcement caused a sensation that was followed in breathtaking fashion by applications to medicine, thousands of serious scientific articles, and quirky, misinformed laws and products. We also explore the appealing relationship between Röntgen and his wife Bertha, how they met, married, and came to Würzburg, how they dealt with his discovery during those breathtaking six weeks, and Bertha's reaction when she saw the X-ray photograph of her hand and wedding ring. We go over the details of the experiment Röntgen was carrying out and how he followed through on it, so that when we visit the Deutsches Röntgen Birthplace and Museum in Lennup, the Röntgen-Memorial in Würzburg, and then his and Bertha's grave in the Alter Friedhoff cemetery in Giessen, we will fully appreciate what we are seeing. At the end of the chapter we describe an added place that all scientific-historical travelers should see if at all possible, the Justus-Liebig-Museum in Giessen.

ADDITIONAL READING

- Q. C. Field-Boden, *A Holiday or a Pilgrimage? Lennup, Wurzburg, and Giessen: From Where it all Began to Where it all Ended*, Spring 1998 newsletter of the Radiology History & Heritage Charitable Trust.
- I. Asimov, Roentgen, Wilhelm Konrad, in *Asimov's Biographical Encyclopedia of Science and Technology*, Doubleday & Company, Inc., Garden City, NY, 2nd edn, 1982, 502–504.

REFERENCES

1. I. James, *Remarkable Physicists: From Galileo to Yukawa*, Cambridge University Press, Cambridge, 2004, Wilhelm Conrad Röntgen, pp. 177–183.
2. O. Sacks, Penetrating Rays in *Uncle Tungsten, Memories of a Chemical Boyhood*, ed. A. A. Knopf, New York and Toronto, 2001.

CHAPTER 14

The Discovery That Atoms "Fly to Bits": Becquerel and the Curies (Paris and Warsaw)

14.1 A QUICK LOOK AT PLACES TO VISIT "TRAVELING WITH THE ATOM" IN PARIS AND WARSAW CONNECTED TO THE DISCOVERY OF RADIOACTIVITY

Becquerel Sites in Paris: Muséum d'Histoire Naturelle ✺✺, *Henri Becquerel plaque* ✺ *near Jardin des Plantes; Curie Sites, Paris: Two Commemorative Plaques* ✺ *marking residences (Marie); Curie Graves in the Panthéon* ✺✺✺*; Bibliothèque Nationale de France (Curie radioactive papers)* ✺✺*; "Parcours des Sciences" Scientific Tours* ✺✺✺*; Sceaux: Pierre Curie Home Commemorative Plaque* ✺*, Marie Curie Home Commemorative Plaque* ✺*, Original Curie Graves,* ✺✺*; Ploubazlanec, l'Arcouest, Brittany: Irène and Frédéric Joliot-Curie Memorial* ✺✺*; Warsaw, Poland: Maria Skłodowska-Curie Museum* ✺✺✺*; Monument to Maria Sklodowska-Curie overlooking Fountain Park* ✺✺*; Mural of Maria Skłodowska-Curie near the Copernicus Science*

Traveling with the Atom: A Scientific Guide to Europe and Beyond
By Glen E. Rodgers
© Glen E. Rodgers 2020
Published by the Royal Society of Chemistry, www.rsc.org

Centre ▓▓; Commemorative Plaque near the site of the "Flying University" ▓; Curie Statue at Warsaw University of Technology ▓; Memorial Plaque at the first radiology laboratory in Poland ▓; Maria Skłodowska-Curie Institute of Oncology ▓▓; Maria Skłodowska-Curie Park and Statue ▓▓; Skłodowski Family tomb at Powązki Cemetery ▓▓; Mural Commemorating the Universe of Maria Skłodowska-Curie ▓▓.

Antoine Becquerel, Marie Sklodowska-Curie, Pierre Curie, Irène and Frédèric Joliot-Curie and the Discovery of Natural and Artificial Radioactivity

- In 1896, Antoine-Henri Becquerel (1852–1908) discovered that uranium salts spontaneously emitted a penetrating radiation that Marie Sklodovska-Curie dubbed "radioactivity".
- Becquerel also discovered that beta radiation was a ray of electrons.
- In 1898, Marie and Pierre Curie announce the discovery of the elements polonium and radium.
- In 1903, Marie receives her PhD and the Curies are awarded the 1903 Nobel Prize in physics.
- In 1906, Pierre is killed in a traffic accident.
- In 1911, Marie's tragic and triumphant year, she received the Nobel Prize in Chemistry.
- In 1932, Irène and Frédéric Joliot-Curie discover artificial radioactivity.
- In 1935, Irène and Frédéric Joliot-Curie share the Nobel Prize in Chemistry.

"So the atoms in turn, we now clearly discern,
Fly to bits with utmost facility;
They wend on their way, and, in splitting, display
An absolute lack of stability."

From the poem, "The Death-knell of the Atom," by Sir William Ramsay[1] written in 1905.

The discovery of radioactivity changed forever the way we view atoms. In Sir William Ramsay's poem, he writes that radioactive atoms "fly to bits with utmost facility". [Recall that Ramsay is honored in Westminster Abbey (see Chapter 7, p. 193) as the winner of the 1904 Nobel Prize in Chemistry, awarded for his role in discovering five of the noble gases.] Before we look in detail at the places in Paris to visit relative to the discovery of radioactivity, let's review what we learned in earlier chapters about Paris (Chapter 8) and radioactivity (Chapter 6). First, recall that in Chapter 8 ("The New French Chemistry and Atomism: Franklin, Lavoisier, Gay-Lussac, Ampère") we highlighted the seminal work of Antoine Lavoiser and his successors who, starting in the late eighteenth century, established the oxygen-based combustion theory (that replaced phlogiston), proposed a new systematic language of chemistry and thereby revolutionized chemistry for all time. In that chapter we discussed striking places to visit in Paris related to Benjamin Franklin, Lavoiser, Claude Berthollet, and Joseph Louis Gay-Lussac, as well as the chemist and microbiologist Louis Pasteur. Readers preparing to visit Paris should consult Chapter 8 as well as this chapter.

We have also discussed radioactivity in a prior chapter. In Chapter 6, "The Brits, Led by the 'Crocodile', Take the Atom Apart", we briefly noted that Wilhelm Röntgen's discovery of X-rays in 1895 in Würzberg, Germany (see Chapter 13) was followed by Antoine Becquerel's discovery of "radioactivity" in Paris in 1896. As briefly noted in Chapter 6, it quickly became apparent that Becquerel's rays were composed of two types. Ernest Rutherford ("the crocodile") named them "alpha" and "beta". He and his co-workers in Montreal and Manchester found that alpha rays were really particles identical to helium nuclei, He^{2+}, that is, helium atoms from which the two electrons had been removed. The trick was that these α- and β- particles were spontaneously and continuously emitted from a "parent" radioactive atom which was then changed into – or "transformed into" as Rutherford liked to say – a new atom, often called the "daughter" atom. Daughter atoms themselves were often intensely radioactive and short-lived so unscrambling all these various radioactive products turned out to be quite the challenge. A decade or so later, Rutherford used Geiger and Marsden's alpha-particle gold-foil experiment (Chapter 6, pp. 155–156) to discover the atomic nucleus. He and Chadwick (see Chapter 6, pp. 162–164) were gradually able to discern that a nucleus was composed of protons and neutrons. It soon became apparent that some atoms of a given element had the same number of protons, but differing numbers of neutrons. These were called "isotopes", a coin termed in 1913 by Frederick Soddy, Rutherford's invaluable co-worker in Montreal but who by then had moved on to the University of Glasgow (see Chapter 6). Rutherford's colleague Francis Aston characterized a number of isotopes using his mass spectrometer. Chapter 6 describes stunning places to visit in New Zealand, Canada and the United Kingdom related to Rutherford and his "boys".

Before we move on to the discovery of radioactivity in Paris, let's take a moment to establish a little nomenclature that some of you are familiar with but others are not. (Here's an opportunity for fellow scientific-historical travelers to help each other learn the language of nuclear chemistry which will, in turn, enhance our understanding of these amazing discoveries and the men and women who first discovered them.) Atomic nuclei are represented by a symbol of general form $^{A}_{Z}X$ where Z is the "atomic number" (that is, the number of protons in the atom), A is the "mass number" defined as the number of protons plus neutrons (collectively known as "nucleons"), and X is the symbol for the element. (Recall from Chapter 6, p. 169, that it was Henry Moseley who had initially established the concept of atomic number.) A little math reveals that $A - Z$ gives the number of neutrons. For example, a helium nucleus could more specifically be represented as $^{4}_{2}He$ indicating each has 2 protons and 2 neutrons. As noted above, an alpha particle is, in fact, a helium nucleus but occasionally it is common to indicate specifically that it has lost its two electrons and therefore has +2 charge, that is, an alpha particle is sometimes but not always represented as $^{4}_{2}He^{2+}$. A uranium atom always has 92 protons and its most common isotope has 146 neutrons. This isotope would be represented as $^{238}_{92}U$ and is referred to as uranium-238. Another less common isotope of uranium is $^{235}_{92}U$, called uranium-235. The latter would have 235 − 92 = 143 neutrons. It was using uranium isotopes that radioactivity was first discovered.

14.2 ANTOINE HENRI BECQUEREL (1852–1908)

Now that we are back in Paris, let's start with the story of how Becquerel discovered radioactivity. He came from a scientific family. For example, his father was particularly interested in fluorescence and phosphorescence, in which materials absorb one wavelength of light but emit another. Phosphorescence is the basis of "black lights", where materials absorb ultraviolet light but re-emit striking visible colors. Recall that Röntgen observed a fluorescent glow (see Figure 13.1) on the wall of his cathode ray tube when it produced X-rays. In light of Röntgen's results, Becquerel wondered if any of his fluorescent materials might also be emitting these mysterious, highly penetrating rays. Working at the Muséum National d'Histoire Naturelle (Museum of Natural History) in Paris, where his father and grandfather had worked before him, he wrapped a photographic plate with black paper and then placed a sample of a fluorescent material on top. He put this little assembly on a window sill where it would be exposed to sunlight. His idea was that sunlight would initiate fluorescence and, he hoped, X-rays. The latter, following Röntgen's discovery, would penetrate through the black paper and expose the photographic film. Sure enough, when he developed the film after only a few hours exposure, it was exposed and Becquerel concluded that the fluorescence was, in fact, producing X rays.

However, there followed a few heavily clouded days and then several more when the sun shone only intermittently. During these days he stored his sample atop the black-paper-covered photographic plate in a drawer awaiting sunnier days. Finally, he set the assembly on the sill for a while and then developed the photographic plate, expecting to find only a faint darkening. He was mightily surprised when he found that it was darkly exposed (see Figure 14.1). Apparently, the darkening of the plate was going on continually all the time it had been sitting under the piece of fluorescent crystal. Its exposure was not connected to the amount of time the assembly had been exposed to the sun. What was happening?

As it turned out, the material he was testing was one his father had been particularly interested in, namely, potassium uranyl sulfate. Its ability to expose the photographic plate was, in fact, not related to fluorescence or its exposure to sunlight. What he had observed were not X-rays but instead another new type of radiation soon to be called "radioactivity". It was the uranium in his crystal that was spontaneously radioactive. He studied the phenomenon carefully. He heated his sample, he chilled it,

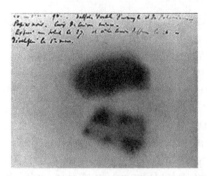

Figure 14.1 Image of Becquerel's photographic plate which has been fogged by exposure to radiation from a uranium salt. The shadow of a metal Maltese Cross placed between the plate and the uranium salt is clearly visible in the lower exposure.

he pulverized it, and he dissolved it in acids, but its activity was undeterred. It looked like the radioactivity was an atomic property having nothing to do with how the atoms were assembled in the crystal. Not only did the radiation from his sample penetrate certain types of matter, but it also ionized the air around it. Today, we know that uranium spontaneously emits both alpha and beta radiation. Uranium-238 is an alpha emitter. U-235 is a strong beta emitter. The alpha particles were prevented from reaching the plate by the black paper, but the beta particles could penetrate right on through. Several years later, in 1899, Becquerel noted that the beta radiation was also a stream of particles that could be deflected by a magnetic field. He determined their charge-to-mass ratio and showed them to be identical to J. J. Thomson's electrons (See Chapter 5). Beta particles, quite remarkably, come from the nucleus but are identical to electrons.

Using our recently introduced notation for atomic nuclei, we can represent the beta particle or electron as $_{-1}^{0}\beta$ or $_{-1}^{0}e$. (It has a negative one charge and contains no nucleons, that is, no protons or neutrons.) We can write a nuclear equation to represent what happens when a uranium-235 nucleus emits a beta particle.

$$_{92}^{235}\text{U} \rightarrow {_{93}^{235}}\text{Np} + {_{-1}^{0}}\beta$$

Note that this is a balanced nuclear equation in that the sum of atomic numbers (the subscripts) on the left side is equal to the sum on the right. The sum of the mass numbers (the superscripts) on each side are also equal. In other words, we say that the parent nucleus, uranium-235, is radioactive and spontaneously emits a beta particle, leaving behind a daughter nucleus, neptunium-235. (For more on the discovery of neptunium, Np, see Chapter 16.)

14.3 TRAVEL SITES RELATED TO HENRI BECQUEREL IN PARIS

14.3.1 Muséum d'Histoire Naturelle

36 rue Geoffroy Saint-Hilaire, Jardin des Plantes, Paris (48.841615°, 2.357066°). Crystals and apparati such as Becquerel's electroscope in the main exhibition hall and in the Trésor. See www.mnhn.fr for further information.

14.3.2 Henri Becquerel Plaque

47 rue Cuvier, Jardin des Plantes, Paris 5 (48.844749°, 2.356940°). A plaque on Georges Cuvier's house denotes the laboratory where Becquerel discovered radioactivity in 1896.

14.4 MARIE SKLODOWSKA-CURIE (1867–1934) AND PIERRE CURIE (1859–1906)

"He caught the habit of speaking to me of his dream of an existence consecrated entirely to the scientific research, and asked me to share that life." Marie Curie

The story of Marie Sklodowska-Curie's work to isolate radium from the uranium ore pitchblende is one of the most remarkable in the history of the atom. Marya Salomea Sklodowska was the youngest of five children living in her native Poland that, officially, did not exist. In 1795 it had been divided up between Austria, Prussia, and Russia. She lived in the section that was ruled by Russia, which severely restricted women from attending university. For three years (1886–1889) she worked as a governess and saved money that she sent to her older sister, Bronya, who was attending medical school at the Sorbonne in Paris. Back in Warsaw for a year, she lived with her widowed father, worked in a chemistry laboratory, and studied in the underground, so-called "Floating University", where both Polish men and women students could study in defiance of the Russian restrictions.

In 1891, now with Bronya's financial help, it was Marya's turn to study in Paris. Now 24 years old, she boarded an unheated, ladies-only, fourth-class, bare-bones carriage for the forty-hour, steam-driven train trip from Poland to study at the Sorbonne. She registered at the Sorbonne under the name Marie. Although Bronya and her husband urged Marie to live with them, Marie found that she was not able to concentrate on her studies well enough and soon chose to live in a sixth-floor, cold, rented room in the Latin Quarter closer to the university. These were difficult years. She had barely enough money for both food and coal for her stove, often having to choose between the two, causing her to faint from hunger in the classroom on several occasions.

Men outnumbered women 100 to 1 at the Sorbonne, her French did not hold up well when professors spoke quickly, and her mathematics background was inadequate. Nevertheless, she prevailed. After two years she took the exam for certification in physics and ranked first among all the candidates. A year later she came in second in a degree in mathematics. Her intention was to earn a teaching degree and return to Poland. However, during that time she met Professor Pierre Curie (who was 8 years older), who was even then an internationally known physicist. Pierre had purposely avoided romantic encounters until he met Marie, whom he found not only charming but easily fluent in physics and mathematics. They were both smitten and after a few hesitant steps, they were married in 1895. It was a perfect match. Pierre wrote that

> "It would ... be a beautiful thing in which I hardly dare believe, to pass through life together hypnotized in our dreams: your dream for your country; our dream for humanity; our dream for science." Pierre Curie in a letter to Marie written in 1894.

Neither of them wanted a church wedding and so were married in a civil ceremony in the town hall of Sceaux (where Pierre's parents lived), just south of Paris (see Chart 14.2). As shown in Figure 14.2, Marie did not wear a traditional wedding gown but

The Discovery That Atoms "Fly to Bits" 335

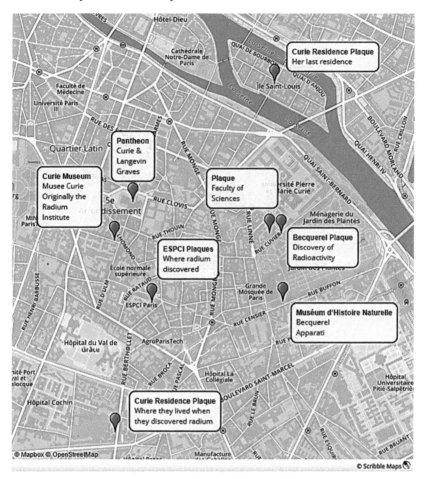

Chart 14.1 Travel sites in downtown Paris related to the discovery of radioactivity by Henri Becquerel and Pierre and Marie Curie and the discovery of artificial radioactivity by Irène and Frédéric Joliot-Curie.

instead a practical dark blue dress that could be worn in the laboratory. There were no wedding rings and the Curies used the money they were given as a wedding present to buy bicycles and, for their honeymoon, toured the French countryside together. Figure 14.3 shows the Curies with their new bicycles.

Pierre also earned his doctoral degree in 1895. His research was on the temperature dependence of various types of magnetism and resulted in what is still known as Curie's Law (after Pierre of course, not Marie). Marie earned her teacher's diploma

Figure 14.2 Pierre and Marie Curie's wedding photograph. Marie wears a simple blue dress.

in 1896 and their first child, Irène, was born in 1897. As head of the laboratory at the School of Industrial Physics and Chemistry (*École de Physique et Chemie*) Pierre arranged for Marie to work there. For her doctoral research, she decided to systematically investigate Becquerel's rays. Becquerel had noted that when his rays passed through a gas, it acquired the ability to conduct electricity. Marie had an electrometer, constructed by Pierre and his brother Jacques, that proved to be most useful in accurately investigating this phenomenon. Almost immediately, she found that, in addition to uranium, the element thorium also emitted Becquerel's rays. Marie turned next to an investigation of ores and minerals containing these elements. One uranium ore, pitchblende, exhibited an activity much larger than could be attributed to the uranium in the sample. It was during this time that they used the term "radioactivity" for the first time. (The name comes from Latin *radius*, meaning "ray".) By then, Pierre had decided to give up his research so that he and Marie could work together.

Figure 14.3 Pierre and Marie Curie toured the French countryside on their bicycles purchased with money given to them for their wedding. This photograph was taken in front of Pierre's family home in Sceaux.

Marie advanced the idea that minerals such as pitchblende must contain very small amounts of an unknown element or elements of extraordinarily intense radioactivity. By 1898, the Curies were able to announce the discovery of two elements, which they called *polonium*, after Marie's homeland, and *radium* (again after *radius*). During this time, with the help of a chemist (Gustave Bemont) and a spectroscopist (Eugène Demarcay), they obtained a clear spectral line due to radium. The previously unknown element had now been "fingerprinted". (See Chapter 11, pp. 271–272 for the role of Robert Bunsen and Gustav Kirchhoff in developing spectroscopy.) To complete their discoveries, they still needed to produce compounds of their new elements.

Seven tons of pitchblende was needed to produce a single gram of radium. Pitchblende itself was too expensive for the Curies to purchase in sufficient quantities, so they had to work with the tailings left over after the ore had been processed at the mine site. After initially paying out of their own pockets to have this brown residue shipped from the mines of St. Joachimsthal (in what was Czechoslovakia, now the Czech Republic), they processed it in a poorly equipped shed – again provided by the *École de Physique et Chemie* – located in a back alley of Paris. The previously unoccupied shed had a bituminous floor, a leaky glass roof and was excruciatingly hot in summer and bitterly cold in winter. This was back-breaking work for Marie, who said that, "Sometimes I had to spend a whole day stirring a boiling mass with a heavy iron rod nearly as big as myself". Working with quantities approaching 100 lb (40 kg), they laboriously reduced the material to insoluble barium and radium carbonates and sulfates until finally, in 1902, they had isolated 0.120 g of nearly pure radium chloride and determined the atomic weight of the radium.

In June, 1903, a standing-room-only crowd gathered for Marie's oral examination for her PhD. The three professors on her examining committee were dressed in the customary white ties and tails. She had a new dress, black so she could wear it in the lab later of course. She was three months pregnant with a child she would later miscarry. Her defense, with drawings on a chalkboard, was flawlessly presented. Her family and friends and the girls from the teacher's college in Sèvres where Marie was teaching physics burst into applause when the favorable result was announced. She was the first woman to win her doctorate at the Sorbonne. In a twist of irony, Marie and Pierre, their families and friends and colleagues, were unexpectedly joined by Ernest Rutherford and his new wife, May, for a celebratory dinner. Pierre produced a small cylinder of a glowing radium salt for their guests' amusement. Rutherford noted that Pierre's hand was so raw and inflamed that it was difficult for him to handle the tube.

In those days, no one knew just how dangerous radioactivity was. Radium is an extremely powerful alpha emitter. It also emits beta particles and gamma rays; its emissions are so intense that it constantly ionizes the air around it, producing a glow. Figure 14.4 shows a photograph of several grams of radium bromide taken by its own light! The Curies have written of returning to their laboratory-shed at night and seeing their various fractions glowing in the dark! Radium remains

The Discovery That Atoms "Fly to Bits"

Figure 14.4 A photograph of 2.7 g of radium bromide, taken by its own light on October 15, 1922. Used by permission from the Musée Curie (coll. ACJC).

perpetually ionized, shoots powerful rays through matter, and burns human flesh without feeling hot to the touch! The Curies were constantly subjected to this ionizing radiation, and Marie's laboratory notebook is to this day unavailable for direct consultation because it remains contaminated by the materials she and her husband worked with on an everyday basis. Even reprints of her thesis, published in Britain in 1903, should be checked for radioactivity since they may have been used in contaminated laboratories all over the world. Not surprisingly, both Marie and Pierre suffered from radiation poisoning all the rest of their lives. They were often in intense pain, suffered with burned, scarred and trembling hands, and were almost constantly fatigued. These problems were not mitigated by Pierre's habit of carrying a sample of radium salt in the pocket of his waistcoat and Marie's having some by her bed because it shone pleasantly in the darkness.

In December of 1903 – the very same year Marie received her PhD – Marie, Pierre, and Becquerel shared the Nobel Prize in Physics for their study of radioactivity. Most likely, you will not be surprised that, originally, only Pierre and Becquerel were nominated! What might surprise you is that four *French* scientists had actively campaigned to exclude her because not only was she a young woman, but she wasn't French! Pierre was told surreptitiously about this campaign and responded that he did not wish to be considered unless it was with Marie. Be that as it may, the Curies could not attend the ceremony due to their poor health due to exposure to radiation.

Pierre, at least, did not have to suffer from radiation sickness for long. Tragically, on April 19, 1906, he was run over and killed by a horse-drawn wagon near the Pont Neuf in Paris (see Chart 14.2). Preoccupied with his work, and walking in the rain with a limp (due to chronic pain), he was instantly killed by a wagon carrying 13 000 pounds of military gear. When his father got the news, he asked "What was he dreaming of this time?"

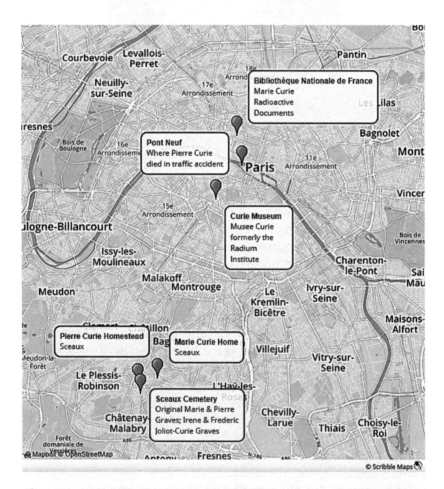

Chart 14.2 Travel sites in greater Paris related to the discovery of radioactivity by Pierre and Marie Curie and the discovery of artificial radioactivity by Irène and Frédéric Joliot-Curie.

Marie was devastated and was left with two daughters, Irène, now 9, and Ève, aged 2. Marie succeeded Pierre as the head of the laboratory at the School of Industrial Physics and Chemistry. She also took over his teaching duties and therefore became the first woman to ever teach at the Sorbonne. When she gave her first lecture in November 1906, the large amphitheater was packed with students, journalists, and photographers. In 1908 she became the first woman appointed as a professor at the Sorbonne. She continued to produce pure radium chloride and with others, was able to isolate metallic radium.

1911 was both a tragic and triumphant year for Marie. It started in late 1910, when Marie and Paul Langevin, one of Pierre's first students, now a prominent physicist in his own right and a good friend of the Curies both before and after Pierre's death, became romantically involved. Also late in 1910, Marie let it be known that she was willing to stand for the public election to *l'Académie des Sciences*, an all-male honorary society. The combination was too much for the French press to resist. The election, normally not note-worthy, drew journalists, photographers, and numerous curious bystanders. It is fair to say that a prejudice against women and immigrants, a combination of anti-science and even anti-Semitic sentiments, and the rumors of the Curie–Langevin affair resulted in Marie losing the election by one vote! Adding insult to injury, the academy then went on to vote to ban women "forever"! ("Forever" turned out to be about half a century.) This was a humiliation that Marie never forgot.

Tragic though the year was shaping up to be, Marie was triumphantly invited to the first Solvay Conference in Brussels. She was the only woman invited to this elite conference that discussed the latest problems in radiation and quantum theory. There were 20 pre-eminent scientists there including Albert Einstein, Max Planck, Ernest Rutherford and, yes, Paul Langevin (see Figure 14.5 for the famous picture of all the participants). During the conference, however, newspaper articles appeared that accused Marie of being an "eloper" and a "husband snatcher", the husband being Paul Langevin. She demanded and received an apology, but the damage was done and she was subsequently hounded by the press for years. In a remarkable coincidence, on the very same day that these articles appeared, Marie received the news that she had been awarded a second Nobel Prize, this

Figure 14.5 Photograph of the first Solvay Conference in 1911 in Brussels. Seated (L–R): W. Nernst, M. Brillouin, E. Solvay, H. Lorentz, E. Warburg, J. Perrin, W. Wien, M. Curie, and H. Poincaré. Standing (L–R): R. Goldschmidt, M. Planck, H. Rubens, A. Sommerfeld, F. Lindemann, M. de Broglie, M. Knudsen, F. Hasenöhrl, G. Hostelet, E. Herzen, J. H. Jeans, E. Rutherford, H. Kamerlingh Onnes, A. Einstein, and P. Langevin. Photography courtesy of Hulton Archive/Stringer/Getty Images.

time in chemistry, for her discovery of the elements polonium and radium and the isolation and characterization of radium and its compounds.

When Marie returned to Paris still under attack from a rabid press, Langevin challenged an acutely aggravating newspaper man to a duel! As you might suspect, scientists don't do duels particularly well; Langevin's second was a mathematician and future Prime Minister who was not well-versed in duel rules; they fought with pistols at a distance of 25 meters but, thankfully, no shots were fired. However, news of the duels caused the Swedish Academy of Science to express reservations about Mme Curie actually attending the ceremonies in Stockholm. Standing their ground and with great difficulty and trepidation, Marie, Bronya, and Irène travelled to Stockholm so Marie could receive her second Nobel Prize in person. Back home again in Paris, she and her daughters, faced with hostile crowds day and night, had to be taken elsewhere for their safety and ultimately moved to an apartment on Île St. Louis. For several years after this, still suffering the ravages of radiation poisoning, Marie kept a low, even anonymous, profile and slowly recovered both mentally and physically.

The Radium Institute was jointly established by the University of Paris, *i.e.*, the Sorbonne, and the Pasteur Institute. Built to honor the memory of Pierre Curie, it was divided into two pavilions or sections; one, to be headed by Marie, for research in radioactivity, and a second for research in medicine and biology. Marie took an active part in the planning and construction of the building and even created gardens for it before the foundations were complete. It was completed in July of 1914, but the outbreak of the First World War delayed its opening. When the war broke out, Irène (17) and Ève (10) were sent to l'Arcouest in Brittany, where they often summered. Marie took a 20 kg lead-lined container full of her radium on a crowded train to Bordeaux. She slept next to it that night and then placed it in a bank vault and returned to Paris. To aid the war effort, she supervised the equipping of 20 vans as mobile field hospitals and established 200 fixed installations with X-ray apparatus. Figure 14.6 shows Marie sitting in the driver's seat of the vans, which were known as "Petite Curies". When Irène turned 18 she too was active in this effort. As a result of their efforts, as many as a million soldiers were X-rayed, and thousands of lives were saved. Not surprisingly, both Marie and her daughter were exposed to large amounts of radiation. At the

Figure 14.6 Marie Curie sitting in the driver's seat of, but not actually driving, one of the radiology cars, a "Petite Curie", in 1917. Used by permission from the Musée Curie (coll. ACJC).

end of the war, Irène received a military medal but Marie did not. The Radium Institute (which had been completed in 1914) was opened in 1918 after the war was over.

Radium became the miracle drug of the early 1900s. Rather incredibly, it was used to treat heart trouble, a variety of cancers, tuberculosis, rheumatism, high blood pressure, asthma, ulcers, impotence, and even some mental illnesses. Radium therapy, although it turned out to be too dangerous, was the forerunner of modern radiation therapy. Radium in various forms was also used for a variety of more trivial and now highly questionable applications including liniments, toothpastes, skin creams, hair tonics, mouthwashes, bottled water (known as "liquid sunshine"), and in a glow-in-the-dark paint for watch dials. This paint contained a small amount of a radium salt and zinc sulfide, a phosphor that glows in the dark when hit by an energized particle, in this case an alpha particle from the decay of radium. (This is the same phosphor used by Rutherford, Geiger and Marsden in their gold-foil, alpha particle experiments that revealed the existence of the nucleus. See Chapter 6.) Unfortunately, the women employed to paint the hands and numerals of the watches with fine brushes (which they inserted in their mouths to obtain a fine point) paid dearly for the ignorance of the effects of radioactivity. (See *The Radium Girls: The Dark Story of America's Shining Women*,[2] by Kate Moore, for the tragic stories surrounding these young girls.)

Because of the applications found for it and its extreme scarcity, radium soon became extraordinarily expensive, on the order of $100 000 per gram. In the early 1920s, Marie Curie had only a gram or so in her laboratory, and could not afford to purchase any more. On the other hand, there were about 50 grams in America! Enter Marie ("Missy") Maloney, a prominent American female journalist. Understandably, Marie distrusted all members of the press and routinely spurned every request for an interview. Missy, however, had some contacts unavailable to most and was able to set up a meeting with, in her words, the "pale, timid little woman in a black cotton dress, with the saddest face I had ever looked upon". When Marie, in her distinct Polish-accented English, lamented that she did not know how to obtain more radium, Missy had an answer: "the Women of America". Quite remarkably, Marie almost immediately agreed that Missy could proceed to set up a campaign to raise the necessary funds and that Marie, Irène (23 years old), and Eve (16 years old) would go to America to receive her radium. Most of the funds came from small donations from ordinary American women. Missy made the necessary arrangements with the press (getting them to agree to never mention the Langevin affair, for example), the American Chemical Society, the National Academy of Sciences, women's colleges like Smith and Mount Holyoke, and the Standard Chemical Company in Canonsburg, near Pittsburgh that was contracted to produce Marie's radium. It was a six-week trip, one that was terribly difficult on Marie, who almost always appeared shy, drawn, tired, and unsmiling. However, as shown in Figure 14.7, on May 20, 1921, she did smile when President Warren G. Harding presented Marie Skladowska Curie, "the radium lady", with one gram of radium paid for by the women of America. (In 1929, again with her now-good-friend Missy's help, she returned to America once again – this time with few ceremonies and no daughters in tow – and received sufficient funds, many from the Polish Women's Alliance of America, to buy more radium, which she donated to the medical institute in Warsaw that she and her sister Bronya had established. She returned home just before the great stock exchange crash in that year.)

Figure 14.7 Marie Curie with President Warren Harding on May 20, 1921, the day he presented her with a gram of radium, a gift of the "women of America". Used with permission of Emilio Segrè Visual Archives/American Institute of Physics/Science Source.

After the 1921 trip, Marie returned triumphantly to Paris. Nanny Fröman says "after being dragged through the mud ten years before, she had become a modern Jeanne d'Arc". Irène, now a doctoral candidate at the Sorbonne, worked with her mother as a technical assistant at the Radium Institute. Specifically, Irène was studying the alpha rays emitted from polonium. In 1924, Marie hired one of Paul Langevin's students as an additional technical assistant at the institute. This was Frédéric Joliot. It didn't take long for Irène and Frédéric to start working together. Marie gave them highly concentrated samples of polonium-210 for their experiments. Irène defended her PhD thesis on March 30, 1925 in front of 1000 people! In 1926 they married and combined their surnames to become Irène and Frédéric Joliot-Curie, although they still signed their scientific papers Irène Curie and Frédéric Joliot. Their combined research was fruitful and so were they. In 1927, their daughter, Hélène, was born, followed in 1932 by Pierre, named for Irène's revered father. Also in 1932, Irène replaced her mother, whose health was rapidly failing, as the director of the Radium Institute.

In the early 1930s, Irène and "Fred" started to use their polonium to bombard beryllium (Be) atoms with alpha particles. They produced a very energetic and penetrating radiation that they misinterpreted as gamma rays. As it turned out, they had just missed discovering neutrons. James Chadwick (see Chapter 6, p. 163), working in Rutherford's laboratory in Cambridge, repeated their experiments, came up with the correct interpretation, and is credited with the discovery of the neutron. He received the Nobel Prize in physics in 1935. The nuclear equations representing these processes are given below. (A neutron is represented as $^1_0$n.) Notice that they are balanced nuclear equations. (See earlier in this chapter for a quick review of how to read nuclear equations and tell that they are balanced.)

$$^{210}_{84}\text{Po} \rightarrow ^{206}_{82}\text{Pb} + ^{4}_{2}\text{He}$$ (alpha particle here not represented with its +2 charge)

$$^{4}_{2}\text{He} + ^{9}_{4}\text{Be} \rightarrow ^{12}_{6}\text{C} + ^{1}_{0}\text{n}$$ (neutron)

They also just missed the discovery of positrons, which are exactly like electrons but with a positive charge rather than negative. The positron, represented as $^{0}_{+1}$e or $^{0}_{+1}\beta$, was the first example of "antimatter", exactly like matter but with all the particles having the opposite charge. The credit for this discovery went to Carl Anderson working at Caltech in California. (He received the Nobel Prize in physics in 1936.) All was not lost, however, as the couple persevered at the Radium Institute. They continued to bombard various light elements like boron and aluminum with alpha particles. For example, when aluminum was bombarded with alpha particles, first a neutron is produced, as well as a brand-new isotope of phosphorus, phosphorus-30. Let's see the balanced nuclear equation for this process.

$$^4_2He + ^{27}_{13}Al \rightarrow ^1_0n + ^{30}_{15}P$$

Phosphorus-30 was the first example of an "artificially produced" radioactive isotope. It decays by emitting a positron, as shown in the following balanced nuclear equation. Phosphorus-30 has a half-life of 150 seconds, so Irène and Fred had to work quickly to chemically separate the aluminum from the newly created and quickly disappearing radioactive phosphorus.

$$^{30}_{15}P \rightarrow ^{30}_{14}Si + ^0_{+1}e$$

This was a remarkable discovery. The Joliot-Curies had discovered how to make brand new radioactive materials. That is, they had produced atoms previously unknown to anyone. They had missed the neutron and the positron, but in 1935 they were jointly awarded the Nobel Prize in Chemistry for their discovery of "artificial radioactivity". Marie, now near death, knew about their discovery and even held the evidence in her quivering and scarred hands. However, she died before the Nobel Prize was awarded to her beloved daughter, Irène, and her husband, Frédéric. Marie died of radiation-induced leukemia at the age of 66. She was buried in the Sceaux cemetery, just above her beloved husband, Pierre. Marie and Pierre had discovered natural radioactivity in 1898. Irène and Fred discovered "artificial radioactivity" in 1932. (Eve was a pianist, journalist, and author. She wrote a definitive first biography of her mother.)

It did not take long to realize that artificially produced radioactive isotopes could be used to follow physiological processes. Frédéric himself used "radioiodine" (I-131) as a tracer to study the thyroid gland. It can also be used to treat hyperthyroidism (enlarged thyroid) and thyroid cancer. When their radiophosphorus (P-32) is incorporated in phosphates, the metabolism of various organisms can be followed in great detail. It also did not take long for others to use radioactivity for nefarious purposes. One of the most famous is using polonium-210, a lethal alpha particle emitter, to assassinate people like Aleksandr Litvinenko, a former KGB officer who had fled Russia and sought political asylum in the United Kingdom. He died of "lethal polonium-210-induced acute radiation syndrome" in November 2006.

14.5 TRAVEL SITES RELATED TO THE CURIES IN PARIS

14.5.1 Curie Residences

Pierre Curie was born at 16 rue Cuvier. Later, at 12 rue Cuvier, the Laboratoire Curie was established by the Sorbonne in 1904. The laboratory remained on that site until 1914, when the Institut du Radium was built on rue Pierre Curie.

Marie Skladowska's first 6th floor flat was at 3 rue Flatters in the Latin Quarter. Her second residence was also on the 6th floor at 11 rue des Feullantines. (Pierre reportedly climbed up to this room to visit Marie during their courtship years.) Marie's sister, a physician, had an office at 37 rue de Chateaudun. Marie lived and studied in a small nearby room during her final year at the Sorbonne.

See Charts 14.1 and 14.2 for the location of many of the following sites.

14.5.1.1 Commemorative Plaque. 24 rue de la Glacière (48.835584°, 2.344926°). Pierre and Marie Curie were living in this building when they discovered radium in 1898 at the School of Industrial Physics and Chemistry (*École de Physique et Chemie*). This is on the other side of the Seine, close to the Paris Observatory. The plaque marks the apartment where Marie and Pierre were living at the time and where their daughter Irène was born in 1897.

14.5.1.2 Commemorative Plaque. 36 Quai de Béthune (48.851217°, 2.356455°). Marie Curie lived in this building on the Île St. Louis from 1912 to her death in 1934. She moved here from Sceaux after her triumphal and tragic year of 1911. During this time, she set up an informal school for children of university faculty. Irène had fond memories of attending this school.

14.5.2 Curie Workplaces

14.5.2.1 Commemorative Plaques. 10 rue Vauquelin (48.841247°, 2.347570°). Designates that radium was discovered here with the assistance of the chemist Gustave Bémont. Just below the commemorative plaque is a new brass plaque designating the location of the ESPCI (School of Industrial Physics and Chemistry). A plaque to left of the vehicle entrance gives a brief history of the school and includes a quotation by Pierre Curie, who worked here for over 20 years before moving to the Sorbonne. The plaque announcing the discovery of radium is to the right of door. Mary Virginia Orna says one needs to get permission to get into the building which houses a collection of the original equipment used by the Curies. Louis Nicolas Vauquelin, for whom this street was named, was a chemist and discovered chromium.

14.5.2.2 Curie Museum (Musée Curie). 1 rue Pierre-et-Marie-Curie (48.844346°, 2.344811°), established in 1934 after Marie's death, located in the Curie Institute (Institut Curie) on the ground floor of the Curie Pavilion. The museum was renovated in 2012, thanks to a donation from Ève Curie (who died in 2007 at the age of 102). On the exterior façade are portraits of Marie, Pierre, Irène, and Frédéric. The institute, originally called the Radium Institute (Institut du Radium), was built between 1911 and 1914 for Marie after Pierre's death. Marie did research here between 1914 and 1934. Irène joined her in the 1920s and Frédéric a little later. Her laboratory and office are preserved including her desk, notebooks, laboratory coats, and some equipment. There is a replica of the Pierre Curie's piezoelectric-electrometer used for the quantitative measurement of ionizing radiation. If the opportunity presents itself, ask the staff for an explanation of how the electrometer works. Briefly, piezoelectric current, developed by putting pressure on a crystal of quartz that is suspended above a weight pan, is used to balance the current of the electrometer which increases as the air around a radioactive substance is ionized. Historical exhibits on radioactivity and its applications at the museum vary over time, but often include various nefarious uses of radium including face cream, fish bait, cigarettes, baby-warmer blankets, atomic perfume, and radium water for both bathing and consumption. The lead-lined mahogany box used to transport the radium presented to her by President Warren Harding is on display. The plaque on the box says it was presented to her for "transcendent service to science and to humanity in the discovery of radium". Captions for the exhibits are in both French and English. Much of the equipment is from Irène and Frèdèric Joliot-Curie's day. Some of the original technical artifacts were highly contaminated and had to be destroyed. There are sculptures of Marie and Pierre in the courtyard, which Marie dearly loved. There are portraits of some of the Curie's associates around the garden area. A giant portrait of Marie on a wall of the Curie Institute can be seen from 20 rue d'Ulm (48.843770°, 2.344813°). For more information, see https://musee.curie.fr/.

The Institut Curie, 26 rue d'Ulm, is just down the street from the museum. Today, the institute is a modern cancer research institution. Rue Gay-Lussac intersects with rue d'Ulm – this is where you will find the Pub Gay-Lussac. This is different from the Café Gay-Lussac, a little farther up the street at 15 rue Royer Collard.

14.5.2.3 Commemorative Plaque. Faculty of Sciences Building, University of Paris, 12 rue Cuvier (48.844771°, 2.356075°). This building was constructed in 1900, the first annex of the Sorbonne. Pierre, and later Marie Curie taught physics here. Behind the lecture hall is a small shed, still radioactive today, where both Marie and Pierre carried out some of their studies on radioactivity. The plaque says *"Dans ce pavilion Pierre et Marie CURIE on poursuivi de 1903 à 1914 leurs recherches fondamentales sur le radium"*. ("In this pavilion Pierre and Marie CURIE pursued their basic research on radium from 1903 to 1914".)

14.5.3 Curie Graves

14.5.3.1 Panthéon (Sorbonne). (48.846182°, 2.346200°). Marie and Pierre's remains were moved here from the cemetery in Sceaux in 1995. Lech Walesa, President of Poland, and François Mitterand, President of France, were present for the ceremonies. There is always a flower on their graves. Other scientists buried in the Pantheon include Louis Braille, Paul Langevin (!), Jean Perrin, Pierre Eugene-Marcellin Berthollet, and his wife Sophie (who died an hour before her husband and is the only other woman buried here). Marie Curie is the only person not born in France who is buried here. Stand here and think about the great trials, tribulations, and triumphs that characterized Marie and Pierre's lives. A visit to the grave of Paul Langevin might also be appropriate.

14.5.4 Curie Artifacts and Tours

14.5.4.1 Bibliothèque Nationale de France. 5 Rue Vivienne (48.867397°, 2.338743°). Some writings and personal documents of Pierre and Marie Curie are held here. They are still intensely radioactive, and one needs to sign a waiver to view them. They are sealed in a lead-lined box.

14.5.4.2 Parcours des Sciences. http://parcoursdessciences.fr/ This organization conducts a variety of guided tours for scientists including the following related to the Curies.

14.5.4.3 "Festival of Science: Pioneers of Radioactivity" (Pionniers de la radioactivité). This tour includes the National Museum of Natural History where Becquerel discovered radioactivity, the Radium Institute, the Industrial School of Physics and Chemistry, and various plaques related to the discovery of radium. It does not include the Curie Museum.

14.5.4.4 "Festival of Science: From Maria Sklodowska to Marie Curie" (De Maria Skłodowska à Marie Curie). This tour follows Marie from her arrival at the Sorbonne, to the Sainte-Geneviève library where she sought a warm reprieve, to the School of Physics and Chemistry, the trace of the shed where she and Pierre discovered radium and polonium, and the Curie Pavilion where she worked toward her second Nobel Prize, this time in chemistry.

14.5.4.5 "Heritage days: the Curia, Street Cuvier Pavilion". Discover the Pavilion built in 1905 by Pierre Curie just before his death and where Marie Curie then went on to perform the work that earned her second Nobel Prize.

14.5.4.6 Sceaux, Paris. Pierre's childhood home, Marie's residence after Pierre died, original burial site of both Marie and Pierre Curie, as well as the permanent burial site of Irène and Frédéric Joliot-Curie.

14.5.4.7 Pierre Curie Home Commemorative Plaque. 9 rue Pierre Curie (48.776373°, 2.285231°). This was the family home of Pierre Curie and his brother Jacques. Dr Eugene Curie and his wife lived here from 1892 to 1900; his son Pierre lived here from 1892 to 1895. A plaque was placed here on the 50th anniversary of Pierre and Maries' discovery of radium. The photograph of the Curies on their bicycles shown in Figure 14.3 was taken in front of this house.

14.5.4.8 Marie Curie Home Commemorative Plaque. 6 rue St. Jean Mascré (48.781579°, 2.294581°). This is only a kilometer or so from 9 Pierre Curie Street. Marie and her two daughters lived here after the accidental death of Pierre in 1906. Dr Eugene Curie, Pierre's father, lived in an outbuilding. A Polish housekeeper took care of the Marie's daughters while she continued her research.

14.5.4.9 Original Curie Graves. Sceaux Cemetery, 174 rue Houdan (48.779769°, 2.283876°). The Curie gravestone appropriately lists Marie on top, Pierre in the middle and Dr Eugene Curie, Pierre's father, on the bottom. Pierre was actually the first one buried here in 1906. When Dr Curie died in 1910, Marie requested that the father be buried deepest, Pierre next, with room for her at the top. She was buried here as she intended in 1934. A plaque in front of the white marble grave notes that Marie and Pierre now repose in the Panthéon. There is not a Memorial number listed for Marie or Pierre since they no longer are interred here. No Memorial number is available for Dr Eugene Curie either. Irène and Frédéric Joliot-Curie are also buried here, Irène at Memorial number 38021075, Frédéric at Memorial number 38021079. Michel Langevin, the Joliot-Curie's son-in-law, is also buried here.

14.6 TRAVEL SITES RELATED TO THE CURIES IN PLOUBAZLANEC, L'ARCOUEST, BRITTANY

14.6.1 Irène and Frédéric Joliot-Curie Memorial

Ploubazlanec, l'Arcouest (approximately 48.821443°, −3.013706°). L'Arcouest was a summer gathering place for a number of Paris academics and became a second home for all of the Curies. Sometimes it was even called the "Sorbonne-by-the-Sea". Marie Curie had a cottage on the Arcouest, and brought in many scientists in the 1920s, which made the place "Sorbonne Plage" or "Fort la Science". Her villa still stands. The Joliot-Curie Memorial here on the north coast of Brittany is situated in a small open enclosure formed by walls built of granite, about 50 cm high. The monument is rectangular and carved in granite. The plaque says "In tribute to Irène and Frédéric Joliot Curie, lives devoted to science and peace. Their friends and the commune of Ploubazlanec, where they liked to stay." Two sculptures carved from a coarse granite, representing human shapes, face themselves behind the monument on the low wall, symbolizing the two scientists linked in life and death by their research.

14.7 TRAVEL SITES RELATED TO THE CURIES IN WARSAW, POLAND

See Chart 14.3 for the location of these sites.

14.7.1 Maria Skłodowska-Curie Museum

ul. Freta 16 (52.251566°, 21.008524°) (Muzeum Marii Skłodowskiej-Curie), the birth place of Marie Skłodowska Curie. The museum includes many personal items like the little elephant given to her by Herbert Hoover, President of the United States, the leather bag in which the Polish Women's Alliance of America gave Marie the money for opening the Radium Institute in Poland. Another exhibit includes replicas of laboratory equipment and a model of her laboratory. Many photographs, postage stamps and postcards are on display. This is also the home of the Polish Chemical Society (Polskie Towarzystwo Chemiczne). In 2011 (the 100th anniversary of her 2nd Nobel Prize), a mural was painted on this building in Warsaw's New Town. The infant Maria holds a test tube from which emanate the elements discovered by her: polonium and radium. During the Warsaw Uprising in 1944, the building was demolished by German forces. However, the plaque survived and was put back in place when the building was reconstructed following the war.

14.7.2 Monument to Maria Sklodowska-Curie

ul. Kościelna (52.254164°, 21.009377°). Maria loved to stroll along the banks of the River Vistula. This monument features a 2.5 m statue depicting Maria sweeping along in a full bronze-colored

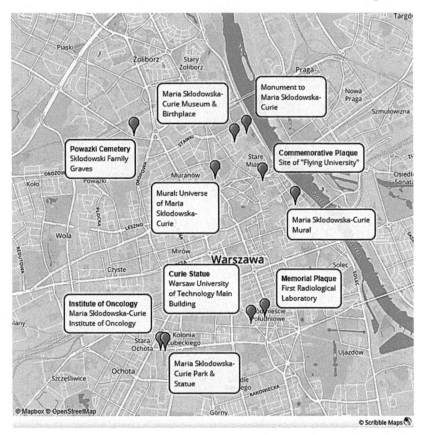

Chart 14.3 Travel sites in Warsaw celebrating the Life and work of Marie Sklodowska Curie.

dress. She holds aloft a box within which is a representation of her element polonium, Po, with its atomic number of 84. There are some benches nearby which afford great views of the River Vistula shore and the Multimedia Park with its synchronized fountain. What a wonderful place to stop and think about Maria. Try to be here at sunset if possible. It is on a hill not far from the Maria Skłodowska-Curie Museum.

14.7.3 Mural of Maria Skłodowska-Curie

ul. Lipowa 3 (52.241559°, 21.024071°) near the Warsaw University Library and the Copernicus Science Centre. Here Maria, her arms tattooed with her two famous elements, is balancing two large beakers. With considerable visible poetic license, the polonium beaker is shown being metaphorically "filled" from a large pipe coming up from the ground and spiraling a radiant lime green liquid into the beaker.

14.7.4 Commemorative Plaque

Central Agricultural Library (*Centralna Biblioteka Rolnicza*) ul. Krakowskie Przedmieście 66 (52.245826°, 21.014242°). The year before Marie went to Paris, she defied Russian prohibitions and studied at the "Flying University" (*Uniwersytet Latający*) that was housed in this building, then the home of the Museum of Industry and Agriculture. In a small outbuilding near the courtyard she did chemical experiments as she prepared to apply to the Sorbonne. Years later, she said "had they not taught me chemical analysis so well in Warsaw .... I would never have isolated radium." A plaque hangs on the wall of the building on the side of Krakowskie Przedmieście Street. The outbuilding still stands and now contains guestrooms.

14.7.5 Curie Statue

Warsaw University of Technology (Politechnika Warszawska) plac Politechniki 1 (52.220531°, 21.010564°). In 2005, a 2 m statue of Maria Skłodowska-Curie was unveiled in the hall of the main building. The artist reportedly wanted to represent Skłodowska as a woman who made difficult choices.

14.7.6 Memorial Plaque

First Radiological Laboratory in Poland, 8 Śniadeckich Street (52.221602°, 21.014796°). The plaque translates to: "In this building was from 1913 till 1939 a radiological laboratory of the Warsaw Scientific Society, the Honorary Director of which was Maria Skłodowska-Curie. This plaque was installed in 1997 to celebrate the centenary of the discovery of polonium and radium and in honor of the discoverers of these elements – Maria Skłodowska-Curie and Pierre Curie. Polish Academy of Science, Warsaw Scientific Society." The building now houses the Institute of Mathematics of the Polish Academy of Science (PAN) and the Warsaw Science Laboratory.

14.7.7 Maria Skłodowska-Curie Institute of Oncology

ul. Wawelska 15 (52.215902°, 20.983628°) (Centrum Onkologii – Instytut im. Marii Skłodowskiej-Curie - dawny Radowy). This is the former Radium Institute which, thanks to Maria's initiative, was opened in Warsaw in 1932. A substantial sum of about $80 000, raised from the Polish women in America, supported the institute. Maria laid the foundation stone in 1925. Seven years later, she planted a gingko tree and presented the institution with of a gram of radium for medical purposes. A large mural dedicated to the elements she founded is on the Wawelska side of the building. In the year after the scientist's death, a statue of Maria Skłodowska-Curie, showing her with bowed head, was unveiled in front of the Institute and the street next to it was named after her. The Institute is now the seat of the Society in Tribute to Maria Skłodowska-Curie. Inside the building are a sculpture of Skłodowska's head and a commemorative plaque dedicated to Franciszek Łukaszczyk, the first director of the Institute, who became famous also for having hidden the radium given by Maria Skłodowska from the Nazis during the occupation. During the occupation, patients were given fake radiation treatments during the day (to fool the Nazis) and then real ones at night.

14.7.8 Maria Skłodowska-Curie Park and Statue

ul. Marii Skłodowskiej-Curie (52.215823°, 20.984639°). This park is just across the street from the Institute of Oncology. It contains a life-sized statue and an educational trail that opened in 2011 featuring several large cubes that share the story of Maria's life and her scientific contributions.

14.7.9 Powązki Cemetery

(Cmentarz Powązkowski) ul. Powązkowska 14 (52.253834°, 20.976064°). Many eminent Poles are buried here. There are many monuments and medallions of artistic value. The Skłodowski Family tomb includes the graves of Maria's parents, her sisters Zofia, Helena, Bronisawa, and her brother Józef. (Section no. 164, row III, entrance through the St. Honorata gate near St. Karol Boromeusz Church).

14.7.10 Mural Commemorating the Universe of Maria Skłodowska-Curie

ul. Nowolipki 11 (52.245944°, 20.996196°). The central point of the mural is an atom, around which the world spins – represented in this painting by plants and animals. This is where the Skłodowskis moved to when Maria was a one-year-old. The tenement house into which they moved no longer exists. This was also where Marie's father taught mathematics and physics at the 2nd Public Secondary School for boys.

14.8 SUMMARY

After an introduction to the symbols commonly used to represent atomic nuclei and the nature of balanced nuclear equations, we see how Antoine Becquerel discovered his "rays" while searching for Röntgen's X-rays in fluorescent materials found at the Museum of Natural History in Paris, where his family had worked for many years. His potassium uranyl sulfate emitted two types of rays and he was able to show that one of them was a small stream of particles identical to J. J. Thomson's electrons. We visit places near the Jardin des Plantes where Becquerel's work is celebrated. Next, we follow how the Warsaw-native Maria Salomea Skłodowski earned her way to the Sorbonne in Paris where she excelled, under extraordinarily impoverished conditions, in her study of physics and mathematics. After she married Pierre Curie, they worked together, again under bleak and dangerous conditions particularly for Marie, to isolate two new elements, polonium and radium, and coin the term "radioactivity". Becquerel and the Curies shared the 1903 Nobel Prize in Physics for these discoveries. After Pierre died in a tragic traffic accident, Marie, now the mother of two young daughters, continued her work and earned a second Nobel Prize in Chemistry. She survived twenty more years of tragedy and triumph, living just long enough to see her daughter Irène, with her husband Frédéric Joliot, discover artificial radioactivity. We visit myriad residences, workplaces, laboratories, hospitals, monuments, statues, museums, and graves all over Paris and Warsaw (and even Brittany) that honor and preserve the memory of the Curies. Several of these are four - and five-atom sites, among the best scientific-historic traveling sites in the world.

ADDITIONAL READING

- L. Redniss, *Radioactive, Marie & Pierre Curie, A Tale of Love & Fallout*, Dey Street, An Imprint of William Morrow Publishers, 2010. For information, address HarperCollins Publishers, New York.
- G. Gablot, A Parisian Walk along the Landmarks of the Discovery of Radioactivity, in *The Physical Tourist: A Science Guide for the Traveler*, ed. J. S. Rigden and R. H. Steuewer, Birkhäuser Verlag AG, Switzerland, 2009.
- N. Fröman, Marie and Pierre Curie and the Discovery of Polonium and Radium, (originally delivered as a lecture at the Royal Swedish Academy of Sciences in Stockholm, Sweden, on February 28, 1996), http://www.nobelprize.org/nobel_prizes/themes/physics/curie/index.html, accessed October 17, 2016.
- M. V. Orna, Paris: A Scientific 'Theme Park', in *Science History: A Traveler's Guide*, ed. M. V. Orna, American Chemical Society, 2014, distributed by Oxford University Press, Oxford, 2014.
- E. Curie (transl. V. Sheean), *Madame Curie: A Biography*, Da Capo Press, Boston, MA, 1937. For further information, contact Doubleday & Co., New York.
- R. Pflaum, *Grand Obsession: Madame Curie and Her World*, Doubleday, New York, 1989.

REFERENCES

1. S. W. Ramsay, The Death-knell of the Atom, *Ind. Eng. Chem., News Ed.*, 1930, **8**, 18.
2. K. Moore, *The Radium Girls: The Dark Story of America's Shining Women*, Simon & Schuster, UK, 2016.

CHAPTER 15

Quantum Mechanics Reluctantly Proposed: Planck and Einstein (Germany and Switzerland)

15.1　A QUICK LOOK AT PLACES TO VISIT "TRAVELING WITH THE ATOM" IN GERMANY AND SWITZERLAND

Max Planck Sites in Berlin: Planck Statue and Plaque at Humboldt University ▓▓; *Berlin-Dahlem: Max Planck residence* ▓; *Archives of the Max Planck Society* ▓▓▓▓; *Göttingen: Max Planck's grave* ▓▓, *I. Physikalisches Institut, Lichtenberg Collection* ▓; *Albert Einstein Sites in Ulm, Germany: Einstein Birthplace Monument and Memorial Plaque* ▓▓, *Einstein Window in Ulm Münster* ▓▓, *Einstein Fountain Sculpture* ▓, *Einstein House* ▓▓; *in Munich, Germany: Memorial Plaque* ▓; *in Bern, Switzerland: Einstein House and Plaque* ▓▓▓; *Einstein Plaque at the former Patent Office* ▓, *Café Bollwerk* ▓, *Einstein Exhibit at Historical Museum of Bern* ▓▓▓; *in Zurich, Switzerland: Einstein Memorial Plaque* ▓, *Bust of Einstein* ▓, *Memorial Plaque to Mileva Marić* ▓; *in Berlin, Germany: Einstein Memorial Plaque* ▓, *Plaque at Einstein Residence* ▓, *Great Synagogue* ▓; *in Caputh, Germany: Einstein's Summer House* ▓▓; *in Brandenberg, Germany: Einstein Tower at Telegraphenberg* ▓.

Traveling with the Atom: A Scientific Guide to Europe and Beyond
By Glen E. Rodgers
© Glen E. Rodgers 2020
Published by the Royal Society of Chemistry, www.rsc.org

> **Max Planck and Alfred Einstein and the Wave-particle Duality of Light**
>
> - In 1900 Max Planck solves the "ultraviolet catastrophe" by reluctantly proposing that light energy comes in "bursts" or "quanta" described by his equation $E = h\nu$. Planck never really accepted the reality of his own proposal.
> - In 1905, in one of three seminal articles in his *annus mirabilis*, Albert Einstein explains Brownian motion in terms of atoms and molecules bombarding small particles in a fluid like water.
> - Also in 1905, Einstein explains the photoelectric effect by proposing that light is particulate, acting like particles, later called "photons".
> - Einstein has proposed the "wave-particle duality of light", one of the most mind-boggling concepts in science, that when extended to the wave-particle duality matter, will help explain the internal structure of the atom.
> - In 1922, Einstein is awarded the Nobel Prize in Physics for his explanation of the photoelectric effect, instead of special theory of relativity.

15.2 MAX PLANCK (1858–1947)

> Planck has come up with a "previously unimagined thought, the atomistic structure of energy" .... "Without this discovery it would not have been possible to establish a workable theory of atoms and molecules and the energetic processes which govern their transformations." Albert Einstein

It's time to go back to a consideration of the inner structure of the atom. In Chapter 12, we explored Niels Bohr's "explanation" of the nature of electrons purportedly orbiting around the nucleus of an atom. In 1913, he had maintained that Rutherford's nuclear atom does not collapse as expected because classical laws, particularly Maxwell's laws of electromagnetic fields, are not valid inside the atom. In that chapter, we noted that Bohr applied the new "quantum ideas" of Planck and Einstein to the nuclear atom. He then went on to assert that the orbits and energies of electrons are "quantized" and that spectral lines are due to electrons moving from one allowed orbit to another.

Quantum Mechanics Reluctantly Proposed

This assertion was mysterious at best and implausible at worst because he maintained that the orbits and energies were quantized because "I say it is true" and "you're too young to know" as the real reason. His colleagues were stunned and amazed, mostly because his audacious quantum assumption worked so well, albeit only for simple hydrogen atoms, and therefore they "wouldn't [*i.e.*, couldn't] say boo to him". In Chapter 12, we left it at that and had a terrific time exploring the Niels Bohr Institute in Copenhagen and all the great sites related to Bohr, the great Dane. Now, however, it is time for us to come to grips with these "quantum ideas" of Planck and Einstein. What are "quantum ideas" anyway? Where did they come from? Who was responsible for first proposing them? And how does a knowledge of them give us any insight into the internal structure of the atom? After that we will discuss the many places in Germany and Switzerland that we can visit related to "the quantum" and "quantum mechanics", one of the most revolutionary theories in the history of the atom. With this background information, we will more fully appreciate what we are seeing. So, let us carry on.

We start with a reluctant quantum warrior, a dubious "quantum mechanic" if you will, Max Planck. Planck was definitely not from a revolutionary background. He came from an upright, conservative German family of ministers and lawyers. His father became a professor of law in Munich when Max was 9. Max himself was a person of extraordinary intellectual integrity and fearlessness. He was rigorous, deductive, highly organized, and steeped in the classics of both physics and, as it turns out, music. He was an accomplished pianist who played only certain "allowed" sonatas, those of Haydn, Mozart, and Beethoven (in that order!) He only played Schubert at Einstein's urging and then considered that a "degradation". He also played the violin, the cello, and the organ and composed songs and operas. He believed in the classics – the classics of music and the classics of physics. Planck even looked the part, as he wore pince-nez glasses, dressed meticulously, and was "proudly German". He was, as Einstein biographer Walter Isaacson puts it, "shy, steely in his resolve, conservative by instinct, and formal in his manner". Figure 15.1 shows the younger Max Planck. What do you think? Do you see signs of a reluctant revolutionary warrior?

Figure 15.1 The "classical" Max Planck as a young man.

As a young man – as young men, but rarely young women, had the chance to do in those days – Planck had many options and debated long and hard about his career plans. Music and linguistics seemed attractive, but physics was the most alluring. He was warned about physics, however, in that it might not provide much of a challenge because, in the words of his thesis advisor, physics in the 1880s was just about finished. All the important discoveries had been made and, just maybe, all that was left was the detail work, suitable only for second-class minds. If only they could have known about the upcoming discoveries of X-rays by Röntgen in 1895, radioactivity by Becquerel in 1896, and the electron by J. J. Thomson in 1897. The next decade was going to be characterized by immense, paradigmatic changes that would challenge the best minds for decades. As it turned out, Max Planck played a significant role in meeting those challenges.

Having tentatively decided on physics, Planck found himself in a continuing dialogue with Ludwig Boltzmann and Ernst Mach. (See Chapter 10 for the intense interpersonal dynamics between those two.) Their debate was whether or not to use "physical atoms" and molecules, still viewed questionably, to explain the behavior of gases and their thermodynamics, that is, how they responded to heat and other types of energy. Again, Planck was dubious – as a young man he did not want to accept the idea of using atoms and molecules or probability to explain thermodynamics. He admitted that he was just not convinced that the atomic theory was going to hold up. So, to avoid this uncomfortable situation, he switched to studying the radiation of hot bodies. The rest is history because, to explain this so-called "black-body radiation", he had to propose the "quanta" that would ultimately help determine how we currently picture the inner structure of those atoms.

The University of Berlin was one of the first universities to establish a special chair and institute for theoretical physics. The first person to hold this chair (1874–1887) was Gustav Kirchhoff, who was one of Planck's professors. (The reader may recall that Kirchhoff's contribution to spectroscopy and his work with Robert Bunsen is described in Chapter 11.) In 1889, Max Planck was appointed to succeed Kirchhoff. It was Kirchhoff who first referred to "black-body" radiation. Admittedly, this is an odd term, but a black-body is defined as an opaque and non-reflective object or "body". When it is heated, it radiates light of various frequencies, that is, black-body radiation. For example, we all know that when a piece of metal (the "body" in this case) is heated in a flame it glows yellow, orange, red, and even white hot. The same thing happens when we observe the glow of the embers (the bodies in this case) of a campfire. How do we explain these colors? It looked to be a fairly simple task at first, but, in the end, it was a most challenging enterprise. Many scientists before Planck had tried using the prevailing idea that light was electromagnetic radiation with characteristic frequencies and wavelengths. Rather amazingly, they had failed, and rather catastrophically. In fact, this failure is sometimes called the "ultraviolet catastrophe". (Classical Newtonian physics predicts that a black-body, for reasons not worth pursuing here, should emit infinite amounts of ultraviolet radiation. Black-bodies don't do this, so the failure to explain this result was called the "ultraviolet catastrophe".)

So how did Max Planck solve this "catastrophic" problem? Forced to give up thinking of light as a continuous stream of energy in the form of electromagnetic radiation, Planck reluctantly proposed that light energy comes in discrete "lumps" or "chunks". Max Planck, who as a young physicist was reluctant to think of matter as coming in discrete packages called "atoms" had to propose that the energy coming from a hot body could only come in certain size packages, that is, it came in "bursts" that could not be made infinitely small. The timing for his announcement marked it as the transition from one era to another. During the last month of the last year of the nineteenth century – it was actually December 14, 1900 (the nineteenth century goes from January 1, 1801 to December 31, 1900) – Planck stepped up to the blackboard at the Christmas Meeting of the German Physical Society in downtown Berlin. This decidedly classically trained physicist reluctantly made an extraordinarily unclassical proposal that the energy of a packet of light was given by the expression,

$$E = h\nu,$$

where E is the amount of energy in the package of light, ν ("nu") is the frequency of the light when it is described as a wave, and h is a constant. Planck called these energy packages "quanta" for the Latin for "how much". Planck's constant, h, turns out to be an incredibly small number, 6.626×10^{-34} m^2-kg s^{-1}, so these quanta, these bursts of energy, are exceedingly small. His constant, h, is now regarded as one of the fundamental constants of the universe. Planck also proposed that the intensity (I) of a beam of light was given by the expression,

$$I = nh\nu,$$

where n is an integer. This means that the light beam is composed of an integral number of tiny "bursts" of light. Note that the higher the frequency of light, the more energy in a "burst" and the more intense the light. For example, purple light is roughly twice as energetic as red light because the frequency of the former is roughly twice the frequency of the latter.

These equations defied common sense and were received skeptically. How could light, which certainly looks smooth and continuous, be made of tiny bursts? Skepticism aside, Planck proposed these ideas and equations because they worked to account for the "ultraviolet catastrophe". Be that as it may, there is plenty of evidence that he never believed they were a true depiction of reality. He suspended his disbelief because they worked but he pretty much spent the rest of his life trying to make it up to his classical colleagues. He hoped that the equations would turn out to be, as Isaac Asimov put it, a piece of "mathematical jugglery without any correspondence to anything in nature".

So, this was the origin of the "quantum ideas" that we referred to earlier and that Niels Bohr applied to the atom in 1912. They were proposed at the turn of the millennium from the nineteenth to the twentieth century. These ideas are so important that the year 1900 is often referred to as the dividing line between "classical physics" and "modern physics". Planck had started what is called the quantum revolution and he did this against his will. He never really believed in the reality of quanta and constantly tried to make up for his "error". Be that as it may, let's step back for a moment and acknowledge that Max Planck dared to step outside of his comfort zone and show an intellectual integrity and fearlessness. This took extraordinary courage.

Planck survived a number of "catastrophes" in his long life. The "ultraviolet" version was mild compared to the many personal and national catastrophes he experienced. His young wife and the mother of four of his children (Karl, Margarete, Emma, and Erwin) died in 1909. Planck remarried two years later and had another son, Hermann. Tragically, Karl was killed in battle at Verdun in 1916 during World War I. Both Grete and Emma, identical twins, died during childbirth in 1917 and 1919, respectively. Both of the children, Max's granddaughters, survived and were named after their mothers. Erwin also fought in WWI and was captured by the French. During the second world war he was accused of participating in the 1944 plot to kill Adolf Hitler and was executed by the Nazis in 1945. The Planck's lived in Grunewald near Berlin. Before the war, that household was the gathering place for many of Planck's students and colleagues including the young Lise Meitner, Otto Hahn, and Albert Einstein. Einstein and Planck became great friends and often played music together, Planck at the piano and Einstein on his violin. Amidst tragedies, there was a triumph as well, when Planck was awarded the Nobel Prize in Physics in 1918 for his discovery of the quanta. (See Chapter 19 for some of the political intrigue involved in finally recognizing the theoretical underpinnings of physics.)

Between the wars, Planck lived through the rise of Hitler and the resulting attacks on Jewish and progressive intellectuals, as well as members of what were termed the cultural elite, such as musicians, artists, and authors. Thousands of academics were dismissed, books by Jewish authors (most likely including Einstein's) were burned in the Bücherverbrennung (May 1933) and Planck, then in his mid-seventies, sought a meeting with the Führer to plead for reason and restraint. You can imagine that that meeting did not go well. Unfortunately, the Grunewald house was destroyed in an air raid at the end of the second world war and all of Planck's possessions were destroyed along with it. Planck was caught in a war zone at the war's end and was rescued by American troops who took him to Göttingen where he remained for the rest of his life. After the war, with Planck's help, the Kaiser Wilhelm Institute was renamed for Planck. Today, the Max Planck Society maintains 80 research institutes both in Germany and abroad. Planck, his second wife, and his son Hermann are buried in the Stadtfriedhof in Göttingen.

15.3 TRAVEL SITES RELATED TO MAX PLANCK IN MUNICH

15.3.1 Planck Family Home

Brienner Strasse 33 (48.145141°, 11.567616°). This is where Planck lived when the family moved to Munich when Max was 9 years old. (No plaque is found here.) The Brienner Straße is one of the four royal avenues including the Ludwigstraße, the Maximilianstraße, and the Prinzregentenstraße. You may recall that we visited the statue of Joseph von Fraunhofer at Maximilianstraße 42 in Chapter 11.

During his five years as a *Privatdozent* in Munich, Planck lived at Barer Strasse 48 (48.149668°, 11.572610°) and became engaged to Marie Merck, his future first wife. (There is not a plaque here.)

15.4 TRAVEL SITES RELATED TO MAX PLANCK IN BERLIN

See Chart 15.1 for the location of the sites in this and the following section.

15.4.1 Planck Statue and Plaque at Humboldt University

(formerly the University of Berlin), Unter den Linden 6 (52.5175°, 13.393°). This bronze statue was created in 1949, but was banished until 2006 because of its modern design. The plaque is on the front of the western wing of the main building

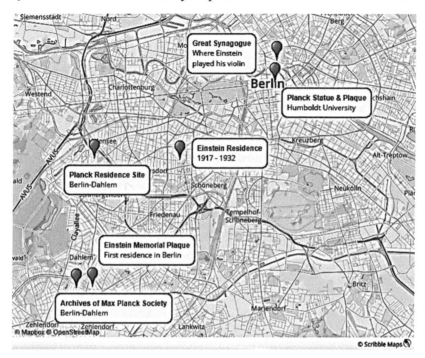

Chart 15.1 Travel sites in Berlin related to Max Planck and Albert Einstein.

of the university and at the entrance to the library. It says "Max Planck, discoverer of the elementary quantum of action h, taught in this building from 1889 to 1928." There is also a statue of Lise Meitner nearby at 52.517793°, 13.394004°. Directly across the Unter den Linden is the Bebelplatz, where the infamous book burning (the Bücherverbrennung) occurred on May 10, 1933. There is a poignant monument there that describes this frightening event. Going about 1 km west along the Unter den Linden quickly brings you to the Brandenburg Gates.

15.5 TRAVEL SITES RELATED TO MAX PLANCK IN BERLIN-DAHLEM

15.5.1 Max Planck Residence

Wangenheimstrasse 21, Berlin-Grunewald (52.491281°, 13.288165°). The Planck family lived in a villa at this address from 1905 to 1944. Max and his first wife, Marie Merck, and their four children lived here, but Marie died in 1909. Planck was a wonderful family man who greatly enjoyed playing with his children in this beautiful house. In 1911 Planck married his second wife, Marga von Hoesslin, and his fifth child, Hermann, was born. Planck would commute from here by train into downtown Berlin for his work. He always rode third-class. As noted above, the house with all his scientific records and correspondence was destroyed by bombs in 1944 at the end of WWII. There is a memorial plaque at this address that says that Max Planck lived in the house that once stood here from 1905 to 1944. Think about the Planck family, their triumphs and tragedies, when you stand in front of this reconstructed villa.

15.5.2 Archives of the Max Planck Society

Boltzmannstrasse 14 (52.446783°, 13.278150°). According to its website, "The Max Planck Society's Archives in Berlin-Dahlem were established in 1975 in order to safeguard the files of the Kaiser Wilhelm/Max Planck Societies for the Advancement of Science (mainly files, photographs, audio records and film) in one central place and to make them accessible to users." Here we find a large, ornate office with a diorama of the area, a room with lighted pictures of scientists associated with the Max Planck Society, and two chairs, Nos. 1 and 2 (taken from the Institute for Physical Chemistry and Electrochemistry) that were reserved for Planck and Einstein. You can sit in these chairs and think about the contributions of these two physicists to our present concept of atomic structure. Nearby, you can find a statue of the great organic chemist Emil Fischer and the Lightning Tower where Heisenberg carried out experiments related to the development of the German atomic bomb (see Chapter 17 for further details). Also here is the Fritz-Haber Institute of the Max-Planck Society, Faradayweg 4–6 (52.448893°, 13.282731°), that was originally founded in 1911 as the Kaiser-Wilhelm Institute for Physical Chemistry and Electrochemistry. Nearby are memorials and plaques in honor of Hahn, Strassman, and Meitner (Chapter 17). While here consider going over to the nearby Harnack-Haus for lunch or dinner. The Harnack-Haus was named after Planck's good friend, Adolf von Harnack, a German Lutheran theologian and prominent church historian.

15.6 TRAVEL SITES RELATED TO MAX PLANCK IN GÖTTINGEN

15.6.1 Max Planck's Grave

in the Rondell Nobel Laureate (Rondell der Nobel preisträger) of Stadtfriedhof Cemetery (51.529739°, 9.911870°). In a scientists' section, there are graves of 8 Nobel Prize winners including Max Planck (1918 Physics). Others include Max Born (1954 Physics), Otto Hahn (1944 Chemistry), Max von Laue (Physics 1914), Walther Nernst (1920 Chemistry), Otto Wallach (1910 Chemistry), Adolf Windaus (1928 Chemistry), and Richard Zsigmondy (1925 Chemistry). Planck's gravestone is very simple: it just says "Max Planck" with the letter h and its value at the bottom for Planck's constant. The three gravestones are for Planck, his second wife, Marga von Hoesslin, and his son Hermann. The contributions of Max Born will be discussed in Chapter 16, whereas those of Otto Hahn are detailed in Chapter 17.

15.6.2 I. Physikalisches Institut

(I. Physics Institute) Georg-August-Universität, Göttingen, Friedrich Hund Platz 1 (51.558062°, 9.946848°). The Physicalisches Cabinet and Lichtenberg Collection contains a collection of historic instruments from the early days of physics in Göttingen. It is located at the entrance of the lecture halls in the new physics building. Guided tours with the curator can be arranged.

15.6.3 Göttingen, the Science City

(*Stadt die Wissenschaft*) There are many streets in Göttingen that are named after scientists including Robert Bunsen, Enrico Fermi, Max Born, Otto Hahn, Werner Heisenberg, and, of course, Max Planck and Albert Einstein. Many of their residences are marked by white plaques as well.

15.7 ALBERT EINSTEIN (1879–1955)

Albert Einstein, of course, is one of the most famous physicists who ever lived. In this brief introduction we will concentrate on his early life history and contributions to the history of the atomic concept, but not on his general and special theories of relativity. As usual, the object here is to better understand and appreciate the Einstein sites we will visit in Germany and Switzerland.

Albert Einstein was born, as anyone who does crosswords knows, in Ulm, Germany. In 1880, his family moved to Munich when he was but one year old. His father and uncle founded a factory (*Electrotechnische Fabrik*) that produced electrical equipment and, early on, flourished partly because they facilitated the electrical illumination for the Oktoberfest. In 1894, however, the business failed after they lost the contract to illuminate the whole city. Einstein was of Jewish heritage, but received his earliest education in a Catholic grammar school. As was the case for Planck, music was a large part of Einstein's life. His mother gave him a violin. She played the piano and they played duets together.

Einstein had no patience for the formal, authoritarian German schools and certainly did not look forward to his required service in the imperial army. In 1894, when his parents moved to Pavia, Italy, he renounced his German citizenship and made plans to enroll in the Zürich Polytechnic or ETH Zürich (ETH stands for *Eidgenössische Technische Hochschule*), now the Swiss Federal Institute of Technology. As part of his preparation, he was encouraged to learn by visualizing images, a practice that led to his life-long love of *Gedankenexperiments* ("thought experiments"). He did not do uniformly well in all subjects and, for example, was required to do remedial work in both French and chemistry. He caused an explosion in a chemistry lab that damaged his right hand and required stitches.

Walter Isaacson, in his excellent biography of Einstein, says he "had developed into a head-turning teenager who possessed, in the words of one woman who knew him, 'masculine good looks of the type that played havoc at the turn of the century' He had heavy dark hair, expressive large bright brown eyes.... and jaunty demeanor." Figure 15.2 shows the younger Albert Einstein. What do you think? Could he play havoc? When it came to physics, he certainly could.

Albert Einstein met Mileva Marić in 1896 when they both enrolled at the Polytechnic. She was the only woman in his classes and together they studied mathematics and physics. Kindred

Figure 15.2 Albert Einstein in the early 1900s, when he was in his twenties.

souls, their mutual attraction was electric. Not concentrating on their studies, neither of them did particularly well. In 1900, Albert graduated near the bottom of his class and Mileva, after failing her final exams, did not graduate at all. Not surprisingly, given his class standing, Einstein had difficulties in securing a teaching position. Incidentally, both Albert and his father wrote to Wilhelm Ostwald (See Chapter 10, pp. 258–260) seeking his help. Ostwald did not reply to either of them! (Nine years later, Ostwald was the first person to nominate Einstein for a Nobel Prize!) Sometime in that time frame, he first heard that a position of "examiner" might become available at the Swiss Patent Office in Bern.

In 1902, Einstein moved to Bern and took the position of "Technical Expert 3 of the Federal Office for Intellectual Property" in the patent office. This work was not particularly difficult and left him plenty of time to think and write about his own ideas. He said that "Whenever anybody would come by, I would cram my notes into my desk drawer and pretend to work on my office work". He called the drawer his "Department of Theoretical Physics". By then, he and Mileva were together. They had already had a daughter, Lieserl, out of wedlock, but she lived only about a year. In early 1903, they were married in a civil ceremony in the

Bern registrar's office. After the ceremony, they went back to his apartment but, as was often the case, he had forgotten his key and had to wake his landlady. In the autumn of 1903 they moved to Kramgasse No. 49, near the famous clock tower, where Hans Albert Einstein was born in May of 1904. Einstein made little toys for his baby son including a cable car constructed from matchboxes and string. Hans still recalled this toy when he was an adult. Scientific/historical travelers will love visiting their apartment and thinking about Albert, Mileva and little Hans.

Einstein's contributions to the history of atoms was part of his *annus mirabilis* of 1905. This might be compared to Newton's *annus mirabilis*, 1666–1667 (See Chapter 3, p. 42) in which he developed calculus, analyzed the visible spectrum of light, and formulated the laws of gravity. Einstein, for his part, wrote three seminal papers, one an explanation of Brownian motion, one about the photoelectric effect, and the third about his special theory of relativity that, in an addendum to the original paper, included his famous equation, $E = mc^2$.

Let's start with his explanation of Brownian motion. Recall that during the decade of 1895–1905, the hotly contested debates between the anti-atomists, Ernest Mach and Wilhelm Ostwald, and atomists like Ludwig Boltzmann raged on (See Chapter 10). It is important to note that Einstein was firmly in the atomist camp. He believed atoms *physically* existed and was particularly enthusiastic about Boltzmann's work on the kinetic-molecular theory of gases that relied on the existence of atoms and molecules hitting the sides of containers, thereby exerting gas pressure. Einstein's explanation of Brownian motion was another major feather in atomists' caps (assuming atomists have caps). Brownian motion, first documented in detail in 1827 by botanist Robert Brown, is the observation that small particles suspended in a fluid like water haphazardly jump around in a jerky, random way. (You may have seen dust particles bouncing around in a beam of sunlight coming in a window– this is akin to Brownian motion.) Einstein noted that this motion was due to the impact of millions of water molecules randomly hitting a given particle. His calculations, which relied on the atomic/molecular structure of nature, mathematically accounted for this haphazard movement. It was his explanation of Brownian motion that played a large role in converting Wilhelm Ostwald over to the atomist side (and convincing the Royal Swedish Academy of Sciences that he was, in fact, a worthy candidate to receive the Nobel Prize in 1909). Many cite Einstein's special theory of relativity as his primary contribution to physics, but we should not overlook this contribution to cementing the idea of the physical existence of atoms.

Turning now to his explanation of the photoelectric effect, recall that we are trying to understand the "quantum ideas" that Niels Bohr used to construct his Bohr–Rutherford atom (Chapter 12, pp. 297–300), characterized by certain allowed orbits and energies of electrons in the nuclear atom. We have noted above that Planck reluctantly proposed that energy came in bursts, that is, in small bundles of magnitude $h\nu$, rather than in continuous waves. What did Einstein do with Planck's radical idea? The answer is that he used it to explain the photoelectric effect. This experimental result had been discovered by Heinrich Hertz in 1887. (Hertz had also discovered radio waves and helped to fully establish that James Clerk Maxwell's theory of electromagnetism was correct.) A quick check with the "Travelers Guide to the History of the Atom" in the Appendix might help to keep all these discoveries in perspective.

Hertz had found that when light of the appropriate frequency is shone on a metal surface, electrons can be knocked loose and fly away, as shown in Figure 15.3b. When the frequency of light is increased, as shown in Figure 15.3c, the electrons fly away from the metal with greater velocity or kinetic energy. Not shown in the figure is the fact that when the intensity of the light at a given frequency is increased, *i.e.*, when it is brighter, more electrons are produced but they have the same velocity or kinetic energy. As shown in Figure 15.3a, if the frequency of the light is decreased, there comes a point when electrons are not dislodged from the atom. This set of experimental results is called the photoelectric effect. How did Einstein explain what was happening?

Einstein explained the photoelectric effect by assuming that light, instead of acting like a wave, acts like a particle or a packet of energy given by Planck's expression, $h\nu$. Nowadays, we call these light particles "photons", but this term was not used until 1926

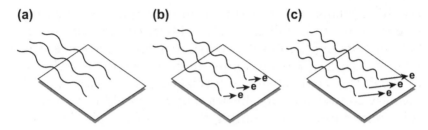

Figure 15.3   The photoelectric effect. In (b), light hits the surface of a metal and knocks loose electrons that fly away. In (c), light of higher frequency also knocks loose electrons, but they fly away with greater velocity or kinetic energy. In (a), light of lower frequency hits the metal, but no electrons are dislodged.

or so. Einstein called these particles *lichtquanten* or "light-quanta" but, for convenience, let's use the word photon. Einstein explained that photons strike electrons of the metal and knock them out or "ionize" them. That is, an electron flies away when it is struck by a particle of light. The effectiveness of a photon in ionizing an electron depends on its frequency: if its frequency is high enough, the photon has enough energy and an electron is ionized and flies away with a certain speed or kinetic energy. If the intensity of the light of that same frequency is increased, that means that more photons strike electrons and ionize more of them. Each one still flies away with the same speed and kinetic energy. If the frequency of the light is not high enough, no matter how many photons you use, *i.e.*, no matter how high an intensity of light you use, no electrons will be knocked off. This was a revolutionary idea. One, in fact, that Planck never accepted. However, Einstein's explanation of the photoelectric effect prevailed and in 1922 he was awarded the Nobel Prize for this work on light-quanta or photons. Some people are surprised that he did not receive the Nobel Prize for his work on the special theory of relativity. See Chapter 19 for the political details on how the Royal Swedish Academy of Sciences made this decision. It is quite fascinating.

There were broader implications to Einstein's work. Essentially, Einstein had proposed a paradoxical approach to the nature of light. For more than 100 years, since James Clerk Maxwell's time (See Chapter 5), light had been best regarded as a wave composed of electric and magnetic fields, *i.e.*, an electromagnetic wave. The Dutch physicist, Christiaan Huygens, in 1678, had also proposed a wave theory of light that defined the paradigm for many years. Prior to Huygens and Clerk Maxwell, Newton's idea that light was best regarded as "corpuscular", that is, particle-like, had held sway. Einstein argued that under certain circumstances (in certain experiments, for example) it was best to "think particle" while under other circumstances, it was best to "think wave". This idea is known as the "wave-particle duality of light". It is one of most mind-boggling ideas ever proposed and still is today. Facetiously, it is almost as if on three alternate days of the week experiments support photon theory but on the other three days evidence supports wave theory. Given this paradox, one could say that one should use the 7th day to pray for divine guidance!

At a conference in Salzburg in 1909, Einstein was worried that his proposal would undermine the work of James Clerk Maxwell and his seminal equations that explained light as electromagnetic waves. Recall from Chapter 5 that Einstein, when asked if he thought of himself as standing on the shoulders of Isaac Newton, replied, to the contrary, "I stand on the shoulders of Maxwell". Recall Ohto Koichi's cartoon shown in Figure 5.5. Einstein would be most pleased to learn that Maxwell's equations have survived intact to this day, despite both the relativity and quantum revolutions. Einstein met Max Planck for the first time at this conference. By now, based on his remarkable year of 1905, he had secured a position as junior professor of theoretical physics at the University of Zurich.

In 1910, Marić gave birth to their second son, Eduard. In 1911, Einstein was invited to the select Solvay Conference that concentrated on "the quantum problem". Einstein was one of the youngest invited and was joined by Max Planck, Marie Curie, Paul Langevin, Ernest Rutherford, among many others. In his lecture, Einstein continued to reconcile his particles of light ideas with Maxwell's equations in which light is pictured as a wave. It was during this conference (See Chapter 14) that the Paul Langevin/Marie Curie affair became public (we would say "went viral" today) and that it was announced that Curie had won her second Nobel Prize, this one in chemistry. Curie was attacked from all sides by the press. Einstein came to Madame Curie's defense.

In 1913, Planck convinced Einstein to return to a position in Germany at the University of Berlin. There's an atomist's irony here. Planck, by then a most respected leader of the physics community, had decided to buy in to the 21 years-younger Einstein's relativistic ideas. Planck was able to convince his colleagues that Einstein was making great contributions to physics, even though "he may sometimes have missed the target in his speculations, as, for example, in his hypothesis of light quanta". Oh well, win some, lose some.

Also, in 1913, Mileva decided to opt out of their marriage. When they separated in 1914, she took up Albert's offer that should he receive the Nobel Prize, she would get the money to support their two sons. Some say that there was some justification in this agreement as Mileva had helped Einstein with some of his mathematics, had proofread his work extensively, and taken care of their family while he wrote his 1905 papers. (See Marie Benedict's *The Other Einstein: A Novel*,[1] for some fascinating insights into Mileva and her contributions to these ideas.). They were officially divorced in 1919 and when Einstein won the 1921 prize (awarded in November 1922), he used part of the money to establish a trust for Mileva and their two sons. Another part went into a more accessible account with the interest designated to go to Mileva herself. In his Nobel Prize lecture given in 1923, he spent most of his time defending his controversial General Theory of Relativity, rather than his explanation of the photoelectric effect, for which he had received the award.

Recall that Planck did not believe his own idea of light quanta. It is natural, then, to wonder if Einstein fully believed the ideas he used to explain the photoelectric effect. He was aware that his proposal of light as particles (later called photons) was inconsistent with the wave theory of light that could explain most of the known characteristics of light, interference being a prime example. Einstein was reluctant to call his idea of light as photons to be a real theory because it could not explain these properties. His 1905 paper is entitled "On a Heuristic Viewpoint Concerning the Production and Transformation of Light". "Heuristic" is a term common in philosophy and denotes something that is "useful for discovering or explaining, but not necessarily to be considered true". We have seen

Quantum Mechanics Reluctantly Proposed

that atoms themselves were classified this way pretty much throughout the entire nineteenth century. Even as late as the last decades of the nineteenth century, atoms were thought to be useful but not necessarily "true". This is the "chemical atom" *versus* the "physical atom" debate that we discussed in Chapter 10, p. 247. Einstein, for many years, was wary of his own proposal about photons and found it difficult to believe that they corresponded to reality. Ultimately, by the mid-1920s or so, he seems to have been able to convince himself that photons were real.

So how did Planck and Einstein's ideas help explain the internal structure of the atom? We have already seen how Niels Bohr used them to advance his picture of the atom. Nevertheless, Bohr still maintained that electrons orbited the nucleus, albeit at only certain allowed radii and energies. It was apparent that such an idea would not stand the test of time. What replaced orbiting electrons? We're now on the right pathway to see the answer to that question. It turns out that the "wave-particle duality of light", proposed by Einstein was soon to be followed by the "wave-particle duality of matter" in which electrons could be pictured as waves. This will lead us to the modern theory of the atom. In the next chapter, we will see how Prince Louis-Victor de Broglie and Erwin Schrödinger led the way to our present ideas of the structure of the atom. Hold on, it will be quite the ride. In the meantime, fellow atomic travelers, enjoy all the places we can visit in Germany and Switzerland that celebrate the heuristic contributions of Max Planck and Albert Einstein.

15.8 TRAVEL SITES RELATED TO ALBERT EINSTEIN IN ULM, GERMANY

See Chart 15.2 for the location of these sites.

15.8.1 Einstein Birthplace Monument and Memorial Plaque

, Bahnhofstrasse 135 (48.399088°, 9.984934°). This building was destroyed in the bombings of 1944. It is now marked by a flat square. The monument was erected in 1979 to mark the 100th anniversary of Einstein's birth. It is made of 24 granite slabs, 12 representing the hours of the day, the other 12 the hours of the night. They symbolize time arranged in space to resemble

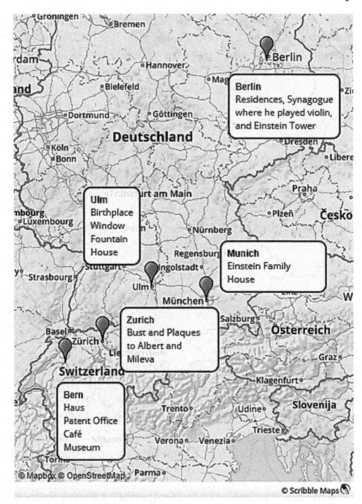

Chart 15.2 Travel sites in Germany and Switzerland where Einstein and his family lived.

a house, presumably a house of birth. Next to this birthplace monument there is a memorial plaque ※ with Einstein's head in bronze relief and the engraving, "a gift from the people of India through Calcutta Art Society".

15.8.2 Einstein Window in Ulm Münster

※※ Münsterplatz 21 (48.398545°, 9.992555°). This is a Gothic-style Lutheran church that boasts one of the tallest steeples in the world. It is rich in art throughout the interior. The window honors Nicolaus Copernicus, Galileo Galilei, Johannes Kepler, and Isaac Newton as well as Albert Einstein. $E = mc^2$ is found there also. It might be advisable to ask a verger for directions to see the window.

15.8.3 Einstein Fountain Sculpture

, Am Zeughaus (48.40047°, 10.00117°). This is a bronze sculpture (only Einstein's eyes are synthetic) created in 1984. The sculpture has three sub-structures. The rocket stands for technology, conquest of space, and the nuclear threat. Atop the rocket sits a large snail, representing nature and wisdom and skepticism about human domination of technology. Out of the shell swells the head of "Albert Einstein", with mischievous eyes and his mocking tongue sticking out.

15.8.4 Einstein House

, Kornhausplatz 5 (48.400132°, 9.994498°) which is home to the Ulm Adult Education Center. Since 1968, a permanent photographic exhibition on the first floor retraces the life of Albert Einstein in a selection of individual photographs. "Alberts Café" is also located here. Cafés are always good places to talk about atomic history.

15.9 TRAVEL SITE RELATED TO ALBERT EINSTEIN IN MUNICH, GERMANY

15.9.1 Memorial Plaque

, Adlzreiterstrasse 12 (48.127022°, 11.555691°). The entire Einstein family lived here. The plaque on the four-story house notes that Einstein spent his childhood and early youth in this house from 1885 to 1894.

15.10 TRAVEL SITES RELATED TO ALBERT EINSTEIN IN BERN, SWITZERLAND

See Chart 15.3 for the location of these sites.

15.10.1 Einstein House and Plaque

Kramgasse 49 (46.947790°, 7.449971°). The Einstein flat, now maintained by the Albert Einstein Society, consists of two rooms located on the second floor. There is a commemorative

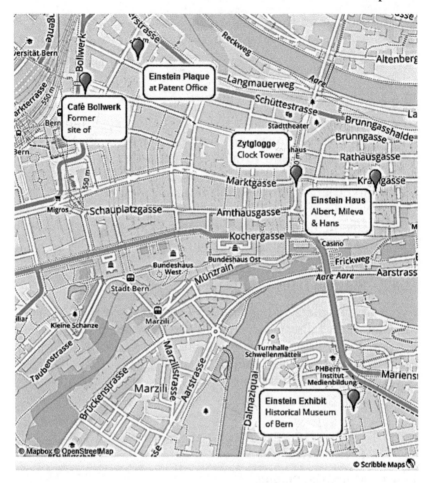

Chart 15.3 Einstein travel sites in Bern, Switzerland.

plaque on the outer façade of the house. The flat is reached by climbing steep stairs, including a tight spiral stairway from the first to the second and the second to the third levels. Albert and Mileva lived here from 1903 to 1905 while he worked at the patent office and published his famous three papers on Brownian motion, the photoelectric effect and the special theory of relativity. Hans Albert Einstein was born here on May 14, 1904. His cradle (with his teddy bear) is in the apartment. Also here is the desk that Einstein used at the Patent Office. One of the drawers in the desk was what Einstein called his "Department of Theoretical Physics". The third floor features an audio-visual presentation of the life and work of Albert Einstein. The nearby clock tower ("Zytglogge") puts on a show every hour, including a dancing jester ringing bells, a parade of bears (Bern means bear), a crowing rooster, an armored knight, followed by an appearance by Father Time with his scepter and hourglass. When Einstein came down the stairs in the morning to go to work, he turned left and walked past the tower every day. When you walk past it, think about Einstein as a young father, working at the Patent Office, but writing these incredible papers that changed physics forever. If you have time, you might stop at the Café Einstein, also at Kramgasse 49, order an "Einstein Bier" and raise a toast to Albert, Mileva, and Hans. Some say that the beer is the best there is, relatively speaking.

15.10.2 Einstein Plaque at the Former Patent Office

Speichergasse 6, corner of Speichergasse and Genfergasse (46.950521°, 7.442454°). This building is now occupied by a Swiss telephone company. Einstein's office has been remodeled and is unrecognizable. The entrance to the building is not public. The plaque on the wall of the building says, in both German and English, "While he reviewed patent requests here, a thought crossed Albert Einstein's mind: 'A man will not feel his body weight during free fall,' a basic idea behind the theory of relativity." You might try walking from the Einstein-haus to the former Patent Office. It is about a 10 minute walk. Think about the young Albert on his way to work.

15.10.3 Café Bollwerk

Bollwerk 21 (46.949789°, 7.440793°). This café, a favorite meeting place for Einstein and his friends, no longer exists. An Italian restaurant called L'Aragosta stands in its place. On the corner, there is a picture of the famous guest and an inscription attesting to the fact that he indeed liked the old café. The café is close to the former Swiss Patent Office. You might stop here on the way from the Einstein-haus to the former Patent Office building.

15.10.4 Einstein Exhibit at Historical Museum of Bern

Helvetiaplatz (46.943059°, 7.449319°). In 2005, the centenary of his *annus mirabilis*, the Historical Museum presented a large exhibition on Einstein's life. Now a permanent exhibition, it occupies all of the second floor and displays 550 items used by Einstein during his life, including books and furniture he once owned. Seventy films and presentations are spread throughout the exhibit, retelling different parts of his biography. All the descriptions are given in English, German and French.

15.11 TRAVEL SITES RELATED TO ALBERT EINSTEIN IN ZURICH, SWITZERLAND

15.11.1 Einstein Memorial Plaque

Unionsstrasse 4 (47.369808°, 8.557657°) (a private residence). The plaque pays tribute to its former famous resident. He lived here twice from 1896 to 1898 and again from 1899–1900. In between he lived at Klosbachstrasse 87, also a private residence. When Planck made his famous proposal in 1900, Einstein lived at Dolderstrasse 17, again a private residence. Upon assuming a junior faculty position at University of Zürich, he lived for a time at Moussonstrasse 10.

15.11.2 Bust of Einstein

ETH Zürich, Campus Hönggerberg, Department of Physics, HPH Building, Entrance hall, level E, Joseph-von-Deschwanden-Platz 1 (47.407596°, 8.509233°). Zürich Polytechnic or ETH Zürich (ETH stands for *Eidgenössische Technische Hochschule*), now the Swiss Federal Institute of Technology. A bronze bust of Einstein is in the main auditorium building of the Department of Physics.

15.11.3 Memorial Plaque to Mileva Marić

Huttenstrasse 62 (47.379348°, 8.552058°). This was Mileva's residence after she and Albert were divorced. The plaque, placed on the house, was sponsored by ETH Zürich and the Fraumünster Society (*Gesellschaft zu Fraumünster*), an organization committed to honoring the role of women in the history of Zürich. She died in Zürich in 1948 and was buried at Friedhof Nordheim, Käferholzstrasse (47.405307°, 8.528333°).

15.12 TRAVEL SITES RELATED TO ALBERT EINSTEIN IN BERLIN, GERMANY

Plaques more related to Einstein's theory of relativity have been omitted here. See Dieter Hoffmann's article in *The Physical Tourist*[2] for details.

See Chart 15.1 on p. 367 for the location of these sites.

15.12.1 Einstein Memorial Plaque

Ehrenbergstrasse 33 (52.447297°, 13.287710°) in Berlin-Dahlem. This was his first flat in Berlin. He worked here from April to November, 1914.

15.12.2 Plaque at Einstein Residence

Haberlandstrasse 5 (52.490716°, 13.337922°) this was his residence from 1917 to 1932. The original building was destroyed in WWII.

15.12.3 Great Synagogue

28–30 Oranienburger Strasse (52.524910°, 13.394179°), where Einstein sometimes played his violin in public concerts.

15.13 TRAVEL SITE RELATED TO ALBERT EINSTEIN IN CAPUTH, GERMANY

15.13.1 Einstein's Summer House

Am Waldrand 15–17 (52.350035°, 13.014053°) about 10 km southwest of Potsdam. Reopened in 2005, this is the small house where Einstein resided from 1929 to 1932. There are few furnishings or exhibits here but the guided tour for the public presents a good overview of Einstein's life.

15.14 TRAVEL SITE RELATED TO ALBERT EINSTEIN IN BRANDENBERG, GERMANY

15.14.1 Einstein Tower at Telegraphenberg

Albert-Einstein-Strasse (52.378859°, 13.063891°). This tower was built in the 1920s as an observatory for a German astronomer who hoped to verify Einstein's general theory of relativity. Located on the historic campus of Telegrafenberg, it is not open to the public. At the entrance gate, just ask for the "Einstein

Tower". The Café Freundlich here is worth a quick visit for lunch. The name Telegrafenberg derives from an "optical telegraph" that was built in 1832 on top of this hill as part of the Prussian semaphore system. The first observatory was built here in 1874 and it is now a principal center for astronomy, physics, and the geosciences.

15.15 SUMMARY

This chapter describes the work of Max Planck and Albert Einstein, who were the first to propose the new "quantum ideas" that Niels Bohr used in his Bohr–Rutherford atom. In 1900, Planck, a classically trained physicist trying to explain the "ultraviolet catastrophe", reluctantly proposed that light energy comes in discrete lumps called "quanta". His equation, $E = h\nu$, related the energy (E) of the quanta to the frequency (ν) of the light when it is described as a wave. Planck's constant (h) is now regarded as one of the fundamental constants of the universe. We briefly describe Planck's family life and his home that was a center of camaraderie for students and colleagues alike, including Albert Einstein. We recall the tragedies and triumphs of his life when, in Berlin and Göttingen, we visit his residences, honorary statues and plaques, the Archives of the Max Planck Society, and his simple grave amongst those of seven other Nobel Laureates.

During his *annus mirabilus* of 1905, while employed in the Bern patent office, Einstein extended Planck's ideas and proposed that there are occasions when light cannot be described as an electromagnetic wave, but rather as a stream of particles called photons, the energy of which is given by Planck's $E = h\nu$. Using this particulate model of light, Einstein successfully explained the photoelectric effect as due to one-on-one collisions between electrons and photons. In another paper, he explained Brownian motion in terms of atoms and molecules colliding with particles suspended in a fluid like air or water. A third paper described his special theory of relativity. We briefly describe Einstein's early years, his education in Zurich, his marriage to Mileva Marić, the birth of their first son Hans and then go on to visit multiple sites in Ulm, Munich, Berlin, Bern, and Zurich, where his life is celebrated with plaques, monuments, statues, elaborate church windows, museums, and in the flat where he and Mileva lived during the most productive time of his life.

ADDITIONAL READING

- R. P. Crease, *The Great Equations: Breakthroughs in science from Pythagoras to Heisenberg*, W.W. Norton & Company, Inc., New York, 2008. Chapters devoted to the equations of Maxwell, Einstein, Schrödinger and Heisenberg are included.
- W. Isaacson, *Einstein, His Life and Universe*, Simon & Schuster, New York, 2007.
- B. R. Brown, *Planck, Driven by Vision, Broken by War*, Oxford University Press, Oxford, 2015.
- J. Teichmann, M. Eckert and S. Wolff, *Physicists and Physics in Munich*, *Phys. Perspect.*, 2002, **4**, 333–359.
- I. Asimov, *Asimov's Biographical Encyclopedia of Science and Technology*, Doubleday & Company, Inc., Garden City, New York, Second Revised Edition, 1982.

REFERENCES

1. M. Benedict, *The Other Einstein: A Novel*, Sourcebooks, Inc., Napierville IL, 2016.
2. D. Hoffmann, Physics in Berlin I: The Historical Center, in *The Physical Tourist: A Science Guide for the Traveler*, ed. J. S. Rigden and R. H. Stuewer, Birkhäuser Verlag AG, 2009.

CHAPTER 16

Quantum Mechanics Brings Uncertainty to the Atom: de Broglie, Schrödinger, Heisenberg, Dirac and Born (France, Switzerland, England, Austria, and Germany)

16.1 A QUICK LOOK AT PLACES TO VISIT "TRAVELING WITH THE ATOM" IN FRANCE, SWITZERLAND, ENGLAND, AUSTRIA, AND GERMANY

Paris: De Broglie Grave and Rue Maurice et Louis de Broglie ; Zurich: Schrödinger residence with life-size cat figure in the garden; Vienna: Schrödinger Akademisches Gymnasium plaque , Institut fur Radiumforschung on Boltzmanngasse, Schrödinger Zimmer (office) , Plaque at Schrödinger's residence on Pasteurgasse, Schrödinger Bust with Equation at the Arcades of the University of Vienna; Alpbach, Austria: Erwin Schrödinger Grave with his wave equation, Schrödinger Hall in Congress Centrum; Göttingen: Max Born grave in Stadtfriedhof Cemetery, The Physicalisches

---

Traveling with the Atom: A Scientific Guide to Europe and Beyond
By Glen E. Rodgers
© Glen E. Rodgers 2020
Published by the Royal Society of Chemistry, www.rsc.org

Cabinet and Lichtenberg Collection ※ in I. Physikalisches Institut at Georg-August-Universität, and Göttingen, the Science City ※※; Bristol, England: Dirac Plaque at his first Childhood Home ※, his second home ※, Dirac Road ※, and "Small Worlds" sculpture ※※ dedicated to Dirac; Cambridge, England: Dirac Residence (no ※); Tallahassee, Florida: The Paul A.M. Dirac Science Library with statue ※※; Dirac Grave in Roselawn Cemetery ※; Paul Dirac Drive ※; London, England: P. A. M. Dirac Memorial Stone in Westminster Abbey ※※※; Island of Heligoland Heisenberg Memorial Stone ※※; Munich: Plaque at Heisenberg youth home ※, Heisenberg Elementary school ※, Maxgymnasium ※; Werner-Heisenberg-Institute of the Max-Planck-Institute of Physics ※; Warsaw: Schrödinger equation in monument at Centre of New Technologies.

Prince Louis de Broglie, Erwin Schrödinger, and Max Born, Based on the Wave-Particle Duality of Matter, Help us Picture the Internal Structure of the Atom

- De Broglie proposes the wave-particle duality of matter summarized by his equation, $\lambda = h/mv$. Electrons in atoms could be regarded as waves, as well as particles.
- Erwin Schrödinger's "wave mechanics" treats electrons in atoms as confined or standing waves. They are confined by the attractive or Coulombic forces between the negative electrons and the positive nucleus.
- Confined waves have certain allowed wave patterns and corresponding energies described by integers called "quantum numbers".
- Max Born proposes that the square of Schrödinger's wave function measures the probability of finding an electron in an atom. Probability can be represented by dot density diagrams.
- When we surround 90 percent of the electron probability, we obtain atomic orbitals. These orbitals replace Bohr's orbits.
- Werner Heisenberg's "matrix mechanics" is found to be equivalent to "wave mechanics"; both yield the "Uncertainty Principle".
- Paul Dirac incorporates the theory of relativity into quantum mechanics and predicts the existence of the positron, the first example of antimatter.
- Schrödinger proposed his cat "thought experiment" to show, in his opinion, how silly quantum mechanics had become.

16.2 PRINCE LOUIS-VICTOR DE BROGLIE (1892–1987)

"After long reflection in solitude and meditation, I suddenly had the idea, during the year 1923, that the discovery made by Einstein in 1905 should be generalised by extending it to all material particles and notably to electrons." Prince Louis-Victor de Broglie, Preface to his re-edited 1924 PhD thesis

Louis-Victor de Broglie is the only prince to contribute to the development of the atomic concept. His father's family had served French monarchs for centuries and was given the hereditary title of Duke by King Louis XIV in 1740. Later, during the Seven Years War (1756–1763), the family acquired the title of Prince. His great-great-grandfather was executed on the guillotine during the French revolution. Studying at the Sorbonne, he originally read history and law, with the aim of pursuing a career in the diplomatic service. He earned a degree in history, but then switched to the sciences, obtaining a second degree in 1913. During the first world war, he served in the wireless telegraphy section of the army and worked in the shadow of the Eiffel Tower. His older brother Maurice was a well-known experimental physicist interested in X-rays and radioactivity. Louis, however, preferred the theoretical side of physics, especially atomic theory.

He presented his idea of the wave-particle duality of matter in his doctoral thesis of 1924. His examiners "liked the math, but did not believe ... [that his 'matter waves'] ... had any physical meaning." His PhD advisor, Paul Langevin, Marie Curie's long-time associate and sometime lover, noted that the thesis was a "brilliant paper", but as philosophy or literature, *not* physics. Maurice challenged Langevin and asked him to send it to Einstein. Einstein said it was completely crazy but original, so they should give him his PhD. Later, Einstein wrote that de Broglie had "lifted a corner of a great veil". Not long afterwards, the paper fell into the hands of Erwin Schrödinger and the rest is (atomic) history. Two weeks later, Schrödinger produced his wave equation that would become the basis for the modern view of the internal structure of the atom. We have discussed Schrödinger's time in Ireland in Chapter 2 (pp. 23–28) and will discuss his "wave mechanics" later in this chapter.

Recall from Chapter 15 that Einstein had proposed that light in certain experiments was best described as an electromagnetic wave, while in other experiments one needed to describe it as particles, ultimately called photons. This mind-boggling circumstance is called the "wave-particle duality of light". To see how de Broglie arrived at his "wave-particle duality of matter", consider the two equations proposed by Planck and Einstein, $E = h\nu$ and $E = mc^2$, respectively. Recalling that the frequency (ν) of light is related to its wavelength (λ) by the equation $\nu = c/\lambda$, where c is the speed of light, we can write the following equality:

$$E = mc^2 = h\nu = h(c/\lambda)$$

Solving for the wavelength, we arrive at the equation $\lambda = h/mc$ or, if we replace the speed of light with a general expression for velocity, v, we obtain the relationship,

$$\lambda = h/mv$$

which is known as the de Broglie equation. This tells us that any particle of mass, m, traveling at a velocity, v, has a corresponding wavelength, λ. These wavelengths are immeasurably small for most common objects. For example, a baseball of mass 145 g thrown at 100 miles/hour would have a wavelength of about 10^{-25} nm (nanometers), a wavelength impossible to measure. For subatomic particles, most notably electrons, however, the wavelengths are significant. An electron (mass 9.11×10^{-28} g) travelling at one tenth the speed of light would have a wavelength of 0.243 nm, a measurable quantity. Prince Louis-Victor de Broglie was awarded the 1929 Nobel Prize in Physics "for his discovery of the wave nature of electrons".

So, to briefly review, de Broglie had asserted that, in theory, electrons should, at times act as waves and that this is an example of the wave-particle duality of matter. This was a truly radical assertion but, astonishingly, as de Broglie had suggested in his thesis, it was soon shown to be experimentally true. In 1927, Clinton Davisson and Lester Germer, at the Bell Labs in the United States, demonstrated that electrons acted like waves and produced an interference pattern when fired at crystals. In 1928, George Paget Thomson, J. J. Thomson's son, working at the University of Aberdeen, demonstrated the same thing. Ironically, J. J. had received the Nobel Prize in 1906 for showing that electrons are *particles*. Incredibly, both father and son were right! Electrons sometimes act as particles but other times they act as waves. G.P. Thomson and Davisson shared the Nobel Prize for Physics in 1937 for demonstrating that electrons undergo diffraction, a "behavior peculiar to waves".

16.3 TRAVEL SITES RELATED TO PRINCE LOUIS-VICTOR DE BROGLIE

Although integral to the development of atomic theory, there are few places to visit where de Broglie is celebrated. Perhaps fellow scientific/historical travelers can find more and let us know.

16.3.1 De Broglie Grave

Cimetière de Neuilly-sur-Seine (Ancien), 3, rue Victor Noir, Île-de-France, Paris (48.88154°, 2.2643°) Memorial ID: 192690358. Readers are encouraged to empathize with de Broglie as they stand beside his grave.

16.3.2 Rue Maurice et Louis de Broglie

, 13th Arrondissement, Paris (48.833419°, 2.369939°), not too far from rue Albert Einstein.

16.4 ERWIN SCHRÖDINGER (1887-1961)

We have already discussed Erwin Schrödinger in Chapter 2, "Bookending the Atom: Boyle and Schrödinger". For a brief overview of his career and contributions to atomic theory, carried out some ten years before Erwin and Anny (and various mistresses and their children!) arrived in Ireland, take a look at the last few pages of that chapter. Now we have the opportunity to discuss in a more detailed, but still appropriately brief manner, Schrödinger's work, so that we can better appreciate the places in Austria and Germany that celebrate his life. It's a whirlwind tour, so hang on tight!

In 1925, after Prince Louis de Broglie had suggested that electrons could be treated as waves and Einstein had endorsed his assertion, Erwin Schrödinger was primed to make his most important contribution to atomic science, the one for which he is most renowned. Schrödinger, who had taken a position as professor of theoretical physics at the University of Zürich in 1921, was a most talented mathematical physicist. Although plagued by ill health in the early 1920s, he had published multiple academic papers on color vision, relativity theory, and atomic structure. He learned of de Broglie's work by reading a reference in one of Einstein's papers. It took him only a week or two to develop the idea that electrons in atoms were "confined waves" and that we should apply what we know about confined or "standing waves" to better understand the internal structure of the atom. He presented these ideas in a seminar to colleagues in Zürich and, in a follow-up question, was asked if he had come up with a "wave equation" yet? He had not. By now it was time for the Christmas break and Schrödinger, most biographers agree, spent it with an unidentified former girlfriend in the ski resort town of Arosa, some 90 miles from Zürich. Evidently, as John Gribben writes, this and other young women "triggered a burst of creative activity [others have called it an 'erotic outburst'!] which carried Schrödinger right through 1926, producing six major scientific papers on what became known as 'wave mechanics'". There are reports that he had the habit of placing pearls in his ears to block out noise and would work for hours uninterrupted (at least most of the time).

So, fellow travelers-with-the-atom, all of the above aside for the moment, let's get back to confined waves! What exactly *do* we know about confined waves and how can we apply – moreover how did Schrödinger and other "wave mechanics" apply – this knowledge to atoms? One place to start are the waves that are set up on a violin string. (Planck and Einstein would fully appreciate this starting point.) We say that the string is a one-dimensional "waving medium" that is confined to wave between two fixed points. When the string is plucked or bowed it produces a characteristic musical tone. Let's say the distance between the two points on the string is L. Figure 16.1 shows the first four confined or standing waves that are set up. Notice that there are only certain allowed wavelengths, λ, for these waves. That is, λ can be $2L$, L, $\frac{2}{3}L$ or $\frac{1}{2}L$. We can express this set of wavelengths using the general formula, $\lambda = 2L/n$, where n is an integer (1, 2, 3, or 4).

We can also produce the same type of result if we use a moderately long jump rope or, as the author did many times in his

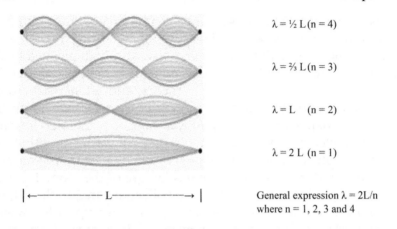

$\lambda = \frac{1}{2}L\,(n = 4)$

$\lambda = \frac{2}{3}L\,(n = 3)$

$\lambda = L \quad (n = 2)$

$\lambda = 2L \quad (n = 1)$

General expression $\lambda = 2L/n$ where n = 1, 2, 3 and 4

Figure 16.1 The allowed wavelengths for a one-dimensional standing or confined wave. The general expression for these wavelengths is $\lambda = 2L/n$ where L equals the distance between the two ends and n is an integer, 1, 2, 3, or 4. Reproduced from ref. 1 [https://courses.lumenlearning.com/chemistryformajors/chapter/electromagnetic-energy-2/] under the terms of a CC BY 4.0 license [https://creativecommons.org/licenses/by/4.0/].

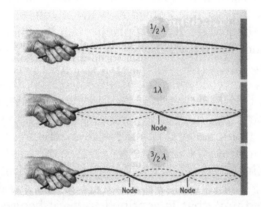

Figure 16.2 Three standing waves produced by shaking a jump rope or telephone cord. Reproduced from ref. 2 with permission from Cengage Learning, Copyright 2011.

classes, a telephone cord. Figure 16.2 shows how this might work. If one end of the cord is held in place and you gently shake the free end, it's fairly easy to produce the $n = 1$ wave (called the fundamental or the first harmonic). However, it quickly becomes evident that it takes a steady hand putting in just the right amount of energy to sustain the pattern. Producing the $n = 2$ (the second harmonic or overtone) is also relatively easy but the energy needed, again nearly a constant, is considerably higher. The $n = 3$ standing wave (the third harmonic) is difficult to achieve because it takes a constant high amount of energy. This might be a fun experiment to try as you discuss this section. Good luck finding a telephone cord nowadays!

To sum up our basic knowledge of confined waves, at least in the "one-dimensional" above case, the energy of the confined wave depends on n, an integer. Low energies correspond to low n's, higher energies correspond to higher n's. The upshot is that there are only certain allowed energy values for these "confined waves" and they depend on the integer, n. We say that the energy is "quantized", that is, it comes in only certain allowed values that depend on a quantum number, n.

These arguments were at the very heart of Schrödinger's reasoning about electrons in atoms. He assumed, using de Broglie's argument, that an orbiting electron acts not as a particle but rather as a wave. Furthermore, he reasoned, the electron wave is "confined" by the attractive force between the negative charge of the electron and the positively charged nucleus. Therefore, it followed, the electron can be pictured as assuming a set of allowed standing or confined wave patterns that correspond to certain allowed energies that, in turn, depend on a quantum number. Each of these confined waves, sometimes called "matter waves", corresponds to an allowed energy in much the same way as in the telephone cord example. However, as we will see in the next few paragraphs, all of this is a vastly oversimplified representation. For now, let's emphasize that we no longer regard an electron in an atom to be a particle orbiting around the nucleus. Rather the electron, confined to the area around the nucleus by attractive forces, is pictured as possessing wave-like properties that give it certain allowed wave-patterns with corresponding energies.

The standing or confined waves set up by a one-dimensional violin string or a telephone cord are one thing. An electron, acting now as a wave and not a particle, confined to the three-dimensional space around a nucleus is another. Things get very complicated very fast. Schrödinger, however, was up to the challenge. He has been described as a mathematical physicist who could write down and solve three-dimensional differential wave equations on the back of a napkin over lunch! With relative ease, Schrodinger produced his "wave equation" for a hydrogen atom with its one electron confined to the volume around the nuclear charge produced by one proton. The confining force is just the plus/minus [sometimes called "Coulombic" after Charles-Augustin de Coulomb (1736–1806)] forces between these two oppositely charged particles. A much-abbreviated version of Schrödinger's wave equation is shown below.

$$i\hbar \frac{\partial}{\partial t}\Psi(r,t) = \left[\frac{-\hbar^2}{2\mu}\nabla^2 + V(r,t)\right]\Psi(r,t)$$

The equation contains information on the masses of the proton and the electron (combined in the symbol μ – pronounced "mu") and the total energy of the system when in motion (shown in square brackets). Notice the equation contains the symbol i (which stands for the square root of -1), the derivative with respect to time, $(\partial/\partial t)$, and Planck's constant, h, here represented in the form $\hbar$ (pronounced "h-bar"), which is h divided by 2π. Ψ is the "wave function" (represented by the Greek letter psi) that depends on position (r) and time (t). Once the wave equation was solved, the resulting wave functions gave information about the possible three-dimensional confined wave patterns and the energies associated with those wave patterns. As in the violin string or telephone cord, there would be many allowed wave patterns not just one, arrayed from lower to higher energies. Somewhat logically, each wave function, Ψ, for this three-dimensional case, was found to be dependent not only on three spatial coordinates (for example, x, y and z) but also on three quantum numbers, commonly given the symbols n, l, and m. (For travelers-in-the-know, the spin quantum number, s, was added later.) There are various abbreviations of the Schrödinger equation. One is

$$i\hbar \frac{d}{dt}|\Psi(t)\rangle = \hat{H}|\Psi(t)\rangle$$

Here $\hat{H}$ is the "Hamiltonian operator", which characterizes the total energy of the system under consideration. When we get to Schrödinger's grave in Alpbach, Austria, we will see his equation further abbreviated to $i\hbar\dot{\psi} = H\psi$. After all, there's only a limited amount space on a gravestone. No matter how abbreviated the equation, this is intricately complicated stuff and Schrödinger was deservedly awarded the 1933 Nobel Prize in Physics for this work.

So, what is the significance of the Schrödinger wave equation? It is safe to say that it is one of the most important equations we have discussed, including Clerk Maxwell's four equations for electrodynamics on his statue in Edinburgh, Boltzmann's $S = k \log W$ on his tombstone in Vienna, and Einstein's $E = mc^2$. Now we have Schrödinger's equation on his epitaph above his tombstone. Solving it produces a set of wave functions that contain *all the information that is possible to obtain* concerning, in our particular case, an electron in a hydrogen atom. Somewhat disconcertingly, this information has a built-in or inherent uncertainty or indeterminism. For example, it cannot give us the precise location of the electron at a particular time. Instead, it provides only the likelihood (or the probability) of an electron being found at a particular point. Yup, it's a whirlwind tour all right! Keep hanging on tightly!

But you say, given the above built-in uncertainty of the information, how are we to represent the nature of electrons in atoms? In other words, how are we going to represent these mysterious, inherently uncertain, wave functions? Schrödinger referred to his confined wave patterns as "Einstein-de Broglie waves" but even Einstein had difficulties picturing what these waves looked like. Where does that leave us considerably more mathematically (and philosophically) challenged beings? In 1926, Max Born, the head of the theoretical physics group at the university in Göttingen, proposed a way of visualizing electrons in atoms. He maintained that Ψ^2, that is, the square of the wave function, was a measure of the *probability* of finding an electron in an atom. It took a while for this assertion to catch on but Born was awarded the Nobel Prize in Physics in 1954 "for his fundamental research in quantum mechanics [a term he invented], especially for his statistical interpretation of the wavefunction". Neither Einstein nor Schrödinger was comfortable with this idea of introducing probability into quantum mechanics. More on their reactions shortly.

OK, now we need to discuss how we can represent probabilities. One way to do this is to use dot density diagrams or, if you like, Stippling diagrams (see Figure 2.1 for an example of stippling art.) We are all familiar with dot density diagrams. For example, Figure 16.3 shows a dot density diagram representing the population of Europe. Most of us know that the closer together the dots, the greater the probability of finding human life. (Note that the diagram tells us nothing about finding a particular human being at a particular time or how a person, perhaps a scientific/ historical traveler, moves from one place to another.) We can do

Figure 16.3 A dot density diagram representing the population of Europe. The closer the dots the greater the population. Used by permission from Migration Watch UK.

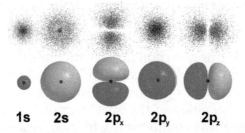

Figure 16.4 Dot density diagrams or orbital representations for electrons.

the same thing for electrons in atoms. Figure 16.4 shows several dot density diagrams (as was the case for the population chart, these diagrams tell us nothing about finding the electron at a given location at a particular time or how an electron moves from one place to another.) As noted in the figure, these probability representations are known as "orbitals". Bohr's well-defined electron orbits are replaced with fuzzy orbitals derived from Schrödinger's wave equation.

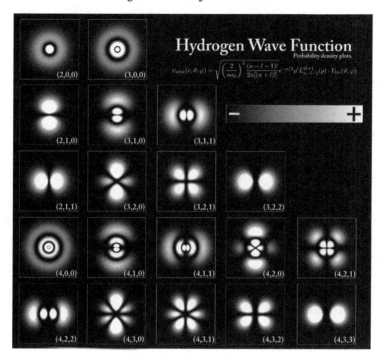

Figure 16.5 Atomic orbitals as boundary surface representations.

A more popular way to represent atomic orbitals is shown in the bottom row of Figure 16.4 and in Figure 16.5. In these diagrams, ninety percent of the electron density (the probability of finding the electron) is surrounded by a definite border. These are sometimes called "boundary surface representations". These are the familiar 1s, 2s, 2p, 3s, 3p, 3d, and so forth, orbitals that many but not all of us have studied at one or another point in our education.

Schrödinger's wave mechanics is the basis of our modern representations of electrons in atoms. Einstein continued to have difficulties with the role of chance and probability in quantum mechanics all the rest of his life. Schrödinger also was not entirely comfortable with the implications of quantum theory. He wrote about Born's probability interpretation of quantum mechanics, saying: "I don't like it, and I'm sorry I ever had anything to do with it." Again, more on his dissatisfaction later.

16.5 TRAVEL SITES RELATED TO ERWIN SCHRÖDINGER IN DUBLIN

See Chapter 2 (pp. 25–27) for his residence, his workplace at the Dublin Institute for Advanced Studies, and the "What Is Life?" aluminum statue in the National Botanic Gardens.

16.6 TRAVEL SITE RELATED TO ERWIN SCHRÖDINGER IN ZURICH

16.6.1 Schrödinger Residence

Huttenstrasse 9, (47.382710°, 8.549026°) There is life-size cat figure in the garden where Erwin Schrödinger lived 1921–1926. Depending on the light conditions, the cat appears either alive or dead. We will discuss the Schrödinger Cat "thought experiment" shortly.

16.7 TRAVEL SITES RELATED TO ERWIN SCHRÖDINGER IN VIENNA

See Chart 16.1 for the location of these sites.

16.7.1 Akademisches Gymnasium Plaque

Innere Stadt, Beethovenpl. 1 (48.20158°, 16.37685°) acknowledges that Schrödinger was a student here 1898–1905 (Lise Meitner also graduated from here; a plaque just below Schrödinger's honors her. See Chapter 17 for more on Meitner.)

16.7.2 Institut fur Radiumforschung

Boltzmanngasse 3, Vienna (48.221319°, 16.356568°); Schrödinger wrote an experimental thesis on electrical-conductivity phenomena and after WWI worked on radioactivity research here.

16.7.3 Schrödinger Zimmer (Office)

Physics Library at Boltzmanngasse 5, Vienna (48.221621°, 16.356293°). This room is actively used by graduate students but it's fun to peek in and ask them about it.

16.7.4 Plaque at Schrödinger's Residence

Pasteurgasse 4 (48.221850°, 16.358463°). This is where he and Anny stayed after they returned from Ireland.

Quantum Mechanics Brings Uncertainty to the Atom

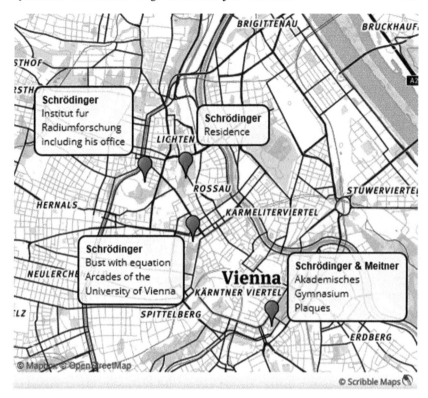

Chart 16.1   Travel sites in Vienna related to Erwin Schrödinger.

16.7.5   Schrödinger Bust with Equation

▓▓▓ The Arcades of the University of Vienna (48.213209°, 16.360051°). This is the same abbreviated wave equation found above his grave in Alpbach. Many other scientists central to the atomic theory are also honored here, including Doppler, Loschmidt, and Boltzmann. See Chapter 10 for details.

16.8   TRAVEL SITES RELATED TO ERWIN SCHRÖDINGER IN ALPBACH, AUSTRIA

16.8.1   Grave of Erwin Schrödinger

▓▓ (47.39852°, 11.94458°). Alpbach is a beautiful mountain town, sometimes called "the village of thinkers". The gravestone has his wave equation as the epitaph. (Mistakenly, a dot was originally put over the right-hand Ψ. One can see that attempts have been made to obliterate this mistake.) It is in the graveyard of the Heiliger Oswald, a Catholic church. Although Schrödinger was not Catholic, the priest in charge of the cemetery permitted the burial after learning Erwin was a member of the Pontifical Academy of Sciences. Although his wife Anny was barren, his three illegitimate children included a daughter, Ruth, who was born to Hilde March, who, you may recall from Chapter 2, lived with Anny and Erwin in the "Clontarf household" in Dublin. At this writing, Ruth and her husband live in Alpbach. Despite all the erotic intrigue in Erwin's life, Anny's name is on his gravestone. A granddaughter of Schrödinger's also lives in Alpbach. She wrote in 2015 that "Annie is not in Schrödinger's grave at all; only her name shows." According to Alpbach tradition, bones are removed after 21 years, but evidently an exception was made for Schrödinger. Given the cat stories, discussed below, some of us would be tempted to open the grave to make the observation for ourselves.

16.8.2 Schrödinger Hall

Congress Centrum, Alpach (47.398080°, 11.947374°). The main hall in this convention center is named for Schrödinger.

16.9 TRAVEL SITE RELATED TO ERWIN SCHRÖDINGER IN WARSAW

16.9.1 Schrödinger Equation

This is part of a monument in front of Warsaw University's Centre of New Technologies. James Clerk Maxwell's four equations for electrodynamics are also shown here (see Chapter 5).

16.10 WERNER HEISENBERG (1901–1976) AND PAUL A. M. DIRAC (1902–1984)

Werner Heisenberg came from an affluent, academic family in a Würzburg suburb. His family moved to Munich when he was nine. He was a good-looking, precocious young man but suffered from debilitating allergies. He played chess at a high level and was an accomplished pianist, playing "master compositions" by the time he was thirteen. During the first World War he was sent to work on a farm and afterwards faced chaotic conditions that made strong impressions on him. He was involved in some street fighting and even led some anti-communist activities against the Munich Soviet Republic. He put these things behind him, however, and by the time he graduated from the Maximilians-Gymnasium in Munich, he had taught himself calculus. He loved anything athletic including mountain climbing, skiing, and camping.

Heisenberg was a student of the great teacher and professor of physics, Arnold Sommerfeld in Munich, the theoretical mathematician, physicist and solid-state chemist Max Born in Göttingen, and the great Dane Niels Bohr in Copenhagen. He did not get along well with Erwin Schrödinger, in large part because they developed competing views of quantum mechanics. With electrons in orbitals around the nucleus, Schrödinger's wave mechanics produced the modern pictorial model of the atom that many physicists and chemists could, and still do, live with. Heisenberg rejected Bohr's fanciful quantum jumps from one allowed orbit to another. Instead, he adopted a Mach-like view (see Chapter 10) of the world in which he only used experimental observable quantities (like spectral line data) and mathematics to develop a complex mathematical version of quantum mechanics called "matrix mechanics". He started working on this in 1925 while studying in Göttingen. Suffering from hayfever, he obtained Born's permission to go to the pollen-free and treeless island of Helgoland in the North Sea off the northwest coast of Germany where, in two weeks, he developed complex mathematics to approach his ideas about electrons in atoms and how to account for spectral lines. When he returned to Göttingen, his mentor Born recognized his strange mathematics to be so-called "matrix algebra". A matrix is mathematical structure in which the numbers are arranged into rows and columns. It did not take long for Born, Heisenberg and Pascual Jordan to publish papers on his non-pictorial, highly theoretical "matrix mechanics" that fully accounted for the spectra of the hydrogen atom without having to refer to electron orbits or even orbitals.

As part of his mathematics, Heisenberg formulated his "Uncertainty Principle". He multiplied the matrix representing the position (p) of the electron times the matrix representing its momentum (q). It turned out that *pq* did not equal *qp*, or as mathematicians say, they are not commutative. Experimentally, this meant that it makes a difference in what order position and momentum are measured. If the position of the electron is determined first, then its momentum cannot be known accurately, and *vice versa*. This result was fully consistent with the results from Schrodinger's "wave mechanics", developed only a short time later. Born and Jordan were the mathematical experts in this work. Born's tombstone in a remote section of the Stadtfriedhof in Göttingen memorializes this foundation of the Uncertainty Principle. Sandwiched between his name and birth and death dates and his wife Hedwig's similar data, is the equation $pq - qp = h/2\pi i$. Another tombstone, another equation! By the way, Max and Hedwig's grand-daughter is the singer, Olivia Newton-John.

Physicists immediately chose sides. Niels Bohr supported Heisenberg's "matrix mechanics", whereas Albert Einstein and Max Planck backed Schrödinger's "wave mechanics". It did not take long, in 1926, for Schrödinger and others to demonstrate that the two approaches were mathematically equivalent. Heisenberg found Schrödinger's version of quantum mechanics to be "disgusting", presumably because it presented a physical model including fanciful orbitals. Schrödinger "believed that to deprive a physicist of the possibility of making space-time models of subatomic phenomena would inhibit further progress".

In yet another extraordinary development, the young Paul Dirac (1902–1984) – sometimes called the "boy wonder" – incorporated Heisenberg's matrix mechanics and Schrödinger's wave equation into one overarching theory. Dirac also incorporated Einstein's theory of relativity into his field theory and, purely based on his mathematics, predicted the existence of the positron, the positively charged particle of the same mass as an electron. The positron was discovered several years later and was the first example of antimatter. You may recall from Chapter 7 that Paul Adrien Maurice (P. A. M.) Dirac is honored in the "Science Corner" of Westminster Abbey. His memorial diamond is only a few meters away from those of J. J. Thomson, Ernest Rutherford, Michael Faraday, James Clerk Maxwell, Charles Darwin, and Isaac Newton. Dirac's "relativistic" equation (reconciling quantum theory with relativity), often referred to as the "Dirac Equation", is engraved on his diamond, which was only put in place in 1995, quite late as these things go. Lately, all these august scientists have been joined by Stephen Hawking who, quite naturally, also has an equation on his gravestone. (It is related to the glow, called Hawking radiation, that is emitted by black holes.) Hawking once commented that: "Dirac has done more than anyone this century, with the exception of Einstein, to advance physics and change our picture of the universe."

Paul Dirac was "the strangest man", as Graham Farmelo discusses in his book, *The Strangest Man: The Hidden Life of Paul Dirac, Mystic of the Atom*.[3] Dirac always kept to himself and hardly ever spoke to anyone unless he was addressed first. This behavior may be attributed in large part to his father, a strict disciplinarian Swiss academic who taught French in Bristol and insisted that Paul address him only in perfect French at the dinner table. (His mother and Paul's siblings spoke English at a different table. Reportedly, this led the young Paul to believe that men spoke French and women spoke English! This might be akin to the metaphor that men are from Mars whereas women are from Venus.) In any case, we can imagine that dinner conversations were rather limited. There is also some evidence that Dirac was autistic, a condition that would certainly affect his behavior and ability to relate to others. As a grownup, he was known among his colleagues for his precise and taciturn nature. His colleagues in Cambridge jokingly spoke of a unit they called the "dirac", defined as a speaking rate of one word per hour! In class, he could speak well and his lectures were always crystal clear. He did not brook questions particularly well, however. When asked a question in class, he would repeat exactly what he had said before, using the very same words. Another student took a different tack and said that he did not understand an equation that Dirac had written on the blackboard. Dirac remarked that that was a comment not a question! Quite an amazing man, Paul Dirac is regarded as one of the most significant physicists of the twentieth century.

The 1932 Nobel Prize in Physics (awarded in 1933) was awarded to Werner Heisenberg. The 1933 Nobel Prize in Physics (also awarded in 1933) was shared between Erwin Schrödinger and Paul Dirac. In a rather stunning development, Max Born and Pascual Jordan were passed over for the Nobel Prize. Jordan had joined the Nazi party in mid-1933 and the Nobel committee did not want to endorse him in any way. Born, who after all was the head of the theoretical physics group at the university in Göttingen and the one who had put Heisenberg on the right track mathematically, was omitted because his and Jordan's work were intimately mixed and could not be separated. Born lived with the sting of this omission for many years but ultimately was given the prize in 1954 for proposing that the square of Schrödinger's wave function, Ψ^2, was a measure of the probability of finding an electron in an atom.

Schrödinger and Einstein just could not stomach the role of chance and probability in quantum mechanics. In attempting to show how ludicrous the situation had become, Schrödinger devised his famous *gedanken* (thought experiment) called "Schrödinger's Cat". In this "experiment", a cat is imprisoned in a steel box with a radioactive source like uranium, a Geiger counter, and a flask of poisonous hydrocyanic acid. Figure 16.6 shows the situation. This is not a real experiment, of course,

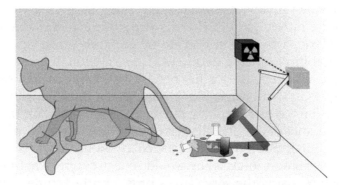

Figure 16.6 Schrodinger's Cat *gedanken* or "thought experiment": a cat, a flask of poison, and a small amount of a radioactive source like uranium are placed in a sealed box. A Geiger counter can detect when a single atom of uranium has decayed. When this happens, a series of events results in the smashing of a flask containing hydrocyanic acid that immediately kills the cat. The probability interpretation of quantum mechanics implies that after a while, the cat is *simultaneously* alive and dead. Yet, when one looks in the box, one sees the cat *either* alive *or* dead not both alive and dead. Reproduced from [https://commons.wikimedia.org/wiki/File:Schrodingers_cat.svg] under the terms of a CC BY-SA 3.0 license [https://creativecommons.org/licenses/by-sa/3.0/deed.en].

merely something that Schrödinger made up to illustrate how silly he thought quantum mechanics had become. Everything is set up so that after one hour, there is a fifty percent chance that an atom of the uranium has decayed, the Geiger counter has detected the decay and a sequence of Rube Goldberg events has broken the flask poisoning the cat. At that point, there is an equal probability that the cat is either alive or dead. In terms of quantum mechanics, the way it was being interpreted in Schrödinger's opinion, the cat's wavefunction would have blended equal parts of the living and the dead cat. The only way we can find out the cat's fate would be to open the box. Only at that point would the combined wavefunction collapse into one of the two possibilities. Schrödinger hoped that anyone considering this thought experiment would realize that the probability interpretation of quantum mechanics was a farce.

In 1927, Heisenberg accepted a position in Leipzig, where he became the youngest full professor of physics in Germany. He established an easy relationship with his students, led weekly seminars preceded by tea, for which he would often go to a local bakery to buy pastries. Afterward, participants would play table tennis, where Heisenberg was most often the victor. He married in 1937, had 7 children but always put his career first. Like Planck (whom he sought for advice), Heisenberg remained in Germany during WWII and was head of the German atomic bomb project. We will discuss this aspect of Heisenberg's life and career in the next chapter. After the war, he was imprisoned in England for 6 months but returned to Germany to reestablish the Kaiser Wilhelm Institute for Physics, which was renamed the Max Planck Institute for Physics.

16.11 TRAVEL SITES RELATED TO MAX BORN

16.11.1 Max Born Grave

Stadtfriedhof Cemetery (Entrance: 51.534273°, 9.910855°). Born is not buried in the Rondell Nobel Laureate (Rondell der Nobelpreisträger) (51.529739°, 9.911870°) section with eight other Nobel Prize winners (See Chapter 15, p. 369). Recall that he was initially omitted from those who received prizes for their parts in the development of quantum mechanics. Born and his wife are buried in a totally different part of the cemetery, near the entrance at

Plot Abt. E7, No. 28, Memorial ID 6447871 (51.533753°, 9.911361°), the family plot of his wife and her forebears. His epitaph, as noted above, is in the form of an equation, a mathematical formula in this case: $pq - qp = h/2\pi i$.

The following two sites were presented in Chapter 15 and are repeated here for convenience.

16.11.2 I. Physikalisches Institut (I. Physics Institute)

Georg-August-Universität, Göttingen, Friedrich Hund Platz 1 (51.558062°, 9.946848°). The Physicalisches Cabinet and Lichtenberg Collection contains a collection of historic instruments from the early days of physics in Göttingen. It is located at the entrance of the lecture halls in the new physics building. Guided tours with the curator can be arranged.

16.11.3 Göttingen, the Science City (Stadt die Wissenschaft)

There are many streets in Göttingen that are named after scientists including Robert Bunsen, Enrico Fermi, Max Born, Otto Hahn, Werner Heisenberg, and, of course, Max Planck and Albert Einstein. Many of their residences are marked by white plaques as well. Born led the theoretical physics section of the university from 1921 to 1933. See the following excellent website on History of Natural Science in Göttingen including a large amount of information and pictures of the cemetery; also listing of Göttingen streets named for scientists http://www.origin-life.gr.jp/3004/3004229/3004229.html.

16.12 TRAVEL SITES RELATED TO PAUL A. M. DIRAC IN BRISTOL, ENGLAND

See Chart 16.2 for the location of these sites.

16.12.1 Plaque at Childhood Home

15 Monk Road Bishopston (51.478543°, −2.595172°). Dirac was born and lived here from 1902 to 1913 although, at this writing, the plaque on the building mistakenly gives the span to be 1902–1923.

16.12.2 Dirac Second Home

6 Julius Road (51.474536°, −2.594870°).

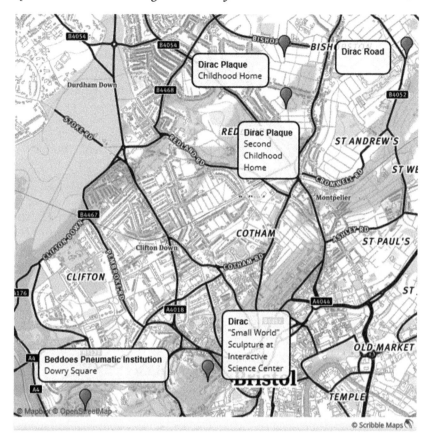

Chart 16.2 Travel sites in Bristol, England related to Paul Dirac (and Humphry Davy).

16.12.3 Dirac Road

(51.477890°, −2.578451°). This rather non-descript little street is about a mile's walk from his home. It is named in recognition of his links with the city. Online one can find a Paul Dirac trail that includes a chart. It includes his two childhood homes, this street, and his primary and secondary schools.

16.12.4 "Small Worlds" Sculpture

Anchor Road (51.450729°, −2.600487°) is dedicated to Paul Dirac. It is near "We the Curious", an interactive Science Center for children and adults at One Millennium Square in the Harbourside area. "Small Worlds" was unveiled in 2001 and consists of concentric cones that create "a scaled path through space pointing towards the ever-smaller worlds studied by Dirac." A plaque in tribute to Dirac is found in the pavement.

While in Bristol travelers might enjoy visiting the Humphry Davy sites associated with Beddoes Pneumatic Institution in Dowry Square (51.451244°, −2.620221°) (See Chapter 3).

16.13   TRAVEL SITE RELATED TO PAUL A. M. DIRAC IN CAMBRIDGE, ENGLAND

16.13.1   Residence

7 Cavendish Avenue (52.184997°, 0.141540°) Dirac held the Lucasian Chair of Mathematics at Cambridge University (the chair held by Isaac Newton) from 1932 to 1969. The chair was held by Stephen Hawking (1980–2009). He and his wife lived here from 1932 to 1969. There is no plaque here.

16.14   TRAVEL SITES RELATED TO PAUL A. M. DIRAC IN TALLAHASSEE, FLORIDA

16.14.1   The Paul A.M. Dirac Science Library

(30.444952°, −84.300190°) at Florida State University (FSU). Dirac was a Professor of Physics here for 12 years after leaving Cambridge. The library is named in his honor and his papers are held here. His daughter Monica has donated materials that will provide researchers further insight into the scope of Dirac's life and work. Outside the library is a statue of him by Gabriella Bollobás. The FSU bookstore is nearby.

16.14.2   Dirac Grave

Tallahassee's Roselawn Cemetery (30.486657°, −84.265754°). Dirac died in Tallahassee and was buried here with his wife of many years.

16.14.3   Paul Dirac Drive

(30.420559°, −84.328093°). This is located southwest of the FSU campus. A good part of it runs along the west side of the Seminole Golf Course and Club.

16.15 TRAVEL SITE RELATED TO PAUL A. M. DIRAC IN LONDON, ENGLAND

16.15.1 Memorial in Westminster Abbey

See Chapter 7 for directions to the honorary diamond placed here in his honor in 1995.

16.16 TRAVEL SITE RELATED TO WERNER HEISENBERG IN HELGOLAND

16.16.1 Heisenberg Memorial Stone

on Klippenrandweg on the island of Helgoland (54.180556°, 7.885278°). In June 1925 he made decisive progress in the creation of "matrix mechanics". He had requested a leave from Göttingen because of a strong attack of hayfever.

16.17 TRAVEL SITES RELATED TO WERNER HEISENBERG IN MUNICH

16.17.1 Plaque at Heisenberg House

Hohenzollernstrasse 110 (48.160872°, 11.571359°). This was Heisenberg's home while a youth.

16.17.2 Elementary School

(Elisabethenschule) Hohenzollernstrasse 140 (48.161142°, 11.566458°).

16.17.3 Maxgymnasium

Karl-Theodor-Strasse 9 (48.165176°, 11.585124°).

16.17.4 Werner-Heisenberg-Institute of the Max-Planck-Institute of Physics

Föhringer Ring 6 (48.184694°, 11.612498°). Located near the northern end of the English Gardens. Formerly known as the Kaiser Wilhelm Society, this is one of the oldest institutes in the Max Planck Society. Albert Einstein was the founding director in 1917 and Werner Heisenberg was its director until 1970.

While in Munich, visitors might also enjoy visiting a few places related to Boltzmann (Chapter 10), Fraunhofer (Chapter 11) and Einstein (Chapter 15). More Heisenberg sites are listed in Chapter 17 where we discuss his contributions to nuclear physics.

16.18 SUMMARY

This chapter completes our description of the role of quantum mechanics in the modern theory of the atom. Working off the wave-particle duality of light proposed by Einstein, Prince Louis-Victor de Broglie proposed the "wave-particle duality of matter" summarized in his equation, $\lambda = h/mv$. The circumstances behind the proposal presented in his doctoral thesis, the skepticism with which it was received by his examiners, and its redeeming reception by Einstein make us appreciate the contribution of the only Prince in atomic history when we visit his grave in Paris. Taking de Broglie's cue, Erwin Schrödinger immediately treated electrons in atoms as confined waves that set up certain allowed wave patterns with corresponding energies described by integers called "quantum numbers". His wave equation, prominently inscribed above his grave in the picturesque village in Alpach and below his bust at the Arcades of the University of Vienna in that elegant city, is one of the most important equations in atomic history. Its importance combined with the "erotic outburst" that produced it make for a truly unique story that travelers will revel in recalling when they visit Schrödinger sites in Austria, Switzerland and Dublin, Ireland.

The role of probability and uncertainty in the interpretation of Schrödinger's "wave mechanics", Werner Heisenberg's "matrix mechanics", and Paul Dirac's quantum mechanics incorporating relativity, is discussed in sufficient detail that travelers can begin to appreciate how they are incorporated into modern atomic theory. The nature of atomic orbitals as depictions of the probability of finding electrons in atoms is explained with the aid of several types of illustrations. Schrödinger's cat, his famous *gedanken* or "thought experiment", is presented as his way of illustrating how preposterous quantum mechanics had become. When we see depictions of "the cat", including at his former residence in Zurich, we will have some appreciation of its role. Stories of Dirac's autism including the "Dirac unit" for rate of speaking,

Heisenberg's retreat to the remote, tree-less island of Helgoland in the North Sea where he invented "matrix mechanics", and Max Born's enigmatic equation on his tombstone all contribute to the fun and adventure of traveling with the atom.

ADDITIONAL READING

- R. P. Crease, *The Great Equations: Breakthroughs in science from Pythagoras to Heisenberg*, W.W. Norton & Company, Inc., New York, 2008. Chapters devoted to the equations of Maxwell, Einstein, Schrödinger and Heisenberg are included.
- K. Hentschel, Some Historical Points of Interest in Göttingen, in *The Physical Tourist: A Science Guide for the Traveler*, ed. J. S. Rigden and R. H. Stuewer, Birkhäuser Verlag AG, Basel, Boston, Berlin, 2009.
- J. Gribbin, *Erwin Schrödinger and the Quantum Revolution*, John Wiley & Sons, Inc., Hoboken, New Jersey, 2013.
- W. Moore, *A Life of Erwin Schrödinger*, Cambridge University Press, Cambridge, 1994.

REFERENCES

1. Electromagnetic Energy, https://courses.lumenlearning.com/chemistryformajors/chapter/electromagnetic-energy-2, accessed June 2019.
2. J. C. Kotz, P. M. Treichel and J. R. Townsend, *From Chemistry & Chemical Reactivity*, Thomson, Brooks/Cole, Belmont CA, 8th edn, 2011.
3. G. Farmelo, *The Strangest Man: The Hidden Life of Paul Dirac, Mystic of the Atom*, Basic Books, New York, 2009.

CHAPTER 17

Nuclear Physics with "the Pope"; Fission and the Hahn/Meitner Controversy: Fermi, Hahn, Meitner, Heisenberg (Italy, Germany, Austria, Sweden, and Norway)

17.1 A QUICK LOOK AT PLACES TO VISIT "TRAVELING WITH THE ATOM" IN ITALY, GERMANY, AUSTRIA, SWEDEN, AND NORWAY RELATED TO NUCLEAR FISSION

Enrico Fermi: Rome: Birthplace Plaque ✤, *Campo de' Fiori* ✤, *Centro Fermi* ✤✤✤, *Fermi Collection, Physics Museum, University of Rome* ✤✤✤; *Florence: Plaque and Engraving at the Basilica di Santa Croce* ✤✤, *Museo Galileo* ✤✤✤; *Pisa: Domus Galilaeana* ✤✤✤, *Collegio Fermi* ✤, *Palazzo della Carovana* ✤; *Chicago: "Nuclear Energy" Sculpture* ✤✤, *Enrico and Laura Fermi "Chicago Tribute" marker* ✤, *Fermi Grave* ✤; *Meitner and Hahn: Vienna: Meitner Birthplace Plaque* ✤ *and Akademisches Gymnasium plaque* ✤; *Mitte-Berlin: Meitner-Hahn Plaque* ✤✤, *Lise Meitner Statue* ✤✤✤; *Dahlem-Berlin: Fritz Haber Institute* ✤✤, *Plaques in honor of the discovery*

Traveling with the Atom: A Scientific Guide to Europe and Beyond
By Glen E. Rodgers
© Glen E. Rodgers 2020
Published by the Royal Society of Chemistry, www.rsc.org

of fission ▓▓, *Hahn Monument* ▓▓, *The Lightning Tower* ▓▓, *Kungälv, Sweden: Plaque honoring Meitner* ▓; *Bramley, Basingstoke and Deane Borough, Hampshire, England: Lise Meitner grave* ▓; *Munich: Deutsches Museum* ▓▓▓▓: *Hahn and Meitner laboratory table* ▓▓, *Meitner Bust* ▓ *in Hall of Fame, Heisenberg Grave* ▓ *in Waldfriedhof; Göttingen: Otto Hahn Tombstone in Stadtfriedhof Cemetery* ▓▓▓; *Rjukan, Norway: Vemork Power Plant* ▓▓▓; *Haigerloch: The Atomkeller-Museum* ▓▓▓.

Probing the Nucleus and the Discovery of Fission

- Enrico Fermi and his "Panisperna boys" in Rome (1) bombard all the known elements with neutrons generated by alpha particles mixed with powdered beryllium; (2) discover the effect of "slow neutrons" and the ability of paraffin wax and water to act as "moderators", and (3) mistakenly believe they have generated transuranic elements but (4) just miss the discovery of nuclear fission. Fermi receives the 1938 Nobel Prize in Physics and immediately emigrates to the United States.
- The German chemist Otto Hahn and the Austrian physicist Lise Meitner collaborate together in Berlin for 31 years, discover a stable isotope of protactinium, and start to follow up on Fermi's work in 1934.
- In 1938, Meitner, a Jewish non-Aryan, is exiled to Sweden; Hahn and Fritz Strassman discover that uranium "bursts" into barium and similar sized nuclei; Meitner and her nephew Otto Frisch explain Hahn's results as due to the splitting of the uranium nucleus, a process Frisch calls "nuclear fission".
- Werner Heisenberg stays in Nazi Germany and is put in charge of the German "uranium project". He attempts to produce a self-sustaining nuclear chain reaction first at the Kaiser Wilhelm Institute of Physics in Berlin and then in a secluded beer cave in Haigerloch. None of these attempts are successful.
- In 1942, Fermi supervises the construction of the first controlled self-sustaining nuclear chain reaction in a squash court at the University of Chicago; he goes on to significantly contribute to the Manhattan Project that produces the first atomic weapons.
- Hahn is solely awarded the 1944 Nobel Prize in Chemistry for his discovery of fission. Lise Meitner, in one of the great injustices of Nobel Prize history, does not receive a Nobel Prize in physics for her explanation of fission.

17.2 A LITTLE REVIEW OF WHAT WE KNOW ABOUT ATOMIC NUCLEI

Now that we have seen how electrons are pictured as atomic orbitals in the modern theory of the atom and, in the last two chapters, visited many sites related to Max Planck, the young Albert Einstein, Prince Louis-Victor de Broglie, Erwin Schrödinger, Max Born, Werner Heisenberg, and Paul Dirac, we can go deeper into a consideration of the atomic nucleus. Rather amazingly for a travel book, we are moving from quantum mechanics to nuclear physics!

Even before we knew how it was discovered, we discussed the inherent instability of the nucleus in Chapter 6 ("The Brits, Led by the 'Crocodile', Take the Atom Apart"). We started with the Kiwi Ernest Rutherford who, soon after emigrating from New Zealand to England in 1895, designated the two types of radioactivity discovered by Antoine Becquerel and the Curies as "alpha" and "beta". He soon was off to McGill University in Montreal where, together with the chemist Frederick Soddy, he studied the complexity of nuclear transformations and defined the term "half-life". During those years, Ernest returned to New Zealand to marry May, brought her back to the new world and then soon after, with their small daughter in tow, was off to Manchester. There, in 1911, after seeing the result of Geiger and Marsden's alpha-particle, gold-foil experiments, he officially announced the discovery of the nucleus. As an aside to his efforts to aid in the war effort, he carried out the first artificial nuclear transformation (bombarding nitrogen atoms with alpha particles producing hydrogen and oxygen atoms). As the director and fourth professor of physics at the Cavendish Laboratory in Cambridge, he supervised the work of John Cockcroft and Ernest Walton, who slammed alpha particles into beryllium nuclei producing what James Chadwick identified as neutrons. In 1913, Henry Moseley, using X-rays, discovered atomic numbers (the number of protons in a nucleus). Soddy, now at the University of Glasgow, coined the term "isotopes" for atoms with the same number of protons but differing numbers of neutrons. As part of these discussions, we have traveled to a large variety of places in New Zealand, Montreal, Cambridge, Manchester, Glasgow, Oxford, Ireland, and London, many of which are among the best traveling-with-the-atom sites in this book.

Chapter 14 ("The Discovery that Atoms Fall Apart") was devoted to the lives and work of Antoine Becquerel, Pierre and Marie Curie, who originally discovered radioactivity in the late 1890s. In that chapter, we introduced balanced nuclear equations to represent these processes. In 1898, the Curies discovered the intensely radioactive elements, radium and polonium. The Curies' daughter Irène and her husband Frédéric Joliot-Curie discovered artificial radioactivity in 1934. During that chapter, we visited residences, workplaces, institutes, statues, museums, graveyards, and tombs, many with ⚛⚛⚛⚛ and ⚛⚛⚛⚛ ratings, all connected to these discoveries. Now, continuing the saga of the nucleus, we come to the remarkable work of Enrico Fermi, who started us on the pathway to the discovery of nuclear fission, the splitting of the atom, that is to say of course, the splitting of the nucleus.

17.3 ENRICO FERMI (1901–1954)

> "I studied mathematics with passion because I considered it necessary for the study of physics, to which I want to dedicate myself exclusively .... I've read all the best-known books of physics." From a letter by Adolfo Amidei, an engineer and colleague of Enrico Fermi's father, quoting Enrico when he was 17 years old.

Fermi was born in Rome, the third child of a lower-middle-class family that highly valued education. Enrico and his brother Giulio, only one year older, were exceptionally close, playing with their older sister, Maria, and building electrical toys together. Laura Fermi, Enrico's wife, who wrote *Atoms in the Family, My Life with Enrico Fermi*,[1] says Enrico and Giulio "completed each other to form a unit, as two atoms unite to form a molecule". In 1915, when Enrico was only 14, Giulio died tragically of complications from a simple operation. Devastated, Enrico became friends with Enrico Persico, one of Giulio's schoolmates. The two Enricos took long walks together, and one day found themselves in Campo de' Fiori, where they searched for physics books in the used book stalls. (See Chapter 9, p. 224 for more about Campo de' Fiori.) Scientific/historical travelers can identify with how difficult it is to find science titles at used book sales, but they discovered a two-volume, 900-page tome with no identifying cover. Some 75 years old and written in Latin, it was entitled *Elementorum Physicae Mathematicae*. The rest, as we have said before, is atomic history. Enrico Fermi, who even at 14 had already completed an intense study of languages, told his sister that he did not even notice that the book was in Latin. He studied it intensely and covered it with extensive, hand-written marginal notes. That was it. Enrico was set on studying physics. He continued studying on his own, particularly during the summer of 1918, when he poured over a five-volume treatise written by the Russian physicist, Orest Khwolson. So, with strong academic abilities, including a prodigious memory, a knowledge of the best-known books in physics, an unswerving devotion to doing well, and an unusual mathematical ability, he had excelled in his secondary education, graduated a year early, and was prepared to go to university.

Fermi was advised to get away from Rome because his home environment was so sad after his brother's death. On the advice of a trusted mentor, Adolfo Amidei, whose letter is quoted above, he applied to be a fellow at the elite *Scuola Normale Superiore* at the University of Pisa, where admission was determined solely by competition. There were three days of eight-hour written exams and an oral exam on the fourth day. Even though he was self-taught in physics he thoroughly impressed the faculty. So, in 1918, he started his studies in the city where Galileo had been born. He was an outstanding student in all his subjects and, by 1920 was far outpacing his professors in his knowledge of physics, including quantum mechanics and relativity. After graduating magna cum laude in 1922, he spent a year studying with Max Born in Göttingen (where he met Werner Heisenberg, but didn't fit in well). In the mid-1920s, he taught mathematical physics at the University of Florence, close to where Galileo spent the last 10 years of life under house arrest. (See Chapter 9 for Galileo sites in Pisa, Florence and Rome.)

In the late 1920s, he courted and married Laura Capon, the daughter of a respected family of non-observant Jews. This was an outstanding love-match that started when they met while Laura was a student in general science at the University of Rome. (One of her mathematics instructors was Enrico Persico!) In 1927, at the age of 25, Fermi had been chosen by a national competition (a *concorso*) to be Chair and Full Professor of Theoretical Physics at the University of Rome. Not surprisingly, older faculty resented his appointment, in a situation much like the way J. J. Thomson had been regarded when he was appointed the third Cavendish professor of physics when he was but 28 (see Chapter 5). Laura and Enrico were married in 1928 and soon she was helping him write a physics textbook for high school students. He dictated the book to her and she commented on its clarity. They had two children: a daughter, Nella, born in 1931 and a son, Giulio, born in 1936 and named after Enrico's much beloved brother.

Fermi was soon regarded as one the upcoming stars of modern physics. He had a full command of both the experimental and theoretical parts of the discipline including relativity, quantum mechanics and wave mechanics. In 1927, he attended the Volta Congress in Como, held to honor the 100th anniversary of the death of Alessandro Volta. (This is when the magnificent Tempio Voltiano in Como was opened. See Chapter 9 for more on the temple and all the Volta sites in that beautiful Italian lakeside town.) The Congress was attended by many of the atomic scientists we have been discussing in the last few chapters including Bohr, Born, Heisenberg, Planck, Rutherford, and Sommerfeld. It was clear that Italy had another major player in physics for the first time since Volta himself. It was soon after this meeting that Fermi decided to move away from exploring spectroscopy and the role of electrons in atoms, and concentrate on the atomic nucleus.

In 1933, he wrote a fundamental theoretical paper on beta decay, which, with only minor changes, continues to be the definitive way we view this process today. This was arguably Fermi's greatest and last contribution to theoretical physics. It was about then that he learned of Irène and Frédéric Joliot-Curie's studies of artificial radioactivity. (See Chapter 14 for the details of their work.) They had used alpha particles to produce new artificial radioactive isotopes, but they could not do this for any element heavier than aluminum. By then, the neutron had been discovered by Chadwick and its ability to penetrate atomic nuclei was becoming apparent. Fermi decided that neutron bombardment was the new wave in physics and he devoted all his time to this exciting experimental line of research. Rutherford wrote to congratulate him on his "successful escape from theoretical physics". He was at the peak of his powers, 33 years old and, as shown in Figure 17.1, had sharp eyes and an easy smile that signified he was ready to go to work.

To carry out this work, Fermi set up his Physics Institute above the Via Panisperna, which winds its way up the Viminal Hill, one of ancient Rome's seven hills. The institute was surrounded by palm trees, thick growths of bamboo, and a goldfish pond in the back. His students included Edoardo Amaldi, Oscar D'Agostino, Emilio Segrè, and Franco Rasetti (who had been a fellow student at Pisa), whom Fermi had appointed as his assistant. D'Agostino was the only chemist among them. Soon nicknamed the "*Via*

Figure 17.1 Enrico Fermi at 33. Reproduced from ref. 2 with permission from Department of Energy. Office of Public Affairs. 10/1/1977–1985.

Panisperna boys", they were a close-knit group, sharing not only a love of physics, but of hiking and skiing and a capacity to have a good time together. Each of them had a nick-name. Fermi's was "The Pope" due, it is said, to his infallibility when it came to quantum mechanics. In many ways, Fermi and Rasetti were the first to organize what today we would call a research group. The Panisperna boys are shown in Figure 17.2.

Fermi and his "boys" set out to bombard every element in the periodic table with neutrons. For a neutron source, they used radon gas, which was steadily produced by the alpha decay of the institute's one gram of radium. The radon had to be used quickly as it has a half-life of only 3.82 days. The radon produced alpha particles that interacted with powdered beryllium to produce about 100 000 neutrons per second. This was a procedure that Rasetti had learned when he worked with Lise Meitner at the Kaiser Wilhelm Institute in Berlin-Dahlem. (We will discuss Meitner's work in this field later in this chapter.) Let's write three balanced nuclear equations that represent the process by which they produced neutrons. The first is the alpha decay (represented by $-\alpha \rightarrow$) of radium into radon gas. Recall that an alpha particle is just a helium nucleus represented by ^4_2He

Figure 17.2 Enrico Fermi and the Panisperna "boys". From left to right, Oscar D'Agostino, Emilio Segrè, Edoardo Amaldi, Franco Rasetti, and Enrico Fermi.

in the equations. The second is the alpha decay of radon-222 gas that produces the Curies' polonium as a by-product. In the third, an alpha particle hits a beryllium-9 nucleus producing a neutron and carbon-12.

$$^{226}_{88}\text{Ra} - \alpha \rightarrow ^{222}_{86}\text{Rn(g)} + ^{4}_{2}\text{He}$$

$$^{222}_{86}\text{Rn(g)} - \alpha \rightarrow ^{218}_{84}\text{Po} + ^{4}_{2}\text{He}$$

$$^{4}_{2}\text{He} + ^{9}_{4}\text{Be} \rightarrow ^{1}_{0}\text{n} + ^{12}_{6}\text{C}$$

A reliable and intense neutron source in hand, Fermi's group started bombarding the lightest elements but did not have much luck in producing any induced artificial radioactivity. However, as they moved into medium-heavy elements like iron and silver, for example, they were successful in producing artificial radioactivity in about 20 cases. It was quite a complex procedure, but by then they had mastered the logistics. The trickiest part was that the irradiated sample had to be quickly removed from the bombardment room because the radon emitted high amounts of "background radiation", consisting of beta particles and gamma rays. In order to see if an irradiated sample was radioactive, it was quickly spirited down a long hallway, where Geiger counters could measure its radioactivity. Fermi and Amaldi were the fastest runners. There's a great story to tell here. One day, a dignified Spanish professor came to call on "His Excellency Fermi". He was told that "the Pope is upstairs" (!) but then was almost knocked flat on his back by Fermi and Amaldi running a sample down the hall. His Excellency Fermi was mighty quick on his feet! By mid-1935 they had published 10 papers, bombarded 62 elements with neutrons and produced some 50 new radioactive elements. Most of the papers had all five of the Panisperna boys as co-authors, all neatly arranged in alphabetical order.

Fermi discovered a most crucial aspect of neutron bombardment, one that had broad implications not only for his work but also for the ultimate production of an atomic bomb. Strangely, he found that for a given element bombarded, the radioactivity produced varied depending upon whether the procedure was carried out on a wooden table or a marble one. You can imagine the dismay and confusion that accompanied this apparent inconsistency. Specifically, they were studying a silver nucleotide (a radioactive isotope of silver) with a half-life of 2.3 minutes. Its activity fluctuated wildly depending on materials close to the neutron source when the silver was bombarded. Fermi, in October 1934, on the spur of the moment, decided to insert a two-inch-thick square of paraffin wax between the neutron source and the target. When he did that, the radioactivity produced increased by more than a hundred-fold! What on earth is going on here, he asked himself. In the end, the explanation had to do with the speed of the bombarding neutrons. When the neutrons collided with the hydrogen atoms of the wax, they were slowed down and now were much more likely to be captured by the target nuclei instead of hitting the nucleus and bouncing off. In the same way, the incoming neutrons had hit the wooden table and had been slowed down; this did not happen with the marble table. This serendipitous discovery in hand, he surmised that water, with its hydrogen nuclei, should have the same effect. This was proven true. Laura Fermi says that the goldfish pond in the back of the institute was used in this experiment. She noted that the goldfish were unharmed. These results were obtained with many elements, in fact almost every element they tested. Substances such as the paraffin wax and water became known as "moderators", as they moderated the speed of the neutrons.

When they finally started bombarding uranium, the heaviest of the known elements, they were convinced (wrongly, it turned out) that they had produced an element with 93 protons, a so-called "transuranic". How did they think this would come about? The reasoning was that the incoming neutron would hit a U-238 nucleus, temporarily producing U-239, which would decay *via* beta emission to produce Element-93 or, with the emission of a second beta particle, Element 94. These proposed reactions are represented in the following balanced nuclear equations. (Yes travelers, here we go again!)

$$_0^1n + {}_{92}^{238}U \rightarrow [{}_{92}^{239}U] - \beta \rightarrow {}_{93}^{239}X + {}_{-1}^{0}\beta - \beta \rightarrow {}_{94}^{239}Y + {}_{-1}^{0}\beta$$

Elements with atomic numbers ninety-three (X) and ninety-four (Y) were unknown at that time. The Fermi group even debated over what to call the new elements 93 and 94. They decided on Ausonium and Hesperium, Greek words for Italy and Land to the West. They ultimately were called neptunium and plutonium after the planets. Some travelers will already know that plutonium turned out to be one of the fissionable nuclei used to produce an atomic bomb in the "Manhattan Project".

It turned out that they had not produced these transuranics. Instead, they had missed one of the all-time important discoveries of the age. Ida Noddack, in a paper published in *Nature* in September 1934, was critical of Fermi's work and maintained that it was perfectly possible that the nucleus could break up into several fairly large fragments, *i.e.*, that the uranium nucleus could be split nearly in two. Her paper was dismissed rather out of hand. Fermi even did some calculations and concluded that Noddack's conclusion could not possibly be correct. Years later, however, it was confirmed that she *was* absolutely right and they had missed the correct interpretation of their results: nuclear fission.

Time was working against "the boys". In Germany, Hitler had taken over in 1933. In early 1935, Mussolini invaded Ethiopia and used "mustard gas". In July 1938, Mussolini, taking a cue from his friend Hitler, launched an anti-Semitic campaign. In October 1938, Fermi visited the Bohr Institute and was told he was likely to receive the 1938 Nobel Prize in Physics. On the way back from Copenhagen, he accepted an offer to join the Department of Physics at Columbia University. On November 10, the phone call from Stockholm came. On November 11, "mixed marriages" between Jews and non-Jews (like Laura and Enrico's) were proclaimed to be forbidden. It was now obvious that the Fermis would have to leave Italy. On December 5, 1938, the Fermi's started the 48 hour train ride to Stockholm. On December 10, Enrico received the Nobel Prize in Physics "for his demonstrations of the existence of new radioactive elements produced by neutron irradiation, and for his related discovery of nuclear reactions brought about by slow neutrons". The chemistry and physiology and medicine prizes were not awarded. Only Fermi and Pearl Buck, who had received the Literature Prize, were on stage. When Fermi backed away from King Gustavus V, who had given him his prize, he almost sat in Pearl Buck's lap. The Italian press criticized Fermi for not wearing a Fascist uniform and for shaking the King's hand instead giving him the fascist salute. On December 24, 1938, the Fermis sailed from Southampton to New York. The short heyday of the Panisperna "boys" had now concluded.

In the United States, Fermi made paramount contributions to the Manhattan Project. As we will see shortly, it did not take long for Otto Hahn, Fritz Strassman, and Lise Meitner to discover

Nuclear Physics with "the Pope"

nuclear fission. Fermi realized that "secondary neutrons" from the fission process would be emitted, and that a nuclear chain reaction was possible. In 1942, Fermi, now at the University of Chicago, supervised the construction of Chicago Pile 1 that produced the first controlled, self-sustaining chain reaction on a squash court underneath Stagg Field, the university's athletic stadium. Fermi went on to help develop the first atomic bombs, including one (called "Fat Man") that used the transuranic plutonium as the nuclear fuel. Fissionable fuels, principally uranium-235 and plutonium-239, became the basis of nuclear power plants.

17.4 TRAVEL SITES RELATED TO ENRICO FERMI IN ROME

See Chart 17.1 for the location of these sites.

17.4.1 Birthplace Plaque

⦿ Via Gaeta 19 (41.904898°, 12.500745°). In 2001, to mark the 100th anniversary of his birth, this plaque was placed on the façade of the house where Fermi was born. It gives the date *and time* of his birth and notes that he was the most celebrated Italian physicist

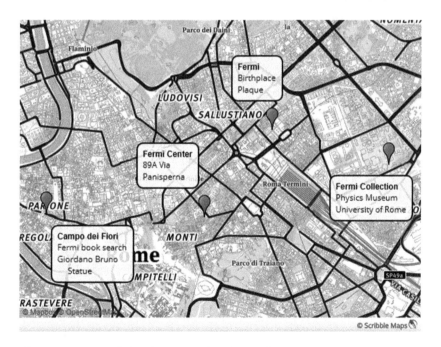

Chart 17.1   Travel sites in Rome related to Enrico Fermi.

of the twentieth century and "His fellow citizens remember him with admiration". In 1908, the family moved to Via Principe Umberto 133 in Rome (41.893951°, 12.510261°). In 1938, Enrico and Laura moved to Villa Borghese Park in Rome (41.912852°, 12.485287°).

17.4.2   Campo de' Fiori

("Field of Flowers") (41.895586°, 12.472160°). This active market area is where Fermi found the physics book that changed his life and, to some degree, all of physics forever. The hooded statue of Giordano Bruno dominates the square. Giordano Bruno was burned at the stake here. This martyr of science believed in many things the church found heretical, including heliocentrism, life on other planets and that matter was composed of discrete atoms. As noted in Chapter 10, this is a most appropriate place to get some coffee and a treat, people-watch, and think about Bruno and Fermi.

17.4.3   Centro Fermi

89A Via Panisperna (41.896793°, 12.492221°). The new *Museo Storico della Fisica e Centro Studi e Richerche "Enrico Fermi"* (Enrico Fermi Centre for Study and Research) was expected to be open to the public in 2019 and will host a permanent exhibit about Fermi. The interior of the building has been reconfigured a little from Fermi's day, but when you are there, ask them about the place where the Spanish professor was almost knocked on his back by "the Pope" rushing a sample down to the Geiger counter room. The goldfish pond (the "garden fountain") where the moderating ability of water (and goldfish) was tested is definitely worth visiting. Contact Prof. Luisa Cifarelli at presidenza@centrofermi.it.

17.4.4   Fermi Collection, Physics Museum of the University of Rome, Enrico Fermi Building, Citta Universitaria

17 *Via* Cesare de Lollis (41.901673°, 12.515826°). The Citta Universitaria is the main campus of the University of Rome or *"La Sapienza"*. The physics museum has a radon-beryllium (radon gas and beryllium dust) neutron generator, Geiger-Müller counters, a lead wedge fashioned by Fermi, and pieces of paraffin that he

inserted when he discovered the effect of slow neutrons. It also has a collection of spectroscopic equipment and some photographic plates of the most significant spectra. Until the end of the 1920s, Fermi and his group focused their research on atomic and molecular spectroscopy. You may also want to visit the entrance to the main physics building (41.901673°, 12.515826°) at 2 Piazza Aldo Moro, where a display cabinet has some good letters to and from Fermi. While you are here, you might go over to the nearby Museum of Chemistry "Primo Levi" (41.902421°, 12.513858°) at 5 Piazzale Aldo Moro which, as noted in Chapter 9, has a good exhibit of scientific equipment and a collection of chemicals and documents that belonged to Stanislao Cannizzaro.

17.5 TRAVEL SITES RELATED TO ENRICO FERMI IN FLORENCE

17.5.1 Enrico Fermi Plaque and Engraving ※※, Basilica di Santa Croce

(43.768550°, 11.262280°). This basilica is also the burial place of Michelangelo and Galileo. It was here that Galileo was finally buried after the church recovered from all the "offenses" he had committed against it. Here also are the graves of Rossini and Machiavelli. There is a statue of Dante Alighieri outside. (See Chapter 10 for more information.)

17.5.2 Museo Galileo

※※※ 1 Piazza Dei Giudici (43.76773°, 11.2559°). Formerly the *Istituto E Museo Di Storia Della Scienza*, it was renamed the Museo Galileo in 2010. Given his status as one of Italy's greatest scientists, it is worth inquiring about temporary exhibits or even new permanent exhibits related to Enrico Fermi. See Chapter 10 for the exhibits here related to Galileo.

17.6 TRAVEL SITES RELATED TO ENRICO FERMI IN PISA

17.6.1 Domus Galilaeana

※※※ 26 Via Santa Maria (43.718457°, 10.397207°). Housed in the Palace of the Observatory of Pisa, this is a specialized library for the history of science. Various sources indicate that twelve of the radon-beryllium sources used by Fermi's group are housed here.

This has also been a depository for Fermi documents including two heavy safes containing original documents relative to the scientific activity carried out by Fermi during his life in Italy. Another source indicates that Fermi's "Sorgenti", used in his experiments on induced radioactivity, is located here.

17.6.2 Collegio Fermi

at the *Scuola Normale Superiore*, 29 *via* S. Apollonia (43.720460°, 10.401880°). This building is dedicated to Enrico Fermi, who studied at the *Scuola Normale* from 1918 to 1922. It opened in 1996. Besides the student rooms, the residence has two classrooms, one used by the Faculty of Arts and one by the Faculty of Sciences.

17.6.3 Palazzo della Carovana

Knights Square (43.719611°, 10.400225°). This ornately decorated palace is the main building of the Scuola Normale Superiore. Stand here and imagine what Enrico Fermi felt when he took the entrance exam to this elite school when he was just 17 years old.

17.7 TRAVEL SITES RELATED TO ENRICO FERMI IN CHICAGO, UNITED STATES

17.7.1 "Nuclear Energy"

a sculpture with four accompanying plaques commemorating Chicago Pile-1. Located on Ellis Avenue, between the Max Palevsky West dormitory and the Mansueto Library on the campus of the University of Chicago (41.792477°, −87.600952°), this bronze sculpture is at the exact location where, under the direction of Enrico Fermi, the pile produced the first sustained nuclear chain reaction under the demolished stands of old Stagg Field. It was dedicated in 1967 on 25th anniversary of the first sustained nuclear chain reaction.

17.7.2 Enrico and Laura Fermi "Chicago Tribute" Marker

5537 South Woodlawn Avenue, Hyde Park neighborhood (41.793964°, −87.596152°). Enrico and Laura lived here from 1946 to his death in 1954. Not open to the public.

17.7.3 Fermi Grave

Oak Woods Cemetery, Chicago (41.7714435°, −87.5965666°). These coordinates give the exact location of his simple gravestone.

17.8 THE DISCOVERY OF NUCLEAR FISSION

17.8.1 Otto Hahn (1879–1968) and Lise Meitner (1878–1968)

"Although I had a very marked bent for mathematics and physics from my early years, I did not begin a life of study immediately.... Thinking back to ... the time of my youth, one realizes with some astonishment how many problems then existed in the lives of ordinary young girls." Lise Meitner

Otto Hahn was an excellent experimental chemist. In 1904, he worked with Sir William Ramsay at University College London the same year the Scottish chemist received the Nobel Prize in Chemistry for his discovery of argon and other noble gases (See Chapter 7). With Ramsay, he developed an interest in pursuing radiochemistry. The next year he worked with Ernest Rutherford at McGill University in Montreal where he succeeded the chemist Frederick Soddy. Subsequently, he went to Berlin-Dahlem to study with Emil Fischer. Here, in 1907, he met and started to work with Lise Meitner. The Kaiser Wilhelm Institute (KWI) of Chemistry was created in 1911 by the Kaiser himself and Hahn became the head of the radiochemistry department. During World War I, he worked with Fritz Haber (1868–1934) who headed the new chemical warfare unit that developed, tested and produced poison gas for military purposes. Hahn had argued against this activity, but was told it would end the war more quickly. It did not, and Hahn was horrified by the results. Others on the German side were as well. One of the most striking examples was Haber's wife, Clara Immerwahr Haber, who was the first woman to receive a PhD in chemistry in Germany. She protested to her husband about using poisonous gases in warfare and when he stridently insisted on continuing, she took Fritz's military pistol and killed herself right outside their residence on the grounds of the KWI. Hahn, by then a lieutenant in the German infantry, participated in the 1914 Christmas Eve fraternization between British and German troops near Ypres, Belgium. In 1915, he helped install the chlorine cylinders used against allied troops in that same area.

Lise Meitner was an excellent physicist, one of the first women to excel in this field. She was born in Vienna to a comparatively well-off Jewish family who valued education for all of their children, both boys and girls. She was inspired to study science when she learned of the Curies' discovery of radium in 1902 (see Chapter 14) but opportunities for women in academia, particularly the sciences, were extremely limited at that time. She studied under Ludwig Boltzmann (See Chapter 10), whom she found to be an inspiring and excellent teacher and earned her doctorate in 1906. In 1907 she visited Berlin to attend lectures by Max Planck (see Chapter 15). From 1912 to 1915 she was an assistant to Planck at the Institute for Theoretical Physics at the University of Berlin. As mentioned in Chapter 15, she often visited the Plancks at their Grünewald house and enjoyed cheerful, informal company and music of all kinds.

Following the example set by Marie Curie, Lise worked as a nurse and an X-ray technician with the Austrian army during WWI. She arranged for her leave to coincide with Hahn's and they worked together at the Kaiser Wilhelm Institute for Chemistry in Berlin-Dahlem. During this time, they discovered a long-lived isotope of protactinium. In 1917, she was appointed head of the Physics Department at the KWI for Chemistry, a position she prized all her life, particularly after she tragically lost it in 1938. After WWI, Lise and Otto continued to collaborate off and on. Figure 17.3 shows the two of them together in the laboratory in 1919. He

Figure 17.3 Otto Hahn and Lise Meitner in his laboratory, 1919. Used with the permission of the Archives of the Max Planck Society, Berlin.

was a handsome, out-going, married man of about her age. She was shy and reserved and never shared a meal with Hahn except on formal occasions. All her life, she had to overcome prejudice against women in science. As just one example, when she first worked in Berlin, Emil Fischer, the great organic chemist, made her promise never to enter the laboratories where men were working. She worked in the basement of the laboratory and could not go upstairs to Hahn's laboratory. The only available toilet was in a restaurant down the street from the laboratory. Despite all this, she persevered to become one of the first women at the KWI to have the title of Professor. She was totally devoted to physics and never married. Between the two wars, now on her own, she became one of the best-known scientists studying radioactivity. Now colleagues, Hahn and Meitner remained close. Lise was the godmother of the Hahn's only child, little Johann Otto.

Einstein called her "the German Madame Curie". In the early 1930s, she had helped Franco Rasetti (Fermi's assistant) learn how to efficiently produce neutrons; he brought this knowledge back to Rome. Fermi and his "Panisperna boys" had been uncertain about what was happening when they bombarded uranium with neutrons. Together with Fritz Strassman, Meitner and Otto Hahn started to investigate products of nuclear bombardment, particularly when uranium was the target. This work was interrupted in the spring of 1938 when Austria was annexed by Germany and Meitner, of Jewish heritage and now suddenly a German citizen, was subject to immediate dismissal from her position. Adding to her dismay, her Austrian passport was invalid, her pension rights were questioned, and she had no funds in reserve. Niels Bohr, consistent with his steadfast efforts to aid scientists escaping Nazi Germany, was able to make some arrangements whereby she could have space to work in the laboratory of Nobel laureate Manne Siegbahn in Stockholm. The problem was how to get there.

The story of Meitner's illegal emigration is harrowing. Ruth Sime, in her biography of Meitner, describes it in riveting detail. (Chart 17.2 shows many of the sites involved in this traumatic journey.) In mid-July, an illegal and perilous scheme to escape by train to Holland (at Nieue Schans, now Bad Nieuweschans) was put together. Nazi authorities knew she was about to flee but missed their chance to intercept her. Meitner spent her last night in Germany at Hahn's house. He gave her his mother's diamond ring to use in an emergency. With only two small suitcases, Lise Meitner, penniless, nearing her 60th birthday, "shaking with fear" as she once wrote, crossed the border without incident.

Hitler was set to invade Czechoslovakia, non-Aryans were pursued throughout Germany and the resulting political tensions had slowed work at the KWI to a crawl. Hahn and Strassman, neither one a Nazi party member and both fearful for their institution and positions, continued to bombard uranium with slow neutrons. It now became apparent that there were unexpected products and that something very mysterious was going on. In early November, 1938, Bohr invited Hahn to give a lecture at his institute and Niels and Margrethe invited Hahn and Meitner and her nephew Otto Robert Frisch to their home in the Carlsberg House of Honor in Copenhagen. This was less than a week after *Kristallnacht*, during which Nazis had savagely attacked Jewish persons and property all over the country. We can only imagine the range of the discussions at the House of Honor, but we do know for sure that Hahn could not reveal that he was in contact with the exiled Meitner and therefore they kept their meeting an absolute secret from anyone outside of Copenhagen. Lise urged Otto to return to the KWI and, with Strassman, redouble their efforts to identify the products of bombarding uranium with neutrons. She returned to Siegbahn's institute in Stockholm where she had only a one-room laboratory but no assistants, only limited equipment and materials, and was not treated as a valued colleague or resource. In December, the Fermis arrived in Stockholm for the Nobel Prize ceremonies (see above) and then immediately started their long trip to New York. The Bohrs attended the ceremonies but then returned to Copenhagen. Lise Meitner was alone.

In Berlin, things came to a head quickly. Strassman and Hahn now produced solid evidence that a radioactive isotope of barium was one of the products. Hahn, incredulous that uranium had apparently "burst apart into Ba", wrote to Meitner, hoping that she "could come up with some fantastic explanation".

Otto Frisch was with his aunt and they were celebrating Christmas together at friends in Kungälv, on the west coast of Sweden. On Christmas Eve they went for a walk in the "snowy Scandinavian woods" to talk about these "startling" results.

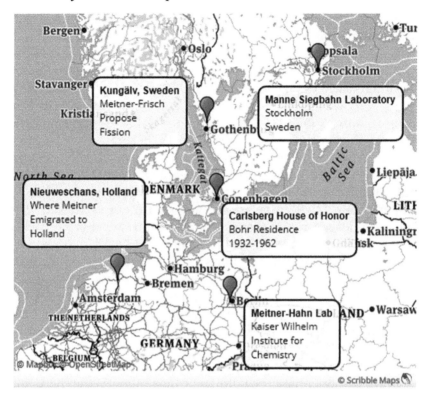

Chart 17.2 Travel sites related to the emigration of Lise Meitner to Sweden.

Using the "liquid drop" model of the nucleus (proposed by Niels Bohr and others), Lise and Otto theorized that when a neutron hits a large uranium nucleus, it is destabilized and splits roughly in two. They estimated the masses of the possible products and used Einstein's $E = mc^2$ to calculate the prodigious amount of energy released by the splitting of the nucleus. Frisch, consulting a biologist friend, called it "nuclear fission".

The discovery of fission by Hahn and Strassman and its explanation by Meitner and Frisch had several repercussions. First, it meant that Fermi and his "Panisperna boys" had been mistaken about producing transuranic elements, that is, elements with atomic number greater than uranium's 92. Instead, nuclear fission had been occurring and they had missed it. Ida Noddack, in fact, had been correct about fission, Fermi's calculations not withstanding. Hahn, Strassman, and Meitner had also been mistaken about the production of transuranic elements. Fermi was embarrassed that he and his group had missed fission and wrote an addendum to his Nobel Prize lecture. One wonders what would have happened if Fermi had recognized the occurrence of fission in early 1935. Is it possible that Fascist Italy and Germany might very well have seized upon the discovery and built the first atomic bomb? One should contemplate that possibility while touring the Fermi Centre at 89A Via Panisperma.

Otto Frisch returned to Copenhagen and consulted Niels Bohr who said "Oh what idiots we have been! Oh, but this is wonderful! This is just as it must be! Have you and Meitner written a paper about it?" Bohr suggested an experiment to measure the energy released when uranium atoms split. He was preparing to travel to the United States but promised to keep these results to himself until Meitner and Frisch could perform a confirming experiment and submit a paper to *Nature*. He was so engrossed in these discussions that he almost missed the ship to New York. Bohr, however, consulted with a colleague on the way to New York and neglected to tell him that the Meitner and Frisch results were a secret. Fission soon became the rage in New York's physics circles.

It did not take long for Bohr to propose that it was the U-235 nucleus that was splitting. With barium as one of the fission products, it soon became apparent that krypton was the other. Soon many research groups were reporting that two or three free neutrons are released during these processes. This meant we could start writing balanced nuclear equations for fission reactions. There are many possible ways for uranium-235 to fission but the one producing barium and krypton could be written as follows:

$$^1_0 n + ^{235}_{92} U \rightarrow ^{141}_{56} Ba + ^{92}_{36} Kr + 3\,^1_0 n$$

With two or three "secondary neutrons" produced, several of these could go on to strike other fissionable U-235 nuclei and quickly create a nuclear chain reaction.

Otto Frisch went on to be active in the Manhattan Project (and the equivalent "Tube Alloys project" in the UK) where he estimated the critical mass, defined as the minimum mass of fissionable material that can sustain a nuclear chain reaction. Once he nearly accidently created a critical mass in his laboratory at Los Alamos.

Lise Meitner stayed in Stockholm for 22 years and became a Swedish citizen. She never got along with Manne Siegbahn (in whose lab she originally worked) who, as it turned out, was known to disapprove of women in the sciences. She was nominated for a Nobel Prize a grand total of 48 times, but never received one due, in large part, to gender and religious discrimination. See Chapter 19 for more details concerning the deliberations of the Royal Swedish Academy of Sciences regarding her nominations. There is some evidence that her personal conflict with Siegbahn and other committee members deprived her of the Nobel Prize in 1946. Things did look up a bit when, in that same year, Meitner (like Marie Curie before her) visited the United States where she was honored as the "Woman of the Year" by the National Press Club and had dinner with President Harry Truman and others. She lectured at Princeton and Harvard and other colleges, who often awarded her honorary doctorates. In 1947, a position was created for her at the University College of Stockholm. She participated in the research that established Sweden's first nuclear reactor. In 1960, she retired and moved to the United Kingdom, where most of her relatives now lived including her nephew, Otto Frisch, whose wife ultimately wore the ring that Hahn had given her for insurance.

The Nobel Prize in Chemistry 1944 (awarded in 1945) was awarded solely to Otto Hahn "for his discovery of the fission of heavy nuclei". He was unable to accept the award because he was interned at Farm Hall, Godmanchester, Huntingdonshire, England, near the water meadows of the River Ouse in Cambridge. At the end of the war, he and other German scientists were suspected of working on the development of a German atomic bomb. However, Hahn's only connection was his discovery of fission. His whereabouts was a secret during that time and the Nobel committee was unable to find or notify him. At Farm Hall, the interned German scientists first learned of the dropping of atomic bombs in Hiroshima and Nagasaki. Hahn felt responsible and contemplated suicide. Some thought that Lise Meitner should have shared in this Nobel Prize but she always maintained that it was strictly a chemical achievement (again, see Chapter 19 for more details.) In 1966, Otto Hahn, Lise Meitner, and Fritz Strassman shared the Enrico Fermi Award, which honors scientists of international stature for their lifetime achievement in the development, use, or production of energy.

Perhaps here at the end of this saga we should note that element 109 is Meitnerium (Mt). It joins Curium (Cm), Einsteinum (Es), Rutherfordium (Rf), Bohrium (Bh), and Fermium (Fm). There is no element Hahnium! There was a proposal to name element 105 Hahnium, but ultimately it was named Dubnium, after the Russian research center in Dubna.

17.9   TRAVEL SITES RELATED TO LISE MEITNER IN VIENNA, AUSTRIA

17.9.1   Meitner Birthplace Plaque

Heinestrasse 27 (48.220205°, 16.386799°). The plaque marks the birthplace of Lise Meitner.

17.9.2   Akademisches Gymnasium Plaque

Innere Stadt, Beethovenpl. 1 (48.20158°, 16.37685°). This acknowledges that Lise Meitner graduated from here in 1901. Meitner's beloved professor, Ludwig Boltzmann (and his daughter Henriette) also graduated from here. There is also a plaque acknowledging that Erwin Schrödinger was a student here 1898–1905.

17.10   TRAVEL SITES RELATED TO LISE MEITNER, OTTO HAHN AND THE DISCOVERY OF NUCLEAR FISSION IN MITTE BERLIN

17.10.1   Meitner-Hahn Plaque

1 Hessischestrasse (52.528401°, 13.380654°). This marks the site where Lise Meitner and Otto Hahn conducted research between 1907 and 1912. Portraits of Meitner and Hahn are displayed on round plaques on either side of the rectangular plaque describing the site.

17.10.2   Statue of Lise Meitner

(52.517793°, 13.394004°). Located in the garden of the Humboldt University of Berlin next to similar statues of Hermann von Helmholtz and Max Planck. This statue, shown in Figure 17.4, was unveiled in July 2014. It was the first monument honoring a female academic in all of Germany. See Chapter 15 for more on the nearby statue of Max Planck.

Figure 17.4 Memorial to the nuclear physicist Lise Meitner (1878–1968) erected in 2014 in the Court of Honor of the Humboldt University in Berlin. Anna Franziska Schwarzbach was the sculptor. Reproduced from ref. 3 under the terms of a CC BY-SA 4.0 license [https://creativecommons.org/licenses/by-sa/4.0/deed.en].

17.11 TRAVEL SITES RELATED TO LISE MEITNER, OTTO HAHN AND FRITZ STRASSMAN AND THE DISCOVERY OF NUCLEAR FISSION IN DAHLEM (BERLIN)

17.11.1 Fritz Haber Institute

Faradayweg 4–6 (52.448588°, 13.282754°). The Fritz Haber Institute of the Max Planck Society was founded in 1911 as the Kaiser Wilhelm Institute (KWI) for Physical Chemistry and Electrochemistry. Otto Hahn was the head of the radiochemistry department. Haber and Carl Bosch devised and perfected a catalytic reaction between nitrogen and hydrogen gases to produce ammonia. Ammonia, through the (Wilhelm) Ostwald process, could be converted to nitrates for use in fertilizers and munitions. It is estimated that the German use of these processes prolonged WWI for about two years. Haber also spearheaded the German chemical warfare unit during the first world war. Otto Hahn objected to the use of chemical warfare, but participated in it nonetheless. The Clara Immerwahr Haber memorial (52.448993°, 13.282872°) marks the spot where she killed herself on the grass near the Haber House after she realized that her husband was going to continue to head the chemical warfare unit. After World

War I, Fritz Haber was regarded by some as a war criminal but, in the end, he was awarded the Nobel Prize in Chemistry in 1918 for his work in devising the Haber-Bosch process.

17.11.2 Plaques in Honor of the Discovery of Fission

Thielallee 63/67(52.447996°, 13.284347°). A large bronze plaque atop the round tower of the old Institute building where the KWI Chemistry was located, now part of the Free University of Berlin. The plaque was put place in 1956 and says that in this building Otto Hahn and Fritz Strassman discovered uranium fission (*"uran-spaltung"*). There is a similar plaque inside the building. Another plaque was added in 1997 that honored Meitner's contribution to the discovery of fission. This plaque also honors Max Delbrück, a pioneer of molecular genetics who was an assistant to Meitner from 1932 to 1937. There is a sculpture of Meitner on the second floor of the building. The tower is attached to the Hahn-Meitner Building formerly known, not surprisingly, as just the Hahn Building.

17.11.3 Hahn Monument

Otto-Hahn-Platz (52.449634°, 13.294361°). The monument is in front of the house where Hahn lived with his family from 1929 to 1944.

17.12 TRAVEL SITE RELATED TO LISE MEITNER AND THE DISCOVERY OF NUCLEAR FISSION IN KUNGÄLV, SWEDEN

17.12.1 Plaque Honoring Meitner

9 Västra gatan (57.865998°, 11.992465°). Located at the Uddmanska house where Meitner was staying when she and Otto Frisch proposed fission to explain Hahn and Strassman's results. The plaque was placed here in December, 2016. It is the second European Physical Society (EPS) historic site. The first was Tycho Brahe's observatory (see Chapter 12) on the island of Ven in the Øresund strait between Denmark and Sweden. At the unveiling of the plaque, there was a guided walk in Lise Meitner's footsteps and a play, "Remembering Miss Meitner".

17.13 TRAVEL SITE RELATED TO LISE MEITNER IN BRAMLEY, BASINGSTOKE AND DEANE BOROUGH, HAMPSHIRE, ENGLAND

17.13.1 Lise Meitner Grave

St. James Churchyard (51.32637°, −1.07572°). Meitner had moved to Cambridge in 1960 to be nearer to her relatives, including Otto Frisch, who was a professor of natural philosophy at the Cavendish. The epitaph, created by Frisch, says "A physics professor who never lost her humanity". A younger brother had been buried here earlier.

17.14 TRAVEL SITES RELATED TO LISE MEITNER, OTTO FRISCH, OTTO HAHN AND FRITZ STRASSMAN AND THE DISCOVERY OF NUCLEAR FISSION IN MUNICH AND GÖTTINGEN, GERMANY

17.14.1 Deutsches Museum

1 Museumsinsel, Munich (48.129904°, 11.583452°).

17.14.2 Hahn and Meitner Laboratory Table

Originally labeled the "Worktable of Otto Hahn", this is a reconstruction of the apparatus used for neutron irradiation. However, the original table had, in fact, been designed by Lise Meitner for the physics section of the Kaiser Wilhelm Institute. In 1989, Meitner's name was added to the inscription on the table here at Deutsches Museum but listed her as an assistant ("Mitarbeiter"). However, the inscription was soon changed to give her full credit. Figure 17.5 shows a replica of the worktable in the Deutsches Museum.

17.14.3 Meitner Bust

Hall of Fame at the museum. Designed to honor the "greatest" German scientists and inventors by means of paintings, busts and stone reliefs. Meitner is now honored with a bust there. Copernicus is honored even though he was a Pole. Albert Einstein is also honored, but he would probably view this as a dubious honor.

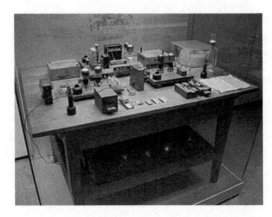

Figure 17.5 The Hahn–Meitner Laboratory Table showing equipment for neutron irradiation and the discovery of nuclear fission experimental setup, reconstructed at the Deutsches Museum, Munich. Reproduced from ref. 4 under the terms of a CC BY-SA 2.0 license [https://creativecommons.org/licenses/by-sa/2.0/deed.en].

17.14.4 Otto Hahn Tombstone

Stadtfriedhof Cemetery ▓▓▓ (51.5325°, 9.909722°) Göttingen. As noted in Chapter 15, in a scientists' section, there are graves of 8 Nobel Prize winners including Max Planck (1918 Physics), Max Born (1954 Physics), Otto Hahn (1944 Chemistry), Max von Laue (Physics 1914), Walther Nernst (1920 Chemistry), Otto Wallach (1910 Chemistry), Adolf Windaus (1928 Chemistry), and Richard Zsigmondy (1925 Chemistry). Otto Hahn's tombstone is unique among these. It simply says Otto Hahn at the top but on the bottom, as shown in Figure 17.6, we see a cryptic $^{92}U + {}^{0}n$ with a double line underneath and then a crooked arrow pointing downward. There's always been a debate as to the meaning of this inscription. Does it mean that he will go to hell for having discovered nuclear fission that led to the atomic bomb? Or perhaps it means that nuclear fission will lead to the end of the world. Horizontal markers in front of the tombstone give his birth and death dates and note he was a professor with a doctor of philosophy (Dr Phil.).

Figure 17.6 A portion of Otto Hahn's gravestone at the Stadtfriedhof in Göttingen. Photo by Glen E. Rodgers.

17.15 WERNER HEISENBERG (1901–1976)

We have already discussed Werner Heisenberg's contribution to quantum mechanics ("matrix mechanics" in 1925 and the "uncertainty principle" in 1927) in Chapter 15. He had been awarded the 1932 Nobel Prize in Physics for his "creation of quantum mechanics". These theories had a direct bearing on how we view the role of electrons in atoms. Now we turn to his contribution to nuclear physics and the German atomic bomb project. He was one of the few top scientists to stay in Germany and work with the Nazis (Max Planck, Hans Geiger, and Otto Hahn were others), but he was constantly subjected to intense scrutiny by the Schutzstaffel (the SS or the "Gestapo") and the Nazi Party. For example, once when he had been lecturing about the theory of relativity, proposed, of course, by the Jewish scientist Albert Einstein, he was attacked by Heinrich Himmler (head of the SS) as a "White Jew" who should be made to "disappear". Such charges were not unusual by the supporters of *Deutsche Physik* or Aryan Physics, who maintained that quantum mechanics and relativity were false theories proposed by Jewish scientists like Albert Einstein and Heisenberg's mentor, Arnold Sommerfeld. In the end Himmler, knowing he needed to keep Heisenberg on as he would be critical for teaching a new generation of German scientists, wrote him a letter minimizing the charges and signing it "Mit freundlichem Gruss und, Heil Hitler!" ("With friendly greetings and, Heil Hitler!") This is sometimes called the "Heisenberg Affair".

Many of his colleagues tried to convince him to emigrate but he always maintained that leaving would be disloyal to his country and to his research students. As the WWII war clouds began to gather, Heisenberg, still based in Leipzig, bought a country home in Urfeld in the Bavarian Alps. When the war began in 1939, German military authorities set up a secret "uranium project" (Uranverein) at the Kaiser Wilhelm Institute for Physics in Berlin (Dahlem). Heisenberg was put in charge of this project. He commuted regularly from Leipzig. He worked in a silo or tower that is still standing today. It was dubbed the "virus house", a name designed to keep away curious visitors. The tower was originally white, but it was thought that the color made it too much of a target of Allied bombers, so they blasted off the white stucco paint down to the red bricks. Heisenberg and his colleagues lived in a nearby underground bunker. By late 1940, things were ready to go. A six-foot deep pit had been built with a brick lining where water could be brought in. Laboratory facilities surrounded the pit. Three atomic pile experiments were carried out here using, among other things, layers of uranium metal powder and paraffin wax. They did not produce a chain reaction.

In Leipzig in September 1941, Heisenberg received a shipment of 40 gallons of heavy water from the Norsk Hydro hydroelectric plant in Norway. The Germans had invaded Norway in 1940 and had quickly established control of this power plant that also produced heavy water. Heavy water, D_2O, containing deuterium, an isotope of hydrogen with one neutron in addition to its proton, was found to be a better moderator than regular water, H_2O. Heisenberg prepared another experiment and this time had evidence that a chain reaction was possible. He now "saw an open road ahead of us, leading to the atomic bomb". At this point, Heisenberg decided he had to talk to his friend and mentor, Niels Bohr.

The discussion with Bohr in 1941 took place at the Carlsberg House of Honor. Bohr had returned to Copenhagen after his trip to New York where he had prematurely revealed the discovery of nuclear fission. Heisenberg was warmly received by Bohr and his wife Margrethe who, most assuredly, were under constant surveillance by the Nazis. The two physicists discussed nuclear fission and its applications but exactly what was said has been subject to much speculation and differences of interpretation. Heisenberg maintained he would be able to keep the German atomic bomb project under control. Bohr was suspicious that Heisenberg was trying to get him to reveal how far the allies had gotten in producing their bomb. Heisenberg gave Bohr a drawing of the heavy-water nuclear reactor that he hoped to build. One of the best representations of this dramatic encounter is the play, "Copenhagen",[5] by Michael Flynn (Winner of the 2000 Tony Award for Best Play). The play has the spirits of Werner Heisenberg, Niels Bohr, and Bohr's wife Margrethe, meeting after their deaths to attempt to answer the question "Why did Heisenberg come to Copenhagen?" Scientific/historical travelers would greatly enjoy reading the play or, better yet, attending a performance of it. In 2002, the play was adapted as a film produced by the BBC and presented on the Public Broadcasting Service (PBS) in the United States.

By August 1943, Dahlem was threatened by ever-increasing allied bombing and the operation was gradually moved to a beer cellar in Hechingen, near the town of Haigerloch, on the edge of the Black Forest. (See Chart 17.3.). Here they built another atomic pile using uranium metal cubes dangled from the aluminum lid of a magnesium cylinder. Figure 17.7 shows a replica of the design. A neutron source was placed in the center and they had the ability to measure the neutron intensity. The experiment resulted in an increase in neutron density, but it was insufficient to produce a sustained chain reaction. The additional uranium and heavy water that was needed was not available. Despite constant air raids, Heisenberg and a colleague travelled, sometimes on bikes, across Germany trying to secure more uranium but time had run out on them. In summary, by the end of the war Germany was on the verge of bringing the first nuclear reactor (at Haigerloch) into its critical state but it was unable to do so.

Heisenberg, like Otto Hahn and other German scientists (10 in all) were interned at Farm Hall, Godmanchester, near Cambridge, England. They were arrested in May and June of 1945 as part of what was known as the Alsos Mission. Heisenberg was cleared and returned to Germany to reestablish the Kaiser Wilhelm Institute for Physics which was renamed the Max Planck Institute for Physics.

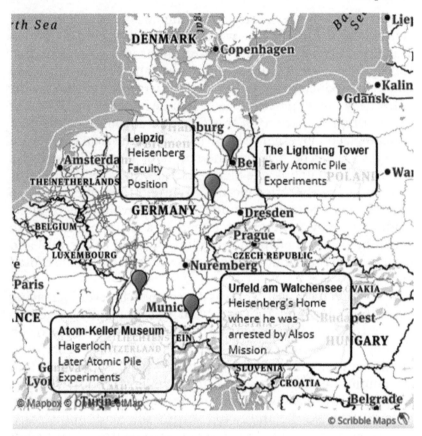

Chart 17.3 Sites involved in the German Nuclear Weapon Project (Berlin-Dahlem, Leipzig, Urfeld, and Haigerloch).

17.16 SCIENTIFIC-HISTORICAL SITES RELATED TO THE GERMAN ATOMIC BOMB PROJECT

See Chart 17.3 for the location of these sites.

17.16.1 The Lightning Tower in Berlin-Dahlem

Berlin-Dahlem ※※ (52.446783°, 13.278150°). This silo-like tower, dubbed the "virus house", is where Heisenberg carried out early experiments related to the development of the German atomic bomb. It was originally white, but the white stucco paint was blasted off to make it a less likely target of allied bomb attacks. A steel square superstructure has been built into the tower to provide for safe storage of materials by the Archives of the Max Planck Society. Consult with the Archives office to visit.

Figure 17.7 A replica of the German nuclear reactor assembly at Haigerloch. Uranium cubes were suspended from the lid of a magnesium vessel and then could be submerged into a bath of heavy water. There was not enough uranium or heavy water to produce a sustained chain reaction. Photograph by G. E. Rodgers.

17.16.2   Vemork Power Plant in Rjukan, Norway

near Rjukan, Norway (59.876910°, 8.588015°). Vemork was a hydroelectric power plant built in 1911 by Norsk Hydro. It was originally designed to electrolyze water into hydrogen gas that in turn would be combined with nitrogen from the air to produce ammonia utilizing the Haber-Bosch process. Later it was the first plant to mass-produce heavy water, D_2O. During World War II, Vemork was the target of multiple allied sabotage operations to prevent the Germans from using the D_2O as a moderator in their atomic bomb project. In 1988, the power station became The Norwegian Industrial Workers Museum and the World Heritage Center. It has an excellent presentation regarding the heavy water production and efforts to sabotage the plant during the war. These sabotage operations were celebrated in the movie "The Heroes of Telemark", starring Richard Harris and Kirk Douglas.

17.16.3   The Atomkeller-Museum at Haigerloch

2–9 Pfluggasse (48.367119°, 8.804038°). This is where Germany, under the direction of Werner Heisenberg, made its final attempt to produce a functioning nuclear reactor. In 1944 their uranium and heavy water were trucked from the KWI Dahlem to Haigerloch. Everything was set up in a beer cellar of a local inn and located directly below a beautiful Schlosskirche (castle church). The final run, the so-called "B-VIII", was carried out in the spring of 1945, but was unsuccessful due to insufficient uranium and heavy water. Soon afterwards, the allied "Alsos Mission" team took over the facility and arrested the nuclear scientists. Heisenberg had continued on to his Bavarian home and was arrested there. The American forces were ordered to blow up the cellar but, fearful that a full-scale explosion would destabilize the church itself, the monsignor in charge of the schlosskirche rushed down and reportedly threw himself in front of the gate. Accordingly, a more limited demolition of the facilities was carried out. Today, the cave is intact and there is a full-size replica of the reactor and even two of the original uranium cubes on display. There are numerous exhibits showing the scientists who worked there and the Alsos soldiers who took it over in 1945. During his time there in Haigerloch, Heisenberg would go up to the church on top of the rock above the cave and play Bach fugues in the organ-loft. He probably used this activity to pause and consider what was happening. Travelers should also take the occasion to do just that.

17.16.4 Heisenberg Grave in Munich

Waldfriedhof, 288 Fürstenrieder Strasse, Munich (48.10278°, 11.49361°). This forest cemetery is on the western outskirts of Munich. He is buried with his parents, August and Annie, and his wife Elizabeth. Memorial ID 61842518.

17.17 SUMMARY

This chapter starts with the great Italian physicist, Enrico Fermi. Soon after the tragic death of his brother Giulio, Enrico, only 14, started to teach himself physics by reading the greatest textbooks available. He received a full scholarship to study in Pisa and soon became a full professor at the University of Rome. His wife, Laura, helped him write a physics textbook, recorded his many accomplishments and explained nuclear physics to the common person. Enrico assembled his Panisperna "boys" and started to bombard all the known elements with neutrons. He discovered that slow neutrons are more effective at producing radioactive isotopes and just missed the discovery of fission. We visit great sites in Rome, Pisa, Florence, and even Chicago in the United States that celebrate the great Enrico Fermi.

Lise Meitner, the German Marie Curie, overcame the extreme gender discrimination of her day and became one of the best-known radiophysicists. For 31 years, this shy and reserved Austrian collaborated with the outgoing German chemist, Otto Hahn, at the Kaiser Wilhelm Institute of Chemistry in Berlin-Dahlem. A non-Aryan Jew, she was tragically exiled to Denmark in 1938 but, with her favorite nephew, Otto Frisch, soon explained that Hahn and Fritz Strassman had split the atomic nucleus, a process that Frisch called fission. Predictions that self-sustained nuclear chain reactions could produce nuclear power and weapons soon followed. At the onset of WWII, Werner Heisenberg stayed in Nazi Germany and headed German efforts to produce an atomic bomb. For Hahn and Meitner, we explore sites in Vienna, Berlin, and Munich, as well as Kungälv, Sweden and Bramley, England. For Heisenberg, we also explore Rjukan, Norway and Haigerloch, Germany.

ADDITIONAL READING

- G. Segrè and B. Hoerlin, *The Pope of Physics, Enrico Fermi and the Birth of the Atomic Age*, Henry Holt and Company, New York, 2016.
- E. Segrè, *Enrico Fermi Physicist*, The University of Chicago Press, Chicago, 1970.
- R. L. Sime, *Lise Meitner, A Life in Physics*, University of California Press, Berkeley and Los Angeles, 1996.

REFERENCES

1. L. Fermi, *Atoms in the Family, My Life with Enrico Fermi*, The University of Chicago Press, Chicago, 1954.
2. https://catalog.archives.gov/id/558578, accessed June 2019.
3. Lise Meitner Denkmal Unter den Linden Berlin,https://commons.wikimedia.org/wiki/File:Lise_Meitner_Denkmal_Unter_den_Linden_Berlin_(3).JPG, accessed June 2019.

4. Nuclear Fission Experimental Apparatus 1938-Deutsches Museum – Munich, https://commons.wikimedia.org/wiki/File:Nuclear_Fission_Experimental_Apparatus_1938_-_Deutsches_Museum_-_Munich.jpg, accessed June 2019.
5. M. Frayn, *Copenhagen*, Anchor Books, New York, 2000.

CHAPTER 18

Mendeleev's and Our Path to the Periodic Table: Mendeleev, Meyer, and Winkler (Russia and Germany)

18.1 A QUICK LOOK AT PLACES TO VISIT "TRAVELING WITH THE ATOM" RELATIVE TO THE DEVELOPMENT OF THE PERIODIC TABLE

Bug, Weißdorf, Germany: Johann Wolfgang Döbereiner Plaque at childhood home ❋; *Jena, Germany: Döbereiner Plaque at residence and workplace* ❋, *Döbereiner Hörsaal, Painting and Statue* ❋❋, *and Döbereiner grave* ❋; *London: John Newlands Royal Society of Chemistry Plaque* ❋ *and Newlands Grave* ❋; *Cognac, France: Paul Émile Lecoq de Boisbaudran Home* ❋ *and Rue Lecoq de Boisbaudran* ❋; *Freiberg, Germany: Clemens Winkler Memorial* ❋❋❋, *Clemens Winkler Collection* ❋❋❋, *Winkler birthplace (Geburtshaus)* ❋, *Winkler Monument* ❋❋, *Terra Mineralia* ❋❋, *Hieronymus Theodor Richter and Ferdinand Reich Monument* ❋❋; *Varel Germany: Lothar Meyer Birthplace Plaque* ❋; *Tübingen, Germany: Lothar-Meyer Bau* ❋; *Tobolsk, Siberia: Monument to Dmitri Mendeleev* ❋❋, *Mendeleev Exhibit at Museum of the Kremlin of Tobolsk* ❋❋, *Mendeleev Mansion* ❋; *St. Petersburg,*

Traveling with the Atom: A Scientific Guide to Europe and Beyond
By Glen E. Rodgers
© Glen E. Rodgers 2020
Published by the Royal Society of Chemistry, www.rsc.org

Russia: Dmitri Mendeleev's Memorial Museum Apartment (D. I. Mendeleev Museum and Archives) ●●●●●, *Rosstandart Metrology Museum at the D.I. Mendeleyev Institute of Metrology* ●●●, *Mendeleev Statue and Three-Story High Periodic Table* ●●●, *Technologichesky Institut (Technology Institute) Metro Stop* ●●; *Monument to Mikhail Lomonosov* ●●, *The Lomonosov Collection in the Kunstkamera* ●●●, *Lomonosov Grave* ●; *Lomonosovo (formerly Oranienbaum), Russia* ●.

The Path to the Periodic Table

- In 1829, Johann Wolfgang Döbereiner proposed the "Law of Triads", sets of three elements with regularly increasing atomic weights and similar properties. Leopold Gmelin added more of these in 1843.
- After the Chemical Congress of 1860 established a commonly accepted set of atomic weights, Alexandre-Émile Béguyer de Chancourtois ordered elements by increasing atomic weight and pointed out that chemically similar elements occurred at regular intervals.
- In 1863, John Newlands proposed his "Law of Octaves", in which elements were ordered by increasing atomic weights and every eighth element showed similar properties. William Odling, who had also proposed a credible periodic system, chaired the meeting where Newlands presented his work and was ridiculed.
- In 1869, Dmitri Mendeleev presented his periodic table of the elements that left room for yet-to-be-discovered elements, for which he predicted sets of properties.
- Three of these elements, eka-aluminum (gallium, discovered in 1875 by Paul Émile Lecoq de Boisbaudran), eka-boron (scandium, discovered in 1879 by Lars Fredrik Nilson), and eka-silicon (germanium, discovered in 1886 by Clemens Winkler) had properties very close to those predicted by Mendeleev.
- Julius Lothar Meyer produced a table very similar to Mendeleev's but publishing delays and Mendeleev's astonishingly bold and accurate predictions established him as the primary creator of the periodic table.
- In the 1890s, the "inert gases" were discovered by William Ramsay and Lord Rayleigh and were soon assimilated as an additional group of the table.

18.2 INTRODUCING DMITRI IVANOVICH MENDELEEV

"[Dmitri Mendeleev]'s name is invariably and justifiably connected with the periodic system, to the same extent ... as Darwin's name is synonymous with the theory of evolution and Einstein's with the theory of relativity." Eric Scerri[1]

The story of Dmitri Mendeleev (1834–1907) and how he came to put together his periodic table of the elements again emphasizes how humble beginnings sometime produce outstanding and unexpected contributions. Mendeleev was born in Tobolsk, in western Siberia, the youngest of 17 children. His father went blind soon after Dmitri's birth and died when his son was only 13. His mother, Mariya, despite the number of children under her care, had become the director of her family's glass factory but, unfortunately, it burned down shortly after her husband died. Together Dmitri and his mother and sister, Elizaveta, left Siberia and set out for Moscow – over the Urals by a horse-drawn four-wheeled cart – and ultimately on to St. Petersburg (about 4000 miles altogether) where Mendeleev, with help from of a friend of his late father, was admitted to the Pedagogical Institute of St. Petersburg. His mother and sister died shortly afterwards.

Mendeleev was not an outstanding high school student but, despite a strong personality and an uncontrollable temper that offended some of his teachers and fellow students, he graduated at the top of his college class. He studied (1859–1861) with Bunsen in Heidelberg (see Chapter 11, where we visited the young Russian's residence) during the time that Bunsen and Kirchhoff discovered the element cesium using spectroscopy. During that time also, still only 26, he was invited to the Karlsruhe Congress of 1860 (Chapter 9) that was convened for the express purpose of resolving the roles of atoms and molecules and determining a system of atomic weights. He was impressed by the Karlsruhe Congress that brought chemists of all stripes together to calmly resolve their disputes. He even wrote a letter to a St. Petersburg newspaper describing the congress as a remarkable event. He returned home soon afterwards and in 1866, after assuming a variety of positions, joined the University of St. Petersburg as

Professor of Chemistry. Reports about his ability as a lecturer varied but, in 1869, he wrote an excellent chemistry textbook (*The Principles of Chemistry*). For the second edition of his textbook, he wanted to present a useful table of the known elements. To devise his table, he is said to have prepared a set of "element cards" (63 then known) and reported that he shuffled them about on the lectern in his office. (He maintained that he would sometimes do this for hours until he fell asleep at his desk.) His final arrangement came to him as he faced a final deadline from his publisher.

18.3 EARLY PROPOSALS LEADING TO MENDELEEV'S PERIODIC TABLE

Before we examine Mendeleev's tables in detail and then explore places that celebrate his life and work, let's back up just a bit and describe some of the early proposals for periodicity that Mendeleev built upon. Eric Scerri, in his comprehensive book, *The Periodic Table: Its Story and Its Significance*,[1] provides an analysis of these early proposals and how Mendeleev's work was the definitive extension of them. In 1829, when about 50 elements were known, Johann Wolfgang Döbereiner (1780–1849) proposed what he called his "analogies" that later became known as his Law of Triads. He pointed out four "triads" of elements as shown in Table 18.1.

In each case, the then-known atomic weight of the middle element was very close to the average of the top and bottom elements. For example, using modern values, the atomic weight of lithium (Li) is 6.941 and that of potassium (K) is 39.10. The average of these is 23.02 whereas the atomic weight of sodium (Na) is 22.99. In addition, the members of each triad showed chemical similarities, for example Ca, Sr, and Ba form similar oxides and Li, Na, and K form similar chlorides. Leopold Gmelin (1788–1853), in 1843, organized the 55

Table 18.1 Four triads of elements listed by Johann Wolfgang Döbereiner.

Triad 1	Triad 2	Triad 3	Triad 4
Li (lithium)	Ca (calcium)	S (sulfur)	Cl (chlorine)
Na (sodium)	Sr (strontium)	Se (selenium)	Br (bromine)
K (potassium)	Ba (barium)	Te (tellurium)	I (iodine)

known elements then known, actually coined the term "triad", and added several more including Mg, Ca, and Ba, and also P, As, and Sb.

The Karlsruhe Congress of 1860, as we have seen in Chapter 9, finally produced a reliable set of atomic weights. It soon became clear that chemists could now proceed to use these weights to organize the 60 elements that were then known. Alexandre-Emile Beguyer de Chancourtois (1820–1886), a geologist, in 1862, was the first to order elements by increasing atomic weight and point out that chemically similar elements occurred at regular intervals of these weights. That quickly led to the work of John Newlands (1838–1898), the son of a Scottish Presbyterian minister, who became the chief chemist at a sugar refinery. As early as 1863, he too ordered elements by increasing atomic weights and found that the properties seemed to repeat themselves in each group of seven elements. The eighth element often had properties akin to the first. Somewhat unfortunately for him, he likened this to the octaves of music. Figure 18.1 shows his table of the elements demonstrating his "Law of Octaves". See if you can spot the various triads (shown horizontally) in his table. During this time, he predicted the existence of two new elements, but neither ever materialized. Also, note that he reversed the positions of several pairs of elements including, most importantly, iodine (I) and tellurium (Te). He did this recognizing that iodine should be in the same group (a row in his table) as chlorine (Cl) and bromine (Br). So, we can see that John Newlands had produced a viable early version of a periodic system.

No.		No.		No.		No.		No.		No.		No.		No.	
H	1	F	8	Cl	15	Co & Ni	22	Br	29	Pd	36	I	42	Pt & Ir	50
Li	2	Na	9	K	16	Cu	23	Rb	30	Ag	37	Cs	44	Os	51
G	3	Mg	10	Ca	17	Zn	24	Sr	31	Cd	38	Ba & V	45	Hg	52
Bo	4	Al	11	Cr	19	Y	25	Ce & La	33	U	40	Ta	46	Tl	53
C	5	Si	12	Ti	18	In	26	Zr	32	Sn	39	W	47	Pb	54
N	6	P	13	Mn	20	As	27	Di & Mo	34	Sb	41	Nb	48	Bi	55
O	7	S	14	Fe	21	Se	28	Ro & Ru	35	Te	43	Au	49	Th	56

Figure 18.1 John Newland's Table of Elements. The numbers are not the atomic weights but rather just give the order of the elements from the lightest (1) to the heaviest (56). The 8th element (F) in the list has similar properties to the 1st element (H) and fifteenth (Cl). Element 3, represented as G, is now called beryllium. G stood for glucinium, from the Greek *glykys* for sweet, due to the sweet taste of its salts, which are toxic to ingest. The symbol of boron is shown as Bo in this table, but now is simply B.

In March of 1865, when he presented his ideas at a meeting of the Chemical Society (a forerunner of the Royal Society of Chemistry) Newlands was greeted with ridicule. For example, someone caustically remarked that he might have produced a superior classification system if he had just ordered the elements *alphabetically*. Others wondered if the elements thus arranged might play a little tune! (Today, we might suggest "Do, a deer, a female deer ....") As Scerri points out, there is some speculation that this derision might have been at least partially motivated by a rivalry between Newlands and William Odling, the chairman of the meeting. Odling, who had succeeded Michael Faraday as the director of the Royal Institution and, like Mendeleev, was invited to the Karlsruhe Congress, had published articles that show he had also designed a credible periodic system of the elements.

See Charts 18.1 and 18.2 for the location of a variety of sites related to the early development of the periodic table.

Chart 18.1 Travel sites related to the development of the periodic table: Jena, Germany (Döbereiner); England (Newlands, Ramsay, and Lord Rayleigh); Cognac, France (de Boisbaudran); Uppsala (Nilson and Cleve); Freiburg, Germany (Winkler); Varel, Germany (Meyer); St. Petersburg (Mendeleev).

Chart 18.2 Travel sites related to Johann Wolfgang Döbereiner in Jena, Germany.

18.4 TRAVEL SITES RELATED TO JOHANN WOLFGANG DÖBEREINER

See *Rediscovery of the Elements: Johann Wolfgang Döbereiner*,[2] by Jim and Virginia Marshall for further information on the life and background of Döbereiner.

18.4.1 Bug, Weißdorf, Germany

18.4.1.1 Döbereiner Plaque at Childhood Home. 17 Bugbergstrasse (50.196141°, 11.847785°). Plaque says "To the memory of the pioneer of experimental chemistry Johann Wolfgang Döbereiner, Professor of Chemistry in Jena. He spent his childhood and youth here as the son of the manor superintendent."

18.4.2 Jena, Germany

18.4.2.1 Döbereiner Plaque at Residence and Workplace. 23 Neugasse (50.924632°, 11.584071°). University of Jena (Friedrich-Schiller-Jena Universität), Germany. The plaque is on the wall of Hellfeld Haus, now a microbiology building, and notes that Döbereiner lived and worked here from 1816 to 1849.

18.4.2.2 Döbereiner Hörsaal, Painting and Statue. 3 Am Steiger (50.932153°, 11.579610°). This building was named after Döbereiner in 1974. The statue, created in 1958, stands in front of the hall. The painting is in the lecture hall, which seats 600 people. There is also a poster exhibit on Döbereiner's life and work.

18.4.2.3 Döbereiner's Grave. Johannisfriedhof (Johannis cemetery), 1 Philosophenweg (50.930714°, 11.582693°). The inscription on the gravestone is difficult to read, but notes that he was "Adviser to Goethe", the "Creater of the Law of Triads", and the "Discoverer of the platinum catalyst". The latter refers to his 1823 discovery that when a stream of hydrogen gas is directed at finely divided platinum powder, it bursts into flame. This became known as the "Döbereiner lighter" which was used for many decades throughout Western Europe until it was replaced by modern safety matches. The platinum was not consumed so this was an early example of a catalytic reaction. The famous poet, author of *Faust*, Johann Wolfgang Goethe, recommended Döbereiner for his position at Jena.

18.5 TRAVEL SITES RELATED TO JOHN NEWLANDS

18.5.1 Newlands Royal Society of Chemistry Plaque

West Square, Lambeth, Surrey, London (51.494944°, −0.106128°) In 1887, the Royal Society awarded Newlands a Davy medal for the same paper that 25 years prior he could not get published. Newlands accepted the medal with grace. In 1998 (100 years after his death), the Royal Society erected a plaque on the side of the house where he was born and raised. The plaque celebrates him as "discoverer of the Periodic Law for the chemical elements" – a far cry from the elements playing a little tune.

18.5.2 Newlands Grave

⬢ West Norwood Cemetery, (51.433080°, −0.097914°). Listed as one of the magnificent cemeteries of London, England.

18.6 MENDELEEV'S PERIODIC TABLE AND HIS FAMOUS PREDICTIONS

This brings us back to Mendeleev. Starting in 1869, Mendeleev, pictured at his writing table in Figure 18.2, published as many as 30 versions of his periodic table. One of the clearer ones, published in 1872, is shown in Figure 18.3. This table does show atomic weights. Elements in the same group ("Gruppe") have similar valences and chemical properties. Valence, first defined by Edward Frankland in 1852, is defined as the capacity for combining with other elements. August Kekulé, one of the organizers of the Karlsruhe Congress also had a hand in clarifying this concept. (You may recall that in Chapter 9 we had the occasion to see a plaque at the site of Kekulé's laboratory in Heidelberg

Figure 18.2 Dmitri Mendeleev standing at his working desk.

Reihen	Gruppe I. R^2O	Gruppe II. RO	Gruppe III. R^2O^3	Gruppe IV. RH^4 RO^2	Gruppe V. RH^3 R^2O^5	Gruppe VI. RH^2 RO^3	Gruppe VII. RH R^2O^7	Gruppe VIII. RO^4
1	H = 1							
2	Li = 7	Be = 9.4	B = 11	C = 12	N = 14	O = 16	F = 19	
3	Na = 23	Mg = 24	Al = 27.3	Si = 28	P = 31	S = 32	Cl = 35.5	Fe = 56, Co = 59,
4	K = 39	Ca = 40	— = 44	Ti = 48	V = 51	Cr = 52	Mn = 55	Ni = 59, Cu = 63,
5	(Cu = 63)	Zn = 65	— = 68	— = 72	As = 75	Se = 78	Br = 80	
6	Rb = 85	Sr = 87	?Yt = 88	Zr = 90	Nb = 94	Mo = 96	— = 100	Ru = 104, Rh = 104, Pd = 106, Ag = 108,
7	(Ag = 108)	Cd = 112	In = 113	Sn = 118	Sb = 122	Te = 125	J = 127	
8	Cs = 133	Ba = 137	?Di = 138	?Ce = 140	—	—	—	— — — —
9	(—)							
10	—	—	?Er = 178	?La = 180	Ta = 182	W = 184	—	Os = 195, Ir = 197, Pt = 198, Au = 199,
11	(Au = 199)	Hg = 200	Ti = 204	Pb = 207	Bi = 208	—	—	
12	—	—	—	Th = 231	—	U = 240	—	— — — —

Figure 18.3 A periodic table published in 1872 by Dmitri Mendeleev. The shaded elements are his eka-boron, eka-aluminum and eka-silicon. Reproduced from ref. 3 with permission from Cengage Learning, Copyright 2012.

that celebrated his proposal that carbon had a valence of four.) The valences start at one for Gruppe I, increase to four at Gruppe IV and then decrease stepwise down to one again at Gruppe VII. That is, the valences of the groups vary periodically, as do their chemical properties. Mendeleev, unlike Newlands, allowed for additional elements in his Gruppe VIII that includes many that we now list as "transition metals". Many travelers will know or recall that the inert or noble gases had not yet been discovered. (As we note later in this chapter, they had a valence of zero.) Most importantly, in his *coup d'etat*, Mendeleev left room for yet undiscovered elements. The most important of these, shaded in the table, were directly under boron (B), aluminum (Al) and silicon (Si). He called these eka-boron, eka-aluminum and eka-silicon where *eka* is the Sanskrit word for "beyond". (Mendeleev actually predicted the existence of many undiscovered elements. Only about half of these materialized.) Mendeleev was confident about the significance of his periodic table. He had some 200 copies printed and sent to chemists throughout Russia and Europe. Even so, his periodic table, like Newlands' without the musical metaphor, was also greeted skeptically. However, Mendeleev, maintaining the strong personality he showed as a youth, was dogged in defending it, constantly improving and refining it for more than 30 years.

What makes Mendeleev's accomplishments so revolutionary were (1) his detailed predictions of the properties of his famous three eka-elements, (2) the subsequent discovery of them within the next 15 years and (3) the amazing agreement between his

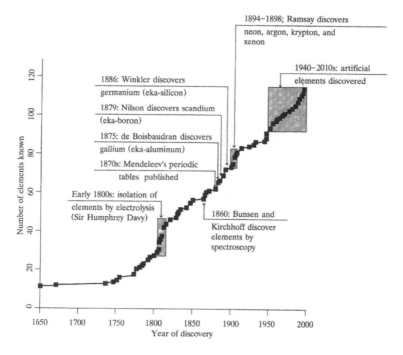

Figure 18.4 A plot of the number of known elements over time. Mendeleev's three most famous eka-elements were discovered by de Boisbaudran (eka-aluminum, gallium), Nilson (eka-boron, scandium) and Winkler (eka-silicon, germanium) in 1875, 1879, and 1886, respectively. Reproduced from ref. 3 with permission from Cengage Learning, Copyright 2012.

predictions and the actual properties. Figure 18.4 shows a plot of the number of elements known *versus* time and the timing for these three famous discoveries.

Travelers-with-the-atom might be curious as to how Mendeleev made his famous predictions. Such considerations might be discussed at a coffee shop near one of the sites recommended below. Look at Figure 18.3, consider the known elements above and below and to the right and the left of the unknown elements, pour another cup, and discuss the possibilities.

18.7 EKA-ALUMINUM AND GALLIUM (PAUL ÉMILE LECOQ DE BOISBAUDRAN)

Paul Émile Lecoq de Boisbaudran (1838–1912) was born in Cognac, France where his family had a wine business. The town gives its name to one of the world's best-known types of brandy. He was independently wealthy and, excited by Bunsen and Kirchhoff's new discovery of spectroscopy, built his own laboratory and began searching for new elements. In 1875, he spectroscopically identified one in a sample of zinc ores. Starting with hundreds of kilograms of ore, he was able to isolate enough of this easily liquefied metal to measure its properties including its atomic weight and even present a tiny amount to the French Academy of Sciences. De Boisbaudran named the new element after *Gallus*, the Latin name for the territory that included what we know as modern France. Of course, *Lecoq* means "rooster" and the Latin equivalent of rooster is *gallus*, so perhaps there was a dual reason that de Boisbaudran chose this name. De Boisbaudran also discovered three additional elements: samarium (1880), dysprosium (1886) and europium (1890).

It was a surprise to de Boisbaudran to learn that Mendeleev was convinced that gallium was the Russian's eka-aluminum. At first, de Boisbaudran even claimed that he'd never heard of Mendeleev's system. Mendeleev, of course, already had his predictions in hand for his eka-aluminum and, as shown in Table 18.2, they were an excellent match for those recorded by de Boisbaudran for his gallium. Mendeleev was aggressive and protective when it came to his predicted elements. For example, de Boisbaudran originally determined the density of the metal to be

Table 18.2 A comparison of the properties of eka-aluminum predicted by Mendeleev with the properties of gallium found by Boisbaudran. Reproduced from ref. 3 with permission from Cengage Learning, Copyright 2012.

	Mendeleev's predictions for eka-aluminum (1871)	De Boisbaudran's observed properties of gallium (1875)
Atomic weight	≈68	69.9
Valence	3	3
Density, g cm^{-1}	5.9	5.93
Melting point	Low	30.1 °C
Volatility	Low	Low
Formula of oxide	M$_2$O$_3$	Ga$_2$O$_3$
Formula of chloride	MCl$_3$	GaCl$_3$
Mode of discovery	Spectroscopy	Spectroscopy

4.7 g cm^{-3}, but after Mendeleev wrote to him to request that he measure the value again (!), the Frenchman then found it to be approximately as given in the table.

18.8 TRAVEL SITES RELATED TO PAUL ÉMILE LECOQ DE BOISBAUDRAN

18.8.1 Boisbaudran Home

- 1 rue de Lasignan, Cognac France (45.694837°, −0.330915°).

18.8.2 Rue Lecoq de Boisbaudran

- Cognac France (45.700543°, −0.336781°).

18.9 EKA-BORON AND SCANDIUM (LARS FREDRIK NILSON AND PER TEODOR CLÈVE)

In 1879 at Uppsala University, Swedish chemist Lars Fredrik Nilson (1840–1899) isolated a new metal from a mineral ore called euxenite. Per Teodor Clève (1840–1905) identified the element as Mendeleev's eka-boron. They named the element scandium after their native Scandinavia. Again, the properties of the scandium were very close to those predicted by Mendeleev for eka-boron. Clève also discovered the elements thulium and holmium. Nilson was born in Skönberga parish in Östergötland (or East Gothland) in southern Sweden. He died in Stockholm. (The famous Viking warrior Beowulf may likely have been from what is now the Östergötland region.).

18.10 EKA-SILICON AND GERMANIUM (CLEMENS WINKLER)

In 1886, Clemens Winkler (1838–1904), a professor of chemical technology and analytical chemistry at the Freiburg University of Mining and Technology, was given a new mineral called argyrodite. It came from the Himmelsfürst mine near the town. When he analyzed the mineral about seven percent of the mass could not be accounted for. After several months of hard work, he was able to isolate the pure element and called it germanium, after his fatherland. Winkler did not identify his element with eka-silicon but rather with eka-stibium, an element that Mendeleev had predicted to lie between antimony and bismuth (see if you can find this in Gruppe V in Table 18.2.) Mendeleev himself thought Winkler's new element to be eka-cadmium, which he had placed between cadmium and mercury (see Gruppe II in Table 18.2). Others (Victor von Richter) correctly identified it as eka-silicon. A comparison of Mendeleev's predictions and the properties of germanium quickly confirmed that identification.

Eka-stibium and eka-cadmium are two of Mendeleev's *not-so-famous blanks*, which we seldom hear about anymore. See Scerri's book for much more on Mendeleev's unsuccessful predictions. He lists 16 major predictions, only eight of which were actually confirmed. Despite this mixed record of correct predictions, Mendeleev was now famous, and his periodic table was well established.

18.11 TRAVEL SITES RELATED TO CLEMENS WINKLER IN FREIBERG, THE SILVER CITY OF SAXONY

See Chart 18.3 for the location of these sites in Freiberg.

18.11.1 The Clemens Winkler Memorial (Gedenkstätte)

5 Brennhausgasse (50.920567°, 13.342240°). This is the former chemical laboratory and residential building of Prof. Winkler. It has exhibits related to Winkler's production of sulfuric acid, analysis of argyrodit (Ag_8GeS_6), isolation of germanium, and equipment for analyzing methane and other gases found in local mines. There are also a few personal items of Winkler's including his pocket watch and a letter opener. The laboratory furniture was installed post-Winkler. A picture of Mendeleev and Winkler when the former visited Freiberg is on display. As they say in this mining town, Gluck Auf (Good Luck!)

18.11.2 Clemens Winkler Collection

Clemens Winkler Bau, Leiziger Strasse 29 (50.925052°, 13.333142°). This is a collection of 1500 inorganic solid and liquid compounds synthesized and collected by Winkler. Many are sealed in glass ampoules and labelled in Winkler's own handwriting. These include the original argyrodite, the first germanium samples, and several germanium compounds synthesized

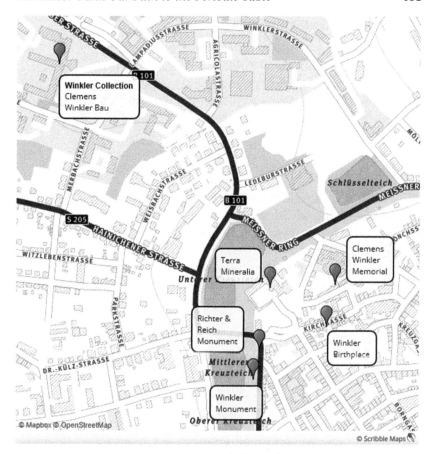

Chart 18.3 Travel sites in Freiberg related to Clemens Winkler.

by Winkler at Mendeleev's request. There is a small piece of gallium sent to Winkler by Lecoq de Boisbaudran in 1886. This collection is used by the inorganic chemistry professors of the TU Bergakademie Freiberg in their courses.

18.11.3 Winkler Birthplace (Geburtshaus)

8 Kirchgasse (50.919744°, 13.341948°) now a private residence. A sign on the side of the building tells the visitor about his discovery of germanium, his invention of the contact process for the production of sulphuric acid, and his work in the analysis of gases.

18.11.4 Winkler Monument

Albert Park, Wallstraße (50.918671°, 13.339516°). The monument to Clemens Winkler is probably the most obvious monument in Albert Park due to its two striking gold leaf reliefs. One, shown below in Figure 18.5, shows the young Winkler being instructed by a Greek muse. Some of his equipment is shown at his feet.

18.11.5 Terra Mineralia

Schloss Freudenstein Schlossplatz 4 (50.920508°, 13.340084°). One of the largest, most beautiful and most modern mineral museums. The memorial plaque in front of the entrance refers to the discovery of germanium and indium, gives their atomic numbers, their discoverers, and dates of discovery. The circles represent the atoms of these elements. This plaque is often removed during the winter months.

18.11.6 Hieronymus Theodor Richter and Ferdinand Reich Monument

Albert Park, Wallstraße (50.919260°, 13.339773°) Richter and Reich discovered indium spectroscopically in 1863. The monument was inaugurated in 2011 at Albertpark.

Figure 18.5 An etching of the young Clemens Winkler being instructed by Calliope, the Greek Muse of Eloquence and Epic Poetry. Winkler's equipment is shown at his feet. Image courtesy of Dr Norman Pohl/TU Bergakademie Freiberg.

18.12 JULIUS LOTHAR MEYER (1830–1895)

Julius Lothar Meyer, a German chemist, had much in common with Mendeleev. He too studied with Bunsen and Kirchhoff in Heidelberg, although Meyer was there five years earlier. Both attended the select Karlsruhe Congress (Chapter 9) and were impressed by Cannizzaro's arguments that clarified atomic weights. Recall that it was Meyer who left the Congress saying "[T]he scales fell from my eyes and my doubts disappeared and were replaced by a feeling of quiet certainty". Meyer also wrote an extremely well-regarded chemistry text and, as early as 1864, produced a periodic table that was organized by increasing atomic weights and recognized the role of valence. (He also found a periodicity in atomic volumes.) He too left a few gaps that he assigned to unknown elements. In short, he was several years ahead of Mendeleev. So what happened? Why is Mendeleev recognized as the primary progenitor of the periodic table? One part of the answer seems to be that Meyer's most compelling table, which he constructed in 1868, a year before Mendeleev's first table, was not published until 1895, some 27 years later! Scerri calls this an "almost comical time delay", and was, in part, the reason for Mendeleev's early and continuing fame at Meyer's expense. Another factor was that Mendeleev was, as we have just seen, much bolder in forcefully predicting the existence of undiscovered elements and listing their properties in striking detail. He constantly monitored and often corrected atomic weights in a timely manner that kept his table current for many years.

18.13 TRAVEL SITES RELATED TO JULIUS LOTHAR MEYER

18.13.1 Meyer Birthplace Plaque

10 Obernstrasse, Varel Germany (now a delicatessen and bakery) (53.395299°, 8.140220°).

18.13.2 Lothar-Meyer Bau

56 Wilhelmstrasse Tübingen, Germany (48.527918°, 9.063668°). Meyers' final university site (now a geology building).

18.14 MENDELEEV: THE LATER YEARS

By 1881, Meyer and Mendeleev found themselves locked in a bitter dispute about priority. In the end, most agree that Mendeleev's system was the more complete and that his bold, detailed and spectacular predictions carried the day and established him as the one who should receive the most credit for the establishment of the periodic table. As a result, Mendeleev quickly became a most celebrated chemist. Nevertheless, in 1882, the Royal Society awarded the gold Davy Medal to him *and* Lothar Meyer.

As we have seen, Mendeleev was strong-headed and now famous. In addition to pursuing his studies of periodic properties, he pursued, throughout his career, the study of a variety of other topics including gases at both high and low pressures as part of his goal to discover the "luminiferous (*i.e.*, light-bearing) aether", the medium in outer space that was proposed as necessary for the transmission of electromagnetic radiation. He also pursued economics, agricultural projects, and coal and oil production studies. As part of latter, he visited the oil fields of western Pennsylvania in 1876. He even designed an icebreaker, the *Yermak*, for use on Russia's frozen seas.

One other issue made life difficult for Dmitri. He had married early (in 1862) and had two children. However, by 1876, now in his mid-forties, he fell in love with 19 year-old Anna Ivanova Popova. He proposed to her in 1881, and threatened suicide if she refused. They were married a month before his first marriage was finalized. The Russian Orthodox Church required seven years before a lawful second marriage could take place, but Mendeleev would not hear of a delay. His violation of the church rule produced a public uproar and, evidently, was a contributing factor to his failure to be elected to the Russian Academy of Science! His marriage to Popova produced four children.

In 1890, he resigned from St. Petersburg University due to a dispute involving oppression of students. Born and raised in Siberia with all those outcasts from Russian society, he was decidedly liberal and often spoke out against the Russian government. Never a particularly good idea, then or now! This also may have contributed to his failure to be elected to the academy. In 1892 he became head of the Archive of Weights and Measures in

Saint Petersburg, and that evolved into a government bureau the following year. He was appointed director of the Central Bureau of Weights and Measures. In this position, Mendeleev is given credit for the introduction of the metric system to the Russian Empire. During that time, he purchased a variety of precision instruments and established a journal of metrology.

18.15 MENDELEEV AND THE DISCOVERY OF THE INERT OR NOBLE GASES – WHERE DO WE PUT THEM IN THE PERIODIC TABLE?

In 1894, William Ramsay (1852–1916) and Lord Rayleigh (John William Strutt) (1842–1919) announced their discovery of a new gaseous constituent of the atmosphere. In 1895, they proposed that it was a new element and named it argon (fr Greek *argos*, for "idle"). Lecoq de Boisbaudran proposed that argon was a member of a new, previously unsuspected, chemical series of elements, later to become known as the noble gases. In 1895, Ramsay (with Frederick Soddy) isolated helium from a uranium mineral, providing evidence that alpha particles are helium nuclei (See Chapter 6). Helium was another noble gas. In 1898, Ramsay and his student Morris Travers announced their discoveries of krypton, neon, and xenon from liquid air. See Chapter 7, p. 193 and p. 181, for a list of places to visit related to Ramsay and Rayleigh. Mendeleev, who was only about 60 at the time, had reservations about these new discoveries but soon recognized them as an additional group in his periodic table that had a valence of zero.

In 1906, just before his death, Mendeleev missed being awarded a Nobel Prize, losing out (by one vote) to Henri Moissan, who had discovered fluorine. See Chapter 19 for more on the politics of this decision. In 1955, element 101 was named Mendeleevium (Md) in his honor. When Mendeleev died in 1907, students carried periodic tables in his funeral procession. To put things in perspective, it was only 63 years before, in 1844, when Mendeleev was a ten-year-old living in Siberia, that similar celebrations had followed the death of John Dalton in Manchester, England. Rather amazingly, only one lifetime separated the first concrete atomic theory from the establishment of the periodic table.

18.16 TRAVEL SITES RELATED TO DMITRI MENDELEEV IN TOBOLSK, SIBERIA

See Chart 18.4 for the location of these sites.

18.16.1 Monument to Mendeleev

🖼️🖼️ Pamyatnik Di Mendeleyevu, corner of Komsomol'skiy Prospekt and Mendeleyeva Prospekt (58.227356°, 68.275966°). This huge monument was unveiled in 1984. Jim and Virginia Marshall's "Rediscovery of the Elements, The Periodic Table"[4] article notes that in Tobolsk there are other historical sites including the church that Mendeleev's family attended, the location of the old Mendeleev home (now gone), and the burial sites of Mendeleev's family. Mendeleev was born and lived as a youth in the village of Verkhnie Aremzyani, near Tobolsk (58.31183°, 68.59083°)

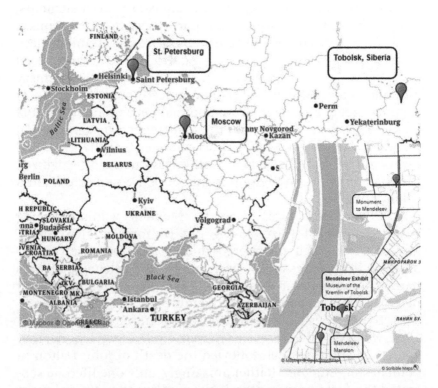

Chart 18.4 Mendeleev sites in Tobolsk, Siberia (with a wider view showing the relative locations of Tobolsk, Moscow and St. Petersburg).

18.16.2 Mendeleev Exhibit

Museum of the Kremlin of Tobolsk, *Tobol'skiy Kreml'* (58.199003°, 68.252724°) overlooking the old town. Designated a national and architectural treasure in 1870, the Tobolsk Kremlin is an elaborate fortress with white walls and towers built high above the Irtysh and Tobol Rivers. Inside the walls is the St. Sophia Cathedral with its gold and blue onion domes. The Mendeleev exhibit includes colored glass items manufactured in the glass factory that Mendeleev's mother managed before it burned in 1848.

18.16.3 Mendeleev Mansion

9 Ulitsa Mira (58.192954°, 68.243609°) (in the old town) where the Mendeleev family once lived. Next door is the Tobolsk Raion Administration Building (10 Ulitsa Mira) where the last Tsar, Nicholas II, and his family were kept for eight months after the revolution. The Museum of the Family of Emperor Nicholas II opened here in 2018.

18.17 TRAVEL SITES RELATED TO DMITRI MENDELEEV IN ST. PETERSBURG

See Chart 18.5 for the location of these sites.

18.17.1 Dmitry Mendeleev's Memorial Museum Apartment (D. I. Mendeleev Museum and Archives)

2 Mendeleevskaya Liniya (59.941654°, 30.299147°) at the "Twelve Colleges Building", the original home of St. Petersburg State University, constructed on the orders of Peter the Great. The long red and white building on Vasilyevsky Island is close to the Neva River, only a short walk from the Hermitage Museum across the Dvortsoviy Bridge (the Palace Bridge). This beautiful apartment has been carefully restored to its original condition. There are poignant photographs including a set showing Mendeleev surrounded by Clemens Winkler, Paul Émile Lecoq de Boisbaudran, and Lars Fredrik Nilson, the three discoverers of his famous eka-elements. Look for his tall "standing desk" and a colorful collection of chemicals, all neatly labelled. The highlight

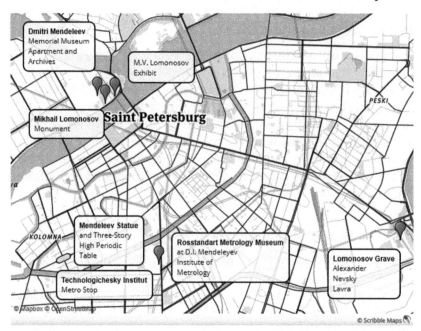

Chart 18.5 Chart of Mendeleev, Lomonosov, and other sites in St. Petersburg.

of the museum is Mendeleev's study. Note the photographs of famous scientists behind his desk and his unusually large, decorative teacup. Nearby is a large collection of instruments he had built for his research including a double-pan, equal-arm balance used to measure small mass differences between two gas samples. A world chart shows the many places that Mendeleev visited over his lifetime including his 1876 trip to Pennsylvania as part of his research on the petroleum industry there. If you are fortunate, you may get a peek into the Mendeleev Archives off the study. It is best to make arrangements in advance in order to visit the Museum. Be sure to bring proof of these arrangements with you.

18.17.2 Rosstandart Metrology Museum at the D. I. Mendeleyev Institute of Metrology

(VNIIM), 19 Moskovsky Prospekt (59.917519°, 30.317341°). This is one of the oldest scientific museums in Russia. From his statue, look for the clock tower that Mendeleev designed. Upon entering the original building, we find a display showing the 131 institutions that awarded Mendeleev honorary degrees. In the Office Room, Mendeleev's desk is set up as he left it as is his chair beside the fireplace. There are exhibits related to Mendeleev's metrological activities, including exquisite balances and timepieces and portraits of Mendeleev by well-known Russian artists. It must be visited by prior arrangement.

18.17.3 Mendeleev Statue and Three-Story-High Periodic Table

(59.917889°, 30.317607°) Mendeleev is shown as an old man smoking a cigarette. A three-story representation of his periodic table is shown behind him on the wall of the building that now houses the Mendeleev All-Russian Institute of Meteorological Research. Installed in 1934 celebrating the 100th anniversary of his birth, elements shown in red on the table were known during his lifetime, whereas those in blue were discovered from 1907–1934. Visiting this scene, shown in Figure 18.6, should be on every scientific/historical travelers' must-do list.

18.17.4 Technologichesky Institut (Technology Institute) Metro Stop

(59.916335°, 30.318417°). This stop is known for its elaborate celebration of Soviet and Russian science. This is the nearest metro stop to the Metrology Museum. The Pushkinskaya Metro station, a 10 minute walk up Zagorodnyy Prospekt is worth stopping to take a look at as well.

18.18 MIKHAIL LOMONOSOV (1711–1765)

Scientific/historical travelers to St. Petersburg should not miss visiting places related to Mikhail Lomonosov, often called the Father of Russian Science. He studied in Moscow, Kiev and St. Petersburg but his advanced education in chemistry was done at the University of Marburg in Germany, where he studied the works of Robert Boyle (1627–1691). He also studied with Johann Friedrich Henckel (physician, chemist, and metallurgist) in

Figure 18.6 The statue of Dmitri Mendeleev in front of the three-story high depiction of his periodic table on the wall of Institute of Meteorological Research building. This display of the periodic table inspired the cover of this book. Photograph by Glen Rodgers.

Freiberg, Saxony, before the establishment of the Bergakademie there. In 1741, he returned to St. Petersburg and became professor of chemistry at the Russian Academy of Science in 1745 and set up the academy's first chemistry laboratory.

Lomonosov anticipated many modern views but is rarely mentioned in traditional histories of chemistry. He expressed strong opposition to the phlogiston theory (see Chapter 8, pp. 200–201) and his views were known, but seemingly ignored by Antoine Lavoisier. Recall that Lavoisier also ignored the work of Priestley and Scheele in their co-discoveries of oxygen. Lomonosov also clearly stated the Law of Conservation of Mass and is even considered to be a co-discoverer of that law, sometimes called the Lomonosov–Lavoisier Law (see Chapter 8). He stated that, "All changes in nature are such that inasmuch is taken from one object insomuch is added to another. So, if the amount of matter decreases in one place, it increases elsewhere." He believed his atomistic views were too revolutionary to publish but anticipated the work of John Dalton. He regarded heat as molecules (he called them corpuscles) in motion and therefore anticipated the work of Count Rumford (See Chapter 11). He and a friend tried to repeat Franklin's kite experiment, but his friend was killed and Lomonosov barely escaped. So, in many ways he was the founder of Russian science. He was also was a poet, astronomer, dramatist, historian, and geographer. He helped found the Russian Academy of Science and the University of Moscow. Given all these accomplishments, it is not clear why we do not know more about Mikhail Lomonosov.

18.19 TRAVEL SITES RELATED TO MIKHAIL LOMONOSOV (ST. PETERSBURG)

18.19.1 Monument to Mikhail Lomonosov

Vasilevsky Island on Mendeleevskaya Liniya (59.940575°, 30.301651°). On the banks of the River Neva not far from the Twelve Colleges and the Mendeleev apartment.

18.19.2 Lomonosov Collection

, an exhibit officially titled "M. V. Lomonosov and the Academy of Sciences of the 18th-century," is located in the dome of the Kunstkamera building (59.941518°, 30.304536°). Lomonosov worked here from 1741 to 1765. In the center is the giant round table where the meetings of the St. Petersburg Academy of Sciences were held. All around the sides are various displays of Lomonosov's personal items as well as his colored glasses, charts, mosaics, portraits, books, and unique scientific equipment, including his collapsible telescope made of paper, wood and leather. Two other exhibits worth taking in are the "First Astronomical Observatory of the Academy of Science" and the Gottorp (Greater Academic) Globe. Lomonosov helped to create this 3.1 meter diameter globe, designed to be a small planetarium holding 10 to 12 people. The outer surface is a chart of the world as known in the 18th century, whereas the inside is a chart of the stars. More information available at www.kunstkamera.ru (English translation available).

18.19.3 Lomonosov Grave

Alexander Nevsky Lavra or Monastery, 179 Nevsky Prospekt (59.921000°, 30.388000°). St. Petersburg's most famous burial ground is divided into two cemeteries. On the left side is the Lazarev Cemetery where Lomonosov and the mathematician Leonhard Euler are buried. The right side is Tikhvinskoye Cemetery that has the elaborate tombstones of novelist Fyodor Dostoyevsky and musicians such as Mikhail Glinka, Pyotr Tchaikovsky, Modest Moussorgsky, Nikolai Rimsky-Korsakov, and Alexander Borodin. Borodin was an amateur musician whose principal profession was chemistry.

18.19.4 Lomonosovo (Formerly Oranienbaum)

(59.915456°, 29.754257°) a town 25 miles west of St. Petersburg, is named in Lomonosov's honor. The town is the site of the 18th-century royal Oranienbaum park and palace complex (a UNESCO World Heritage Site), notable as being the only Imperial Palace in the vicinity of Saint Petersburg that was not captured by Nazi Germany during World War II.

18.19.5 "Akademik Lomonosov"

This is a vessel to be operated as the first Russian floating nuclear power station. It was scheduled to be delivered to Pevek in Russia's far east and become operational in late 2019.

18.20 SUMMARY

This chapter discusses the development of the periodic table of the elements and the places we can visit that celebrate it, some of them among the best scientific/historical sites in the world. The contributions of Johann Wolfgang Döbereiner (the "Law of Triads"), Leopold Gmelin, Alexandre-Émile Béguyer de Chancourtois, John Newlands (the "Law of Octaves"), William Odling, and Julius Lothar Meyer preceded Dmitri Mendeleev's first publication of his periodic table in 1869, about 150 years ago. Döbereiner and Meyer sites in Germany and Newlands sites in London are noted. What established Mendeleev as the primary originator of the table was his prediction of the existence of then unknown elements that he designated eka-boron, eka-aluminum, and eka-silicon. These elements, now called gallium, scandium, and germanium, were subsequently discovered by Paul Émile Lecoq de Boisbaudran (1875), Lars Fredrik Nilson and Per Teodor Clève (1879), and Clemens Winkler (1886), respectively. Each of the elements was named to honor the countries or regions (France, Scandinavia, and Germany) of the discoverers. The de Boisbaudran sites in Cognac, France and the striking Winkler sites in Freiberg, Germany are described. Mendeleev also accurately predicted the properties of his eka-elements and when these were found to be astonishingly accurate, he became world famous. We explore Mendeleev sites in Siberia and St. Petersburg, the latter including two of highest-rated sites in this book. We briefly discuss how Mendeleev handled the discovery of the "inert gases" and then finish the chapter with sites related to Mikhail Lomonosov, the father of Russian science who goes largely unrecognized in the history of science.

ADDITIONAL READING

- J. L. Marshall, Rediscovery of the Elements, Germanium: Freiberg, Germany, *The Hexagon*, Spring 2001, pp. 20–22.
- J. L. Marshall, Rediscovery of the Elements, Gallium, *The Hexagon*, Winter 2002, pp. 78–81.
- M. V. Orna, *Science History: A Traveler's Guide*, ACS Symposium Series, American Chemical Society, Washington, DC, 2014, vol. 1179, pp. 311–313.
- I. S. Dmitriev, Liberty, Labor and Duty, *Russian Journal of General Chemistry*, 2007, 77, 159–172.

REFERENCES

1. E. R. Scerri, *The Periodic Table: Its Story and its Significance*, Oxford University Press, 2007.
2. J. L. Marshall and V. R. Marshall, Rediscovery of the Elements, Johann Wolfgang Döbereiner, *The Hexagon*, Fall 2007, pp. 50–55.
3. G. E. Rodgers, *Descriptive Inorganic, Coordination, and Solid-State Chemistry*, Brooks/Cole, Cengage Learning, 3rd edn, 2012.
4. J. L. Marshall and V. R. Marshall, Rediscovery of the Elements, The Periodic Table, *The Hexagon*, Summer 2007, pp. 23 – 29.

CHAPTER 19

Stockholm, the Atom, and the Nobel Prizes: Berzelius, Scheele, Arrhenius, and the Atomic Nobel Prizes (Sweden)

19.1　A QUICK LOOK AT PLACES TO VISIT "TRAVELING WITH THE ATOM" IN STOCKHOLM

Berzelius Museum (no rating, formerly ●●●●*); Berzelius Statue Berzelius Square (Berzelii Park)* ●●●*; Berzelius Mosaic* ●● *in Stockholm City Hall; Berzelius Grave* ●*; Berzelius Bust* ●● *Uppsala University; Berzelius Laboratory at Gripsholm* ●*; Berzelius School with bust in front* ●● *in Linköping; Scheele Statue* ●● *Flora's Hill in Hops Garden (Humlegården); Scheele Pharmacy* ● *Gamla Stan; The Crown Pharmacy (Apoteket Kronan) in Skansen Park* ●●●*; Arrhenius Bust* ●● *near the Arrhenius Laboratory; Arrhenius Grave* ● *Uppsala gamla kyrkogård; Stockholm Concert Hall (Stockholms Konserthus)* ●●●*; Stockholm City Hall (Stockholms Stadhus)* ●●●●*; The Nobel Museum (Nobelmuseet)* ●●●●*; Alfred Nobel grave* ●● *in Norra begravningsplatsen; Alfred Nobels Björkborn* ●●● *Karlskoga.*

Swedish Atomic Scientists, the Nobel Laureates who contributed to the Atomic Concept and the Politics of the Nobel Prizes

- Jöns Jakob Berzelius devised the modern symbols for the elements, determined many atomic weights and provided numerous examples of the Law of Definite Proportions. He supported Dalton's atomic theory, but rejected the ideas of Gay-Lussac, Avogadro, and Ampère that would have advanced chemistry significantly had they been immediately accepted.
- Carl Wilhelm Scheele, along with Joseph Priestley, is considered a co-discoverer of oxygen. He had a hand in the discovery of many other elements and was primarily responsible for the discovery of chlorine.
- Svante Arrhenius proposed that many solids dissociate to form ions in aqueous solution and that acids produce hydrogen ions and bases produce hydroxide ions. His dissociation theory was initially ridiculed, and he never forgave those who rejected it.
- Arrhenius, Wilhelm Ostwald, and Jacobus van't Hoff were among the pioneers of physical chemistry.
- Alfred Nobel, the inventor of dynamite, established his Nobel Prizes when he died in 1896. The first prizes were awarded in 1901.
- Svante Arrhenius won the Nobel Prize in Chemistry in 1903 and then went on to help prevent prizes for Dmitri Mendeleev and Ludwig Boltzmann. He played strong hands in the politics of the prizes awarded to Ernest Rutherford, Wilhelm Ostwald, Max Planck, Albert Einstein, and the second prize for Marie Curie.
- The bias against theory including the delay in recognizing quantum mechanics, the debate about awarding prizes during WWI, the "Einstein Dilemma", and the "Hahn/Meitner Debacle" are all examples of the role of politics in awarding the Nobel Prizes.

We have been traveling throughout Europe and beyond, visiting sites related to the history of the atomic concept. Now we are ready to wrap up our travels with particular emphasis on the relevant Nobel Prizes in Physics and Chemistry. This brings us to Stockholm, where these prizes are determined and presented. Before we do that, however, we should discuss and visit places in Stockholm related to three important Swedish atomic scientists, Jöns Jakob Berzelius, Carl Wilhelm Scheele, and Svante Arrhenius.

19.2 JÖNS JAKOB BERZELIUS (1779–1848)

> "Chemical signs ought to be letters, for the greater facility of writing, and not to disfigure a printed book ....I shall take therefore for the chemical sign, the *initial letter of the Latin name of each elementary substance* ... [and add a second distinguishing letter when needed.]" Jöns Jakob Berzelius

Jakob Berzelius was the son of a school teacher, but was orphaned before the age of 10. An uncle took him in but soon sent him to a boarding school where, as a gifted student, he tutored private students to make extra money. Sadly, he developed a reputation as a bad-mannered young man who did not apply himself well to his studies. Nevertheless, he started to study medicine at Uppsala University but was set back again when he lost his scholarship. Apprenticed first to a pharmacist and then to a physician who worked with the poor at the Medevi mineral springs, he took up the quantitative analysis of the spring water and started to prove himself proficient in analytical chemistry. At Medevi, he built a Voltaic pile (just invented in 1799, see Chapter 9) and tried, mostly unsuccessfully, to use it to cure various diseases. Back at Uppsala, he earned his doctor's degree with a dissertation on galvanotherapy. In 1802 he moved to Stockholm, where he was appointed as a physician, a *medicus pauperum*, helping the poor and indigent. Impressed with his analytical abilities, Wilhelm Hisinger, a wealthy mine-owner, mineralogist, and chemist, provided Berzelius with laboratory facilities and access to the largest Voltaic pile in Sweden. Together, they electrochemically decomposed many compounds and, in 1803, discovered the element cerium. Within five years, he was appointed a professor of chemistry at a surgical institute. He was now on a path that made him one of the early founders of modern chemistry, right up there with Robert Boyle (Chapters 2 and 3), John Dalton (Chapter 4), and Antoine Lavoisier (Chapter 8).

Back in Chapter 4, we discussed the first concrete atomic theory, proposed by John Dalton in 1803 and published in 1808. At that time, we noted that we were, for convenience, using the modern symbols for atoms/elements that had been devised by Berzelius. Dalton himself had drawn up rather enigmatic, two-dimensional, circular symbols for atoms that he called "pictographs". They are shown in Figures 4.4 and 4.5. Berzelius found these symbols laborious to work with and therefore, in 1813, proposed that we represent each element (and therefore the atoms of that element) using a symbol composed of the initial letter and perhaps a second letter when needed from the Latin name of the element. So, oxygen was represented by O, hydrogen by H, carbon as C, chlorine as Cl, calcium as Ca, but copper was Cu (for the Latin, *cuprum*), gold was Au (for *aurum*), tin was Sn (for *stannum*), sodium was Na (for *natrium*, a Latin name that Berzelius invented), potassium was K (for *kalium*, another Latin name Berzelius coined), and so forth. His system brought order out of chaos. A molecule could be represented by indicating the number of component atoms it contained. For example, the two molecules containing carbon and oxygen would be CO and CO_2, what we now call carbon monoxide and carbon dioxide. Incidentally, Berzelius actually used superscripts, not subscripts, so his representation of carbon dioxide would have been CO^2. Dalton was, in his usual stubborn way, unalterably opposed to Berzelius's symbols for atoms and molecules. He evidently found them "horrifying". Others were openly hostile as well, but by mid-century the symbols had become internationally accepted.

Berzelius went on to make many contributions to chemistry and the atomic concept, including the first reasonably accurate determinations of atomic weights and many examples of the Law of Definite Proportions. Eventually, Berzelius listed the atomic weights of 46 elements comprising some 2000 compounds, all of which he had analyzed himself. Whereas Berzelius had initially been somewhat skeptical regarding Dalton's atomic theory, his own results led him to become one of its most important supporters. He correctly maintained that Dalton's theory would be the greatest advance in chemistry made to that point in history. His work thoroughly supported Dalton's assertion that compounds were composed of atoms combined in whole number ratios. Dalton had assumed that physical atoms were spherical in shape and were most readily distinguished by their different masses. Berzelius, on the other hand, would ascribe neither shape nor dimension to atoms and thought it was their electrical properties that distinguished them. He believed that every atom had either a positive or a negative charge and that compounds were held together by these electrostatic forces. In that sense, he anticipated what we would call ionic bonds. By 1830, he had become one of the primary authorities in chemistry. He wrote the leading textbook of his day entitled *Lärbok i Kemien* (*Textbook of Chemistry*) that had six volumes and 6000 pages and went through five editions. For 27 years (1821 to 1848), he also published an annual survey of the progress in the sciences. Not surprisingly, when Berzelius passed judgement on a new experiment or theory, what he said was considered nearly the last word. Chemists would always wait anxiously to see what Berzelius would say about their work. His opinion could make or break careers and they knew it.

Unfortunately, particularly in his old age, Berzelius, like Dalton as he aged, was on the wrong side of many issues. Perhaps his most egregious error was rejecting the ideas of Joseph Louis Gay-Lussac (his Law of Combining Volumes, 1809), Amedeo Avogadro (his hypothesis, called "EVEN", that equal volumes of gases contain equal numbers of particles, 1811) and similar conclusions by André-Marie Ampère (1814) that would have advanced chemistry significantly had they been immediately accepted. (See Chapter 4, pp. 91–92, Chapter 8, p. 216 and Chapter 9, p. 236.) Their results implied the existence of diatomic molecules like H_2, O_2, and Cl_2 but if, as Berzelius believed, hydrogen atoms were all positively charged, they would repel each other and could not bond together to form H_2. Similarly, for oxygen atoms and chlorine atoms as they were both negative. Today, we know that these molecules are held together by covalent bonds, a type of bonding that was not yet invented.

Another example of Berzelius being on the wrong side of things was his disagreement with Humphry Davy regarding the nature of chlorine, an element that had been discovered by Carl Wilhelm Scheele in 1774. Davy's work had convinced him that chlorine, originally called oxymuriatic acid, did not contain oxygen and was an element. Berzelius believed that chlorine was oxidized hydrochloric (muriatic) acid, that is, that it did contain oxygen. When the element bromine was discovered in 1811, it, like chlorine, was consistent with Davy's conclusions and not Berzelius's. Berzelius wrote forcefully to advocate his viewpoint and when he was proven wrong, his reputation suffered. Nevertheless, he apparently remained friendly with Davy even under those difficult circumstances. (For more on Davy, see Chapter 5).

Berzelius also discovered the elements cerium (with Hisinger in 1803, named after the recently discovered asteroid Ceres), selenium (1817, named after the moon), and thorium (1815, named after the Norse god Thor). He also was the first to isolate elemental silicon and zirconium. He introduced many modern terms including catalysis, isomerism, allotropy, and protein.

Life was not easy for Berzelius. He was plagued by debts incurred when he bought into "get-rich-quick" schemes. A laboratory explosion nearly blinded him. He had migraine headaches that some say were of psychosomatic origin. They were so powerful that he ascribed them as due to the phases of the moon. His last few years were better, however, as he married (1835) a young woman and was made a baron as a wedding gift on the morning of his marriage. He was a great traveler and, together with his accurate experimental work and his many seminal publications, the personal contacts he made traveling resulted in him being one of the most important European chemists for many years. He was a member of 94 academies and scientific societies in many different countries. Figure 19.1 shows Baron Jöns Jakob

Figure 19.1   Baron Jöns Jakob Berzelius wearing his medals. © Shutterstock.

Berzelius wearing two medals. The one in the middle is the Wasa Orden, whereas the one on the bottom is the Nordstjerneordern medal. What looks to be a medal on the top is the emblem on his robe that symbolized the Royal Swedish Academy of Sciences.

19.3 TRAVEL SITES IN STOCKHOLM RELATED TO JÖNS JAKOB BERZELIUS

See Chart 19.1 for the location of these sites.

19.3.1 Berzelius Museum (No Rating)

This museum was inaugurated in 1898 on the 50th anniversary of Berzelius's death. Housed in or near the Royal Swedish Academy of Sciences, it displayed various citations, paintings, furniture, his

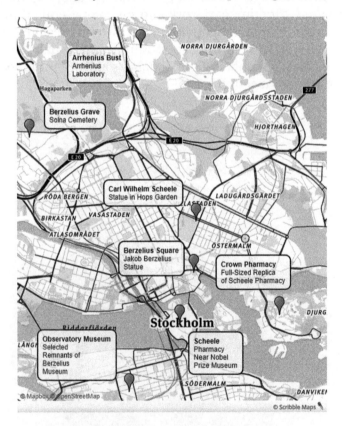

Chart 19.1 Travel sites in Stockholm related to Berzelius, Scheele, and Arrhenius.

wheelchair, the various editions of his textbook, as well as equipment, chemical samples. It was easily a "▓▓▓▓▓" place as late as 2007, but then, mysteriously, was scheduled to be moved to the Observatory Museum (Observatorie Museet), Dottninggatan 120 (59.3146°, 18.055°), at the top of the Observatoriekullen in the city of Vaasa in Stockholm. In 2010, an exhibit about Berzelius was created but, at this writing, neither the former Berzelius Museum nor the Observatory Museum exhibit is available to the public. The website at the Stockholms Gamla Observatorium shows pictures of temporary exhibitions of Berzelius's desktop and his "experimental table". There are a few Berzelius items on display (his blow pipe and blow pipe table) at the Gustavianum Museum (59.857952°, 17.631820°) in Uppsala, about 70 km north of Stockholm.

19.3.2 Berzelius Statue in Berzelius Park (Berzelii Park)

▓▓▓ 2 Nybroplan (59.332586°, 18.074591°). This is easily one of the most beautiful settings honoring a scientist that a scientific/historical traveler will encounter. It is only a short distance from the location of the so-called "German Baker's House" owned by Wilhelm Hisinger, where Berzelius lived and had laboratory space from 1809 to 1819. The building was torn down in 1907.

19.3.3 Berzelius Grave

▓ Solna cemetery (Solna Kyrkogård), Stockholm (59.35295°, 18.02441°) Plot SO 02 30. Both he and his wife, Elizabeth Johanna Poppius Berzelius (1811–1884), are buried here. Note that she was 32 years his junior, was only 37 when he died, and outlived him by 36 years! When Berzelius' wife was herself an old woman, she was asked by a young man at a banquet what her husband's profession had been. She said her husband was a statue in Nybroviken!

19.4 OTHER BERZELIUS TRAVEL SITES

19.4.1 Berzelius Bust

▓▓ Thunbergsvägen 5 (59.8542°, 17.6273°). Above the entrance doorway to the chemistry building at Uppsala University are the busts of Berzelius, Scheele, and Tobern Bergman, an early chair of the department who, it is said, "discovered" Scheele.

19.4.2 Berzelius Laboratory at Gripsholm

(59.2567°, 17.2125°). This simple block building is where Berzelius discovered selenium in the sludge from a sulfuric acid plant which he jointly owned.

19.4.3 Berzelius School with Bust in Front

Cathedral School (Katedralskolan) in Linköping (58.4108°, 15.6183°). Berzelius attended this school and his father was principal there. The bust of Berzelius in front of the school shows him as a youth.

19.5 CARL WILHELM SCHEELE (1742–1786)

Joseph Priestley and the Swedish-German Carl Wilhelm Scheele are usually listed as independent co-discoverers of oxygen. Scheele, a pharmacist who concentrated on research, had a hand in the discovery of a variety of elements (chlorine, manganese, barium, molybdenum, tungsten, and nitrogen in addition to oxygen) but does not receive the sole credit for the discovery of any of them. He also was the first to prepare a number of acids and, astoundingly, actually documented the taste (resembling rotten almonds) of the deadly poisonous hydrogen cyanide gas! As recorded in his laboratory notebook in 1771, Scheele was able to isolate reasonably pure oxygen (what he called "fire air") from various compounds and went on to characterize it well enough that, under normal circumstances, he might very well be listed as the sole discoverer. Tragically, but through no fault of his own, there was a delay in the publication of Scheele's work, so that when it did appear in 1777, Priestley had beaten him into print by about 3 years. (Can you imagine the agony and disappointment Scheele must have felt when he saw Priestley's article?) Due to his careful documentation, however, Scheele is usually listed as the co-discoverer of the element.

So, both Scheele and Priestley (see Chapter 3, pp. 60–62) prepared and characterized a gas that Priestley called *dephlogisticated air*. As we know (from Chapter 8, p. 202), it was Antoine Lavoisier who recognized that the phlogiston theory was inadequate to explain combustion and that Priestley's (and Scheele's) air, which Lavoisier called *oxygen*, was a new element that combined with materials during the combustion process. Travelers-with-the-atom would enjoy the play, *Oxygen*,[1] by Carl Djerassi and Roald Hoffmann. In it, Scheele, Lavoisier, and Priestley and their significant others debate which man should receive a "retro-Nobel" prize for the discovery of oxygen.

Whereas Scheele lost out in his bid to be the sole discoverer of oxygen, when it came to chlorine, he won the day. He isolated it by treating aqueous sodium chloride with sulfuric acid and manganese oxide. This is still a common method for producing small quantities of chlorine. As discussed above, there was a controversy between Humphry Davy and Jakob Berzelius about whether chlorine was, in fact, an element or just oxidized hydrochloric (or muriatic) acid. Davy won that battle and declared the new pale-green gas an element, which he named *chlorine* for the Greek *chloros,* meaning "pale green" or "greenish yellow". This is a highly irritating and toxic gas. Scheele noted that it could decolorize flowers and green leaves. It also went on to be the first war gas, released by the Germans (under the direction of Fritz Haber) against the British in 1915. (See Chapter 17, p. 427 and p. 435).

19.6 TRAVEL SITES IN STOCKHOLM RELATED TO CARL WILHELM SCHEELE

19.6.1 Scheele Statue

Flora's Hill in Humlegården ("Hops Garden"), 22 Sturegatan (59.340249°, 18.075219°). This over-sized, bronze statue in the northeast corner of the garden shows Scheele sitting on a chair with his retort and oven at his feet. He wears a simple suit and a long cloak over his shoulders. The major statue in the center of the park is of Carl von Linné, better known as Linnaeus who, between 1735 and 1768, published the *Systema Naturae* that classified the animal, vegetable and mineral kingdoms using a binary nomenclature composed of two parts, the *Genus* and *species*, both using Latin grammatical forms. We are *Homo sapiens* according to Linnaeus's system. The elemental nomenclature, using abbreviated Latin names, introduced by Berzelius was modeled after the Linnaean system. The Humlegården is a major public park in Östermalm. The National Library of Sweden is located here. Since this is the Hops Garden, scientific/historical travelers should retire to the fine, small restaurant there and lift a beer in honor of Scheele, Berzelius, and Linnaeus. *Homo sapiens bibere cervisiam!*

19.6.2 Scheele Pharmacy

Stortorget 16, Gamla Stan (59.3250°, 18.0703°). Scheele worked as a pharmacist here from 1768 to 1770. Stortorget is the "Grand Square" and is the oldest square in Stockholm, with a historic fountain in the center. This building is now an Italian restaurant and coffee shop. According to Jim Marshall, the "owners swear that the basement is exactly the same as it was in Scheele's day, with masonry dating back to the 1600s." During Scheele's days, the pharmacy was called the "Gilded Raven" (Förgyhllda Korpen). In 1924 it was relocated to 16 Vasterlanggatan (59.3257°, 18.0688°), only a few steps from Stortorget 16. Now called the Raven apothecary (Apoteket Korpen), this modern pharmacy until recently had a sign over the door that said "Apothecary on duty, C. W. Scheele" but this referred not to our Carl Wilhelm Scheele, but a present-day pharmacist. These two sites are very close to the Nobel Prize Museum located in the former Stock Exchange Building (Börshuset) at Stortorget 2.

19.6.3 The Crown Pharmacy in Skansen Park

(59.326188°, 18.100482°) Skansen Park is an open-air museum featuring full-size replicas of buildings of all types. The Crown Pharmacy (Apoteket Kronan) is a full-sized version of Carl Wilhelm Scheele's pharmacy with equipment from historic apothecaries at Drottningholm and Köping, the latter run by Scheele. It is hosted by multi-lingual guides all in costume.

For a more complete guide to scientific sites in Stockholm, Uppsala and, indeed, southern Sweden, Finland and Norway, see "Northern Scandinavia: An Elemental Treasure Trove" by Jim and Virginia Marshall in *Science History: A Traveler's Guide*,[2] edited by Mary Virginia Orna.

19.7 SVANTE ARRHENIUS (1859–1927)

Arrhenius has a duel role to play here in our last chapter. First, let us consider his contributions to chemistry and the atomic concept, his appointment as the first Swede to win the Nobel Prize in Chemistry, and then we can turn to his out-sized, some would say outrageous, role in determining who should be future Nobel laureates. First, his contributions to the atomic concept.

Michael Faraday (see Chapter 5, p. 107), in about 1833, had established the rules of electrolysis and had spoken of "ions" (from the Greek word for "wanderers"), particles that carried electricity. Arrhenius proposed that when salts ("electrolytes") are dissolved in water, even though they are not subjected to an electrical current, they produce ions with varying charges. The atoms making up these salts separated into not neutral atoms but, rather, charged particles, cations and anions. These conclusions were based on accurate measurements of electrical conductivity. The more ions and the more highly charged they are results in greater conductivity. Scientists like Dalton (Chapter 4), for example, would have never accepted this idea. Atoms are hard spheres and don't break up into charged ions!

This idea of dissociation in aqueous solution was revolutionary when Arrhenius proposed it in his doctoral thesis (1884). Per Teodor Cleve was one of Arrhenius's doctoral advisers. (Recall from Chapter 18 that he had helped associate Lars Nilson's new element, scandium, with Mendeleev's eka-boron.) When Arrhenius tried to explain his dissociation theory to Prof. Cleve, he was summarily dismissed, shown out of the office so to speak. Nevertheless, Arrhenius persisted in his premise and even extended his ideas to propose that acids and bases produced ions in solution. Acids, he said, produced hydrogen ions whereas bases produced hydroxide ions. (Arrhenius acids and bases are still considered fundamental concepts in chemistry today.) His examiners were decidedly unimpressed and his thesis, not particularly well-written, was given the lowest possible passing grade – *"non sine laude"* ("not without praise") – and this, some say, may have been an upgrade. Arrhenius, undeterred, responded by sending his dissertation to scientists outside Sweden including Wilhelm Ostwald. Ostwald *was* impressed, so much so that he even traveled up to Uppsala to invite Arrhenius to join his research team in Riga, Latvia. Arrhenius declined because his father was close to death.

Ostwald's trip and his offer took Arrhenius's examiners a bit by surprise and they responded by awarding him a five-year travel-grant which, after his father's death in 1885, Arrhenius was able to use to study with Ostwald in Riga and later Leipzig;

Ludwig Boltzmann in Graz, Austria, and Jacobus Henricus van't Hoff in Amsterdam, and others. Ostwald and van't Hoff were the founders of the new sub-discipline of physical chemistry that emphasized the relationship between the physical structure of a compound, the amount of energy it has and the way it reacts with other compounds. (Recall that Ostwald pictured energy, not atoms, as the chief component of the universe. We visited the Wilhelm Ostwald Park and Museum, with its centerpiece, the "Energiehaus", back in Chapter 10.) Ostwald and van't Hoff (and Walther Nernst) agreed with Arrhenius about dissociation and were sometimes referred to as the "wild army of Ionists". (Did they rule with an Ion Hand?) They quickly convinced other chemists that the ionization theory was correct.

Working with Ostwald and others helped Arrhenius elaborate on his dissociation theory and its application to acids and bases. Dmitri Mendeleev, on the other hand, was adamantly opposed to the idea of dissociation into ions, strongly advocating his own ideas instead. In 1889, Arrhenius added to his reputation by explaining that chemical reactions required an energy of activation to get them going and put this on a quantitative basis in his Arrhenius equation. By 1891, Arrhenius was able to work his way back home and became a lecturer and then, despite substantial opposition, a professor at Stockholm University.

Now the tide of atomic history started to turn in Arrhenius's favor. Things were beginning to happen that showed that atoms had internal structure. Röntgen discovered X-rays in 1895, Becquerel discovered radioactivity in 1896, and J. J. Thomson discovered the electron in 1897. It was now becoming apparent that atoms were, in fact, composed of charged particles. Perhaps dissociation was not such a crazy idea after all. In 1898, Alfred Nobel died, and in his will he set up the new Nobel Prizes. Arrhenius was elected to the Royal Swedish Academy of Sciences in 1901, received the Davy Medal in 1902, and, amazingly, his dissociation theory now fully vindicated, was awarded the 1903 Nobel Prize in Chemistry for the "very same thesis" that had been so severely castigated in 1884. His former adviser Per Cleve, recognizing the error of his ways, strongly advocated for Arrhenius. Arrhenius was the first Swede to be awarded a Nobel Prize.

Arrhenius, then, was a pioneering physical chemist. He was also active in applying it to other disciplines. For example, he was one of the first to predict global warming based on carbon dioxide emissions into the atmosphere. He also wrote a textbook on cosmic physics, in which he discussed phenomena such as ball lightning and thunderstorms, as well as the solar corona and the northern lights, all on the basis of charged particles radiating through space and the earth's atmosphere. He also advocated applying physical chemistry to biochemical topics. He published a book entitled *Immunochemistry* in which he approached toxin-antitoxin reactions, principally diphtheria reactions, using equilibrium concepts and methods common to physical chemistry.

19.8 TRAVEL SITES RELATED TO SVANTE ARRHENIUS

19.8.1 Arrhenius Bust

near the Arrhenius Laboratory, 16C Svante Arrhenius väg (59.365270°, 18.059135°) University of Stockholm. Arrhenius is identified as a *fysiker-kemist* (physical chemist). The Arrhenius Laboratory is located on the main Frescati campus of Stockholm University just north of Stockholm city center.

19.8.2 Arrhenius Grave

Uppsala gamla kyrkogård, Uppsala kommun, Uppsala Iän (59.8549°, 17.62516°). This cemetery is reportedly only a "stone's throw from the house where he spent his childhood and youth".

Arrhenius was born in Wik Castle (or Slott), Sweden (also known as Vik or Wijk) (59.736048°, 17.461568°) near Uppsala. Vik is generally considered to be the best-preserved medieval castle in Sweden. Unlike Robert Boyle, his father did not own the castle but rather had been a land surveyor for Uppsala University, moving up to a supervisory position. Unfortunately, there does not seem to be a marker about the birthplace of Arrhenius.

19.9 THE NOBEL PRIZES

Alfred Nobel (1833–1896) was born in Stockholm, the third of four surviving brothers. His father's business involved explosives and mining machines, so Alfred knew about such matters starting at an early age. When Alfred was just four years old, his father moved to St. Petersburg, Russia where he was able to set up a successful business. The whole family moved there when Alfred was nine and his parents were now able to send him to private tutors.

By the age of 16 he was fluent in five languages: English, French, German, Russian, and Swedish. In 1850, at the age of 17, he went to Paris to study chemistry and then went to the United States for four years, where he worked for John Ericsson, an engineer who, most famously, would build the ironclad warship *Monitor* in 1862. Alfred returned to St. Petersburg, but his father's factory, after the Crimean War had ended, did not fare well and they were bankrupt by 1859.

Alfred and his parents returned to Stockholm where, in 1862, he built a small factory to manufacture the newly discovered nitroglycerin. This oily liquid was exceedingly dangerous to work with, in that it spontaneously exploded when subjected to heat, flame, or even the slightest physical shock. Alfred started researching ways to control its capricious tendency to explode but, unfortunately and tragically, his factory blew up in 1864, killing his younger brother Emil and several others. Alfred responded by creating the Alfred Nobel & Company and moving the whole operation to an isolated site near Hamburg, Germany. This factory also blew up several times. He invented a blasting cap in 1865 and then in 1867 combined nitroglycerin with a porous siliceous earth called *Kieselguhr*, which he found near his factory. He called this formulation "dynamite" (from Greek *dynamis*, "power"). Shaped into rods that could be inserted into predrilled holes, it was much safer to use and easier to handle than nitroglycerin alone. Nobel patented dynamite and similar products all over the world and was soon world famous for inventing safe explosives that could be used to create tunnels, quarries, canals, railways, roads, *etc.* It also could be used to produce more powerful and lethal weaponry, but Alfred thought (or perhaps hoped) that such weapons would be "harbingers of peace", that is, they would be so lethal that no one would venture to use them and there would be no more wars.

In 1873, Nobel moved his headquarters to Paris and continued to set up companies all over the world. They were enormously profitable, but he continued to have misgivings about the potential destructive use of dynamite. A key incident occurred in 1889, when his brother Ludwig died, and a journalist confused Alfred with his brother. Suddenly, Alfred was reading his own obituary, in which he was called the "merchant of death" because of all the money he had made in improving weapons of mass destruction. This was unsettling to say the least, and he resolved to remedy the situation. He rewrote his will such that nearly everything he owned would be used to set up his Nobel Prizes, including the Nobel Peace Prize. Alfred was never of robust health. He suffered from constant headaches, breathing difficulties and heart problems. Ironically, nitroglycerin was found in 1878 to be effective at treating attacks of *angina pectoris*, the medical term for chest pain due to coronary heart disease. (It does this by releasing nitric oxide, NO, which relaxes blood vessels relieving pressure. So, as the saying goes, NO news is good news.) A few months before his death in 1896, Nobel's physicians prescribed nitroglycerin for this heart condition. He declined to take it, writing to a friend: "Isn't it the irony of fate that I have been prescribed [nitroglycerin], to be taken internally! They call it trinitrin, so as not to scare the chemist and the public." Chemists [druggists] are no longer scared of trinitrin. A century later they know it to be a lifesaving compound.

Alfred Nobel, pictured in Figure 19.2, was a loner. He was a retiring person who loved his solitude. He never married and had no children. When he died, his family, Sweden and Norway

Figure 19.2 Alfred Nobel at the age of 50.

(then joined together), and the world were caught by surprise by his magnanimous bequest supported by his enormous estate. As we know, he set up five prizes in physics, chemistry, physiology, or medicine, literature, and the peace prize. He designated which institutions would determine the winners of the prizes but, amazingly, had never consulted with these agencies beforehand or established criteria for how candidates would be nominated or evaluated. He did specify that the winners would be drawn from world-wide nominations and not confined just to Swedish scientists or writers.

19.10 THE EARLY DAYS OF THE NOBEL PRIZES IN CHEMISTRY: ARRHENIUS CARRIES THE DAY

It took five years to start to work out the logistics of awarding the Nobel Prizes. For those in physics and chemistry, nominations would be made by all members of the Royal Swedish Academy of Sciences, the Nobel Committees for Physics and Chemistry, prior winners of the prizes, selected physics and chemistry professors at Swedish and other Nordic universities and technical colleges, and faculty in physics and chemistry at other institutions designated by the Academy of Sciences. As Robert Friedman points out in his comprehensive and nerd-captivating book, *The Politics of Excellence: Behind the Nobel Prize in Science*,[3] decisions were at first difficult because Nobel had specified that winners should be "those who, during the preceding year, shall have conferred the greatest benefit on mankind". Eventually, this was interpreted with a great deal of latitude but, still, most of the living chemists of the day had made their major contributions years if not decades before 1900 so a given committee had to balance specific, recent discoveries against contributions decades in the making. Politics always complicates things and whereas people were judging people as well as their scientific accomplishments, personal feelings and international politics often had a larger than expected effect in the deliberations.

To see these factors in action, let's look at the list of some of the winners of the Nobel Prize in Chemistry for the first decade of these awards. These are included in Table 19.1. The Dutch physical chemist Van't Hoff won the first Nobel Prize in Chemistry in 1901, and the great German organic chemist Emil Fischer won in

Table 19.1 Nobel Prizes in Chemistry related to Sites in *Traveling with the Atom* (most others omitted). (F = France, G = Germany, N = Netherlands, P = Poland, S = Sweden, NZ = New Zealand, UK = United Kingdom, US = United States).

Year	Nobel laureate	Brief description
1901	Jacobus Henricus Van't Hoff (N)	Topics in physical chemistry
1902	Hermann Emil Fischer (G)	Sugar and purine syntheses
1903	Svante August Arrhenius (S)	Dissociation theory
1904	Sir William Ramsay (UK)	Discovery of the inert gases
1906	Henri Moissan (F)	Isolation of fluorine
1908	Ernest Rutherford (UK, NZ)	Disintegration of elements, Chemistry of radioactive substances
1909	Wilhelm Ostwald (G)	Catalysis and rates of reaction
1911	Marie Curie *nee* Sklodowska (P)	Discovery of radium and polonium
1914	Theodore William Richards (US)	Determination of atomic weights
1915	Richard Martin Willstätter (G)	Investigation of plant pigments including chlorophyll
1916	No Prize Awarded	
1917	No Prize Awarded	
1918	Fritz Haber (G)	Synthesis of ammonia from elements
1921	Frederick Soddy (UK)	Chemistry of radioactive substances and isotopes
1922	Francis William Aston (UK)	Mass spectroscopy of isotopes
1935	Frédéric Joliot (F) Irène Joliot-Curie (F)	Artificial radioactive elements
1943	George de Hevesy (G)	Isotopes as tracers
1944	Otto Hahn (G)	Discovery of fission

1902 even though his award was based on decades of work going back many years. (Recall that in Chapter 15 we have seen the great statue of Fischer near the Archives of the Max Planck Society in Berlin-Dahlem. In Chapter 17, we discussed how Fischer had prohibited Lise Meitner from ever visiting a laboratory where men were working.) Who would be next? Despite years of ridicule and disparagement by his peers and examiners, Svante Arrhenius's reputation abroad had grown by leaps and bounds. He was back in Sweden, and young chemists from around the world were coming to work with him. He had many strong nominations, Van't Hoff and Fischer both supported him and Per Cleve, his erstwhile examiner, had recognized the error of his ways and strongly advocated for his former student. So, Svante

Arrhenius, his dissociation theory vindicated, was awarded the 1903 Nobel Prize in Chemistry for the very same thesis for which he had been so severely castigated in 1884. The 1904 prize went to the British Sir William Ramsay, whose striking discovery of the noble gases had been made, not within the past year, but at least within the past decade (see Chapters 7 and 18). A less rigid interpretation of Nobel's original specifications for eligibility for the award was gradually being applied.

But now things were becoming more difficult. Many of the great nineteenth-century chemists were still alive, including one who had contributed so significantly to the history of the atom, Dmitri Mendeleev. Ramsay's discovery of the noble gases and the relative ease with which Mendeleev's periodic table accommodated them seemed to indicate that the periodic law was currently imparting a great benefit to "mankind". The Chemistry Section of the Swedish Academy supported Mendeleev and it was expected that the whole Academy would follow suit. The elderly Russian seemed well on his way to receiving the 1906 prize but now, unfortunately, things suddenly became acutely personal. Arrhenius, now a former winner with an acerbic personality, could never forgive the Russian for not supporting his dissociation theory and vehemently opposed his nomination. Tempers flew and some of Mendeleev's supporters, incensed by Arrhenius's undue opposition, attacked Arrhenius personally. These *ad hominem* attacks backfired and suddenly Henri Moissan's name sprang forward as an alternative. Certainly, his investigation and isolation of the highly corrosive and dangerous fluorine gas was Nobel-prize-worthy but up until this point many assumed that he would easily be in line for future prizes. When the votes were tallied, Moissan had received one more when it counted the most. The grudge that Arrhenius held against Mendeleev had carried the day. Mendeleev was nominated again in 1907 but again Arrhenius objected and, as a result, the Russian never received the Nobel Prize in Chemistry. Some say this was a severe miscarriage of justice. On Feb 20, 1907, two months after receiving his Nobel award, Moissan was stricken with appendicitis and died. Mendeleev died 18 days later.

Looking again at Table 19.1, we are struck once again by seeing Ernest Rutherford listed as the winner of the 1908 Nobel Prize in *Chemistry*. Certainly, the "Crocodile", the archetypal physicist, was flabbergasted. After all, he had once said that "physics is the only real science. The rest [including chemistry] are just stamp collecting". Recall from Chapter 6 that starting in 1902, he and the young chemist Frederick Soddy had worked together at McGill University in Montreal to characterize nuclear transformations and define the half-life of a radioactive element. Rutherford was careful to characterize the processes they were investigating as transformations not transmutations. He told Soddy that if they called them transmutations, "They'll have our heads off as alchemists"! This is the work that earned Rutherford a Nobel Prize. As we noted in Chapter 6, when he learned that the Nobel committee had unexpectedly classified him as a chemist, he joked (most likely in his loud voice) that it was a "transformation" of unparalleled magnitude. So, what happened? Again, Arrhenius exerted a large influence. He had opined that there was little consensus for a chemistry nominee in 1908. That said, he and a few others argued that, whereas Rutherford had shaken the foundations of chemistry by challenging the assumption that chemical elements were not immutable, he should receive the chemistry prize, rather than the one in physics. So, there it was, Rutherford was transformed (but not transmuted!) into a chemist. Soddy had to wait until 1921 to receive his Nobel Prize and it was, quite properly, in chemistry.

Then there was Wilhelm Ostwald! Up until 1908, Ostwald did not even believe in atoms (see Chapter 10). He had said that "what we call matter is only a complex of energies which we find together in the same place". Arrhenius, on the other hand was an avowed atomist and, as far he was concerned, no one, not even his erstwhile champion back in the mid-1880s, was going to receive a Nobel Prize in Chemistry without believing in atoms. This circumstance was partially solved when Ostwald was converted (transmuted?) to atomism in large part due to Einstein's 1905 paper explaining Brownian motion in terms of atoms and molecules colliding with macroscopic particles (see Chapter 15). But still, much of Ostwald's work was decades old. How could he be honored for his lifetime achievements in physical chemistry while still adhering to Nobel's rule that winners of his prizes should have conferred "the greatest benefit on mankind during the preceding year"? The answer was to concentrate on Ostwald's comparatively recent work about rates of reaction and catalysis even though these were not his most important contributions to chemistry. The latter involved his work in analytical chemistry, centering on chemical reactions and equilibrium. Arrhenius wrote a compelling recommendation emphasizing catalysis and once again this carried the day. Wilhelm Ostwald won the 1909 Nobel Prize in Chemistry!

Arrhenius's undue influence in both the physics and chemistry deliberations continued to carry the day for many years. He was vengeful and manipulative and often feuded with other committee members. He vehemently denied or delayed awards for worthy candidates, particularly those with mathematical or theoretical leanings. In 1910 and again in 1911, he strongly opposed the nomination of the French mathematical physicist, Henri Poincaré for a physics prize. French scientists were dismayed and characterized the Nobel Prize selection process as *la comedie humaine*, "the human comedy". Poincaré had received 34 nominations for the physics prize in 1910 and more in 1911. (He died in 1912 without having received his just due.) Parisians, in particular, were upset and when Arrhenius traveled to Paris, he devised a plan to soothe their bruised feelings by advocating Marie Curie for a second prize, this one in chemistry. Recall that she and Pierre (and Antoine Becquerel) had shared the 1903 physics prize for their discovery of radioactivity. Pierre had been killed in a tragic traffic accident in 1906. Then in late 1910, Marie, now involved romantically with Paul Langevin, had allowed her name to be put forward for election to the all-male French Academy of Sciences. In a bitter contest, she lost by one vote and the academy went on to ban women "forever"! Arrhenius knew he could curry favor and re-establish his reputation with the French if he could arrange a second Nobel Prize for Marie, this one for the separation and production of polonium and radium. He did just that and Marie, who was attending the 1911 Solvay Conference, found out about her second Nobel prize on the very same day that newspapers carried shocking headlines about the Curie–Langevin affair (see Chapter 14).

19.11 THE EARLY DAYS OF THE NOBEL PRIZES IN PHYSICS: THE BIAS AGAINST THEORY

Now let's turn to the early Nobel Prizes in Physics. Those relative to our travels are shown in Table 19.2. Again, the first prizes are what we would expect. The revolutions in physics are reflected in

Table 19.2 Nobel Prizes in Physics related to sites in *Traveling with the Atom* (most others omitted) (A = Austria, A-H = Austria-Hungary, D = Denmark, F = France, G = Germany, I = Italy, P = Poland, S = Sweden, NZ = New Zealand, UK = United Kingdom, US = United States, WG = West Germany).

Year	Nobel laureate	Brief description
1901	Wilhelm Conrad Röntgen (G)	Discovery of x-rays
1903	Antoine Henri Becquerel (F)	Discovery of radioactivity
	Pierre Curie (F)	
	Maria Sklodowska-Curie (P)	
1904	Lord Rayleigh (UK)	Discovery of argon
1905	Phillipp Eduard Anton von Lenard (A-H, G)	Work on cathode rays
1906	Joseph John Thomson (UK)	Discovery of the Electron
1914	Max von Laue (G)	Diffraction of x-rays by crystals
1915	William Henry Bragg (UK)	Analysis of crystal structure by x-rays
	William Lawrence Bragg (Australia, UK)	
1916	No Prize Awarded	
1917	Charles Glover Barkla (UK)	X-ray cpectroscopy of the elements
1918	Max Planck (G)	Discovery of energy quanta
1921	Albert Einstein (G, Switzerland)	Explanation of photoelectric effect
1922	Niels Bohr (D)	Investigation of the structure of atoms
1924	Manne Siegbahn (S)	Work in x-ray spectroscopy
1926	Jean Perrin (F)	Verifying Einstein's explanation of Brownian motion
1929	Louis Victor Pierre Raymond, 7th Duc de Broglie (F)	Discovery of the wave nature of electrons
1932	Werner Heisenberg (G)	Creation of matrix mechanics
1933	Erwin Schrödinger (A)	Creation of wave mechanics (Schrödinger) and relativistic quantum theory (Dirac)
	Paul Dirac (UK)	
1934	Harold Urey (US)	Discovery and isolation of heavy hydrogen (Deuterium)
1935	James Chadwick (UK)	Discovery of the neutron
1937	Clinton Joseph Davisson (US)	Experimental discovery of the diffraction of electrons by crystals
	George Paget Thomson (UK)	
1938	Enrico Fermi (I)	Neutron irradiation to produce new radioactive elements; Effect of slow neutrons
1951	John Douglas Cockcroft (UK)	Transmutation of atomic nuclei by artificially accelerated atomic particles
	Ernest Thomas Sinton Walton (UK, Ireland)	
1954	Max Born (WG)	Statistical interpretation of wavefunction
	Walther Bothe (WG)	

the prizes: the German Röntgen wins in 1901 for his discovery of X-rays, the French Becquerel and the Curies win in 1903 for their discovery of radioactivity, and the British J. J. Thomson wins in 1906 for his discovery of the electron.

Fairly early on, however, it became apparent that mathematical and theoretical physics, even though recognized internationally, was getting short-shrift in the awarding of the Nobel Prizes. Arguably, Ludwig Boltzmann, the Austrian mathematical physicist nearly on a par with the Scottish James Clerk Maxwell (see Chapter 5), was one of the first victims of this bias against theory. In Chapter 10, we discussed how Boltzmann extended Maxwell's statistical analysis of the kinetic-molecular theory (KMT) of gases in such detail that we still refer to the Maxwell–Boltzmann distribution of energies in a given gas. This work was perhaps too early to be considered for a Nobel Prize, but Boltzmann went on to explain entropy using his famous equation, $S = k \log W$, which appears on his tombstone. Originally formulated in the late 1870s, this mathematics depended on the physical reality of atoms. Boltzmann, sometimes referred to as the "martyr of atomism", faced fierce opposition from anti-atomist Ernest Mach, was passed over several times and then, after his suicide in 1906, never received a Nobel Prize. Many regard this as an injustice.

In Chapter 15, we discussed Max Planck's reluctant proposal in 1900 that in order to understand black-body radiation, light energy should be viewed as coming in bursts or "quanta". In 1905, Einstein built on this proposal and explained the photoelectric effect in terms of light being composed of photons. Planck was first nominated for a Nobel Prize in Physics as early as 1908 and continued to be nominated on a regular basis. The Solvay Conference of 1911, with its emphasis on quantum theory, had established Planck and Einstein's work as Nobel-Prize-worthy. But there were still those who sarcastically and inaccurately characterized Planck's quanta as "hypothetical molecules of energy" and he was soundly rejected each time around. He did not help his cause when he signed the 1914 "Manifesto of the Ninety-Three", in which 93 leading German academics and artists declared their support for the German War Machine. This pronouncement did not sit well with the international community including Sweden, which hoped to remain neutral and impartial. Wilhelm Röntgen (physics 1901) and Wilhelm Ostwald (chemistry 1909), both

Germans, had signed as well, but they already had their prizes. Ostwald even went as far as equating "the large number of Nobel prizes with Germany's cultural superiority." After the prizes were awarded in 1915, the prospects for Max Planck, both a German and a theoretician, to receive a prize was not looking good as the war started.

The Swedes, as the guardians of Alfred Nobel's legacy, found themselves in a difficult predicament. During the first world war, after heated discussions, the chemistry prizes were not awarded in 1916 and 1917. The physics prize was not given in 1916. In 1918, the prizes were awarded but the presentations were deferred to 1919. Arrhenius, although he favored not making awards during the war, wanted the Nobel committees to be ready to help re-establish harmonious relations and international cooperation amongst those in the scientific community after the war. It helped that Planck was one of the few German physicists who expressed regret for signing the manifesto and corresponded with Arrhenius about maintaining the international spirit of the sciences. Planck's quantum theory remained controversial, but he had become the undisputed leader of German physics and his theory had stimulated research in these matters all over the world. With the war winding to a close, the log jam was broken in 1918 when Germans received both the physics and chemistry prizes. Max Planck won in physics and Fritz Haber in chemistry. (For more on Haber, see Chapter 17. Recall that he was considered by some to be a war criminal for spear-heading the infamous German chemical warfare unit but instead received the Nobel Prize in Chemistry for his role in devising the Haber-Bosch process that used a catalytic reaction between nitrogen and hydrogen gases to produce ammonia that, in turn, could be converted to nitrates used for both fertilizers and munitions.)

19.12 THE EINSTEIN DILEMMA

Albert Einstein was first nominated for a Nobel prize in 1910 and by 1918 his nominations were numerous and well-supported. Unlike most German professors, Einstein had not signed the Manifesto of the Ninety-Three and, in fact, had consistently and ardently opposed German militaristic tendencies. As a result of his *annus mirabilis* in 1905, he had explained the photoelectric effect in terms of particles of light eventually called photons, showed that Brownian motion was due to molecular motions, and posited the theory of special relativity including the equivalence of mass and energy as expressed in his equation, $E = mc^2$ (see Chapter 15). In 1919, in November after Planck received the deferred 1918 Nobel Prize in Physics, at a joint meeting of the Royal Society and the Royal Astronomical Society, J. J. Thomson, 1906 Nobel Laureate and the president of the Royal Society, announced that the observations of the recent solar eclipse had proven that Einstein's theory of general relativity was correct. Suddenly, Einstein was famous. What could stop him from almost immediately receiving a Nobel Prize in Physics?

The answer was virulent, vehement, and vitriolic anti-Semitism combined with a continuing bias against theoretical physics. Much of this was spearheaded by Phillip Lenard, who had been awarded the 1905 Nobel Prize in Physics for his work on cathode rays, but also had had a hand, along with his mentor Heinrich Hertz, in discovering the photoelectric effect that Einstein had explained in terms of photons in his 1905 paper. According to Phillip Ball, "There is no better example than Lenard to show that a Nobel Prize is no guarantee of wisdom, humanity or greatness of any sort ...." Evidently, when Einstein explained the photoelectric effect (using Planck's quanta), Lenard seemed to think that his discovery had been stolen. Lenard denounced what he called "Jewish Physics" as opposed to Aryan Physics (or *Deutsche Physik*). He maintained that "there was a Jewish way of doing science, which involved spinning webs of abstract theory that lacked any roots in the firm and fertile soil of experimental work." In 1920, nominations for Einstein dominated the list being considered by the physics committee. Svante Arrhenius – here we go again – prepared a negative report that included objections from Lenard and discredited the accuracy of the eclipse observations. As a result, Einstein was passed over for the 1920 prize that, incredulously, was given to Charles Guillaume for his discoveries of anomalies in nickel steel alloys!

In 1921, there were 14 nominations for Einstein, but reactionary members of the Nobel Committee for Physics again blocked the awarding of the prize to him. Many of them just could not imagine, as Friedman notes, "this bushy-haired ... radical standing at a Nobel ceremony ... receiving a prize from their king". Some maintained that Einstein *must never* receive a Nobel Prize, even though he was praised throughout Europe as the greatest living representative of physics. They were particularly opposed to the "outlandish" relativity theory which was labeled by some as a "diseased movement". The Academy voted not to award a Nobel Prize in Physics in 1921.

In 1922, the log jam was broken. Einstein again had overwhelming support but now there was also a growing movement to support the candidacy of Niels Bohr for his theory of the atom. Recall from Chapter 12 that Bohr, in 1913, had applied the quantum ideas of Planck and Einstein to explain the stability of the nuclear atom. Relativity theory was still opposed, but Einstein was the recipient of the deferred 1921 Nobel Prize in Physics for "his Services to Theoretical Physics and especially for his discovery of the law of the photoelectric effect". There was to be no mention of relativity on Einstein's Nobel certificate or at the ceremony. Bohr received the 1922 Nobel Prize "for his services in the investigation of the structure of atoms and of the radiation emanating from them".

19.13 MORE NOBEL PRIZES FOR ATOMIC SCIENTISTS

In chemistry, it was time to right the injustice whereby the work of Frederick Soddy, Rutherford's colleague in Montreal, had been ignored and slighted. Soddy and Rutherford, as noted above, had investigated radioactive decay or "transformations", but only Rutherford had received the 1908 prize. Soddy had moved to Glasgow and in 1913 introduced the idea of radioactive isotopes at a dinner party (see Chapter 6, p. 169). In 1922, Francis William Aston, working at Cambridge, had developed his mass spectrometer and found that ordinary neon gas was composed of two isotopes, Ne-20 and Ne-22. This reinforced the importance of Soddy's work. There was also sentiment that it was time to recognize British scientists who had not received an award since Rutherford's in 1908. In addition, Germans had been recognized steadily starting in 1918. In 1920, the German physical chemist, Walther Nernst, despite the usual protestations of Arrhenius driven by decades of interpersonal issues, had received a chemistry prize for his work in thermochemistry. Soddy and Aston were awarded the 1921 and 1922 Nobel Prizes in Chemistry, respectively, in recognition of their work.

In 1926, Jean Perrin (1870–1942) was awarded a Nobel Prize in Physics for his experimental work in verifying Einstein's ideas on Brownian motion as due to collisions of atoms and molecules with small particles in air. This resolved forever that atoms and molecules physically exist and made for a grand celebration to honor the 25th anniversary of the Nobel Prizes. Perrin also had a long history of significant contributions to atomic research. In 1895, very early in his career, he had identified cathode rays as negative particles, anticipating J. J. Thomson's discovery of the electron in 1897 (see Chapter 5). In 1919, he had proposed that nuclear fusion processes were responsible for the illumination of the stars. Some said that Perrin represented the "glory of French science" and was a hero in the tradition of Antoine Lavoisier. He spent the early World War II years in the United States and died there in 1942. After the war, his ashes were returned to France in the cruiser *Jeanne d'Arc* and placed in the Panthéon. How many scientists do you know whose ashes were carried in such a vessel? The scientific/historical traveler should find his grave when looking for those of the Curies (see Chapter 14).

Charles Thomson Rees (C. T. R.) Wilson (1869–1959), a Scottish physicist, won the Nobel Prize in Physics in 1927. By 1912 or so, Wilson had perfected his cloud chamber, in which a supersaturated fog is formed in a dust-free chamber. When an electronically charged particle passes through the fog, it ionizes the molecules in its path. Water droplets then condense on these ions showing the path of the original charged particle. The cloud chamber was used to study radioactivity (alpha and beta particles, for example), X-rays, cosmic rays, *etc*. In 1924 it was used in Manchester to confirm Rutherford's 1919 claim that he had produced the first artificial nuclear transformation (see Chapter 6).

19.14 QUANTUM MECHANICS FINALLY HONORED

Recall from Chapter 16 that Prince Louis-Victor de Broglie, in 1924, had extended Einstein's wave-particle duality of light to suggest that electrons could be treated as waves. De Broglie's equation, $\lambda = h/mv$, had shown that any particle of mass, m, traveling at a velocity, v, has a corresponding wavelength, λ. It did not take long for Erwin Schrödinger to produce his wave equation that would become the basis for the modern view of the internal structure of the atom often called "wave mechanics". Werner Heisenberg developed the competing theory of "matrix mechanics" about the same time. By the late 1920s, nominations for de Broglie, Schrödinger, and Heisenberg started rolling into Stockholm but, as usual, as we have come to expect, they faced resistance from the Swedish Royal Academy of Science. However, by then, the log jam preventing prizes for theoretical physics was starting to break up. Einstein had received the 1921 physics prize and Bohr had won in 1922. De Broglie's extension to the wave-particle of matter was relatively tame. It helped that Davisson and Germer, in 1927, demonstrated that electrons acted like waves and produced an interference pattern when electrons were fired at crystals. In 1928, George Paget Thomson demonstrated the same thing. Given these discoveries, it is not surprising that de Broglie won the 1929 Nobel Prize in Physics "for his discovery of the wave nature of electrons". G. P. Thomson and Davisson shared the Nobel Prize for Physics in 1937 for demonstrating that electrons undergo diffraction, a "behavior peculiar to waves".

With de Broglie the winner in 1929, the stage was set to honor quantum mechanics. Arrhenius had died in 1927 but others were successful in delaying the honors a bit longer. By 1933, nominations for Paul Dirac's relativistic quantum mechanics had been added to those of Heisenberg and Schrödinger. Finally, there was a way forward. Werner Heisenberg won the 1932 prize (which had been deferred and awarded in 1933), whereas Erwin Schrödinger and Paul A. M. Dirac shared the award in 1933. As noted in Chapter 16 (pp. 403–404), Max Born and Pascual Jordan were passed over for any of these prizes because Jordan had joined the Nazi party in 1933 and Born and Jordan's work could not be separated. Heisenberg was dismayed that Born, without whom he could not have developed "matrix mechanics", had not shared the prize with him. Born was finally recognized in 1954.

19.15 NUCLEAR PHYSICS AND THE HAHN/MEITNER DEBACLE

Nuclear physics did not encounter the same resistance. James Chadwick won the Nobel Prize in Physics in 1935 for his discovery of the neutron, which had just been missed by Frédéric Joliot and Irène Joliot-Curie (see Chapter 6, p. 163). In that same year, the Joliot-Curies won the Nobel Prize in Chemistry for their discovery of artificially induced radioactivity (see Chapter 14). Enrico Fermi won in Physics in 1938 (see Chapter 16) for his discovery that fission is promoted by using slow neutrons. He also claimed that he had produced elements of atomic number greater than 92 (the *trans*-uranic elements), but this was later mostly refuted.

These were fairly straightforward awards but now we come to what is sometimes called the Hahn/Meitner debacle. Recall from Chapter 17 the story of the discovery of fission (in 1938–1939) involving Otto Hahn, Fritz Strassman, Lise Meitner, and Robert Frisch. As we saw then, Meitner played a crucial role in this discovery and, indeed, was the intellectual leader of the Kaiser Wilhelm Institute group until, in 1938, she had been forced to emigrate to Sweden. Here, she wound up in the Stockholm laboratory of Karl Manne Siegbahn, who had won the Nobel Prize in Physics in 1924. Siegbahn had been pressured by Bohr and others to accommodate Meitner but, according to Friedman, threatened by Meitner's superior knowledge of nuclear physics, he had "treated her shabbily and without respect".

From 1939 to 1946, there were numerous nominations for both Meitner and Hahn (as well as Strassman and Frisch) to share various combinations of the physics and chemistry prizes. In 1939, they were nominated to share the chemistry award. In 1940, it was for physics. One nominator, a former prize winner, noted that "As I understand the matter, Professor Hahn and Fräulein Meitner should be included in the award for their work respectively in identifying the fission process and in showing the tremendous energy liberated when the fission occurs." Again in 1941, nominators for both the physics and chemistry prizes emphasized that Hahn and Meitner should be honored together. This continued in 1942, 1943, and 1944. In 1945, Bohr expressed the opinion that Hahn and Strassman should share the chemistry prize, while Meitner and Frisch should share the physics prize. This is exactly the way it should have come out. Unfortunately, information regarding Meitner and Frisch's contributions was slow to emerge in the post-war years. In addition, the physics and chemistry Nobel committees could not seem to agree as to which one had jurisdiction to evaluate Meitner's contribution to the discovery of fission. Indeed, there was some mistaken sentiment that Hahn and Strassman had discovered fission after

Meitner had emigrated to Sweden and that they would not have discovered it if she had remained in Berlin. Then there are the unknown effects of gender-bias, anti-Semitism and Siegbahn's fear that giving Meitner the prize would leave him with a diminished role in Swedish scientific circles. In any case, Meitner's role was devalued and, in 1945, Hahn was declared the sole winner of the chemistry prize. In 1946, Meitner and Frisch were nominated for the physics award but did not receive it. At the end of the day, we cannot escape the conclusion that Lise Meitner was unjustly denied a Nobel Prize in Physics for her work in identifying nuclear fission.

19.16 TRAVEL SITES IN OR NEAR STOCKHOLM RELATED TO THE NOBEL PRIZES

The Science Awards are made in Stockholm, whereas the Peace Prize is given in Oslo. The Prize Award Ceremony in Stockholm has, almost without exception, taken place at the Stockholm Concert Hall (Stockholms Konserthus) since 1926. Since 1901, the Nobel Prizes have been presented to the Laureates at ceremonies on 10 December, the anniversary of Alfred Nobel's death. By tradition, established early on, the Swedish monarch gives each laureate a diploma and a gold medal. With few exceptions, the Nobel Banquet has taken place in Stockholm City Hall (Stockholms Stadhus) since 1930. Friedman notes that the "modest and retiring Alfred Nobel never foresaw the hallmark of his legacy: an opulent ceremony and banquet". Scientific/historical travelers should be sure to visit these sites in Stockholm. See Chart 19.2 for the location of these sites.

19.16.1 Stockholm Concert Hall (Stockholms Konserthus)

Hötorget 8 (59.335073°, 18.063130°). The Nobel Prize Award Ceremony has taken place here, almost without exception, since 1926. After speeches extolling the prize winners and their discoveries, the Swedish monarch, currently King Carl XVI Gustaf of Sweden, gives each laureate a diploma and a 200 g, 18 karat gold medal. Both sides of a Nobel gold medal are shown in Figure 19.3. The Nobel Lectures are given by the laureates in the days preceding the award ceremony. Built in the neoclassical style, this blue building housing the concert hall is reminiscent of a classical Greek temple and is considered a

Chart 19.2 Travel sites in Stockholm related to the Nobel Prizes.

Figure 19.3 Official photographs of a Nobel Prize gold medal with the picture of Alfred Nobel on the front with his birth and death dates in Roman numerals and the name of the winner on the back. © ® The Nobel Foundation.

Swedish architectural masterpiece. It is the home of the Royal Stockholm Philharmonic Orchestra. Travelers might get tickets to a concert and consider it a double-treat. Outside the main stairs of the building is the famous bronze statuary fountain, the Orfeus-brunnen.

19.16.2 Stockholm City Hall (Stockholms Stadhus)

Hantverkargatan 1, on the eastern tip of Kungsholmen island, (59.327413°, 18.054795°). The City Hall, built with eight million red bricks, is an active political office building. It is also the site of the Nobel Banquet held each year in the "Blue Hall". 1300 people are served, including 250 students selected by lottery. The organ in the Blue Hall has 10 270 pipes and is the largest in Scandinavia. The staircase leading from the upper balcony down to the main hall is constructed for ladies wearing high-heel shoes and long gowns. Above the Blue Hall is the Golden Hall (*Gyllene Salen*), named after the decorative mosaics made of more than 18 million tiles. This is where the laureates, members of the Royal family, and other dignitaries dance after the banquet. Look for the mosaic of Berzelius in this amazing room. The hall can only be visited on 50-minute guided tours, available in both English and Swedish. The main legislative room for the Stockholm City Council is where the Nobel laureates and their spouses dine. Afterwards, climb up the City Hall Tower with its three golden crowns on top that reflect the Swedish coat of arms. An elevator goes up half-way and from there a helical stairway traces its way up inside the square tower where one is rewarded with breathtaking views of the city. The small park, the Stadshusparken, between the building and Lake Mälaren's shore is adorned with several stunning sculptures.

19.16.3 The Nobel Museum (Nobelmuseet)

Börshuset, Stortorget 2 (59.325353°, 18.070846°), Gamla Stan, Stockholm. Housed in an impressive eighteenth-century building on the edge of Stortorget, this museum was founded in 2001, the same year that the Nobel Prizes celebrated their 100th anniversary. Here you can learn about the Nobel Laureates and the founder of the prizes, Alfred Nobel. Visit the small theater with a panoramic set of three screens that shows films associated with past winners. Suspended from the ceiling are more than 900 banners that move around the museum and show the Nobel Laureates in no particular order. If you miss a favorite laureate, curse and lament, because you will have to wait more than six hours for him or her to reappear. There are more than 38 stops to visit and hear an explanation using an audio guide. Many of these organize the prizes by decades: Stop 13, prizes from 1901 to 1910; Stop 14, 1911–1920, Stop 15, 1921–1930, Stop 16, 1931–1940. Other stops detail the Nobel Prize Awarding Ceremony and the Nobel Banquet, the life of Alfred Nobel, his invention of dynamite, his journeys all over the world, his last will and testament, a description of more 2000 objects owned by the museum, Albert Einstein's letter detailing that his wife, Mileva Marić, was to get his prize money, and so forth. Walk through the museum and experience decades of developments and triumphs since 1901. Visit the Bistro Nobel, have a coffee and a snack and talk about your favorite winners. Be sure to turn your chair upside down and see if it is signed by a Nobel Laureate. Guided tours are available.

19.16.4 Alfred Nobel Grave

Norra begravningsplatsen (Northern burial ground), Stockholm Alfred Nobels allé, Section Kv 4A, grave 170 ($59.356092°$, $18.032557°$).

19.16.5 Alfred Nobels Björkborn

10 Bjoerkbornsvaegen, Karlskoga, Sweden. ($59.340354°$, $14.534645°$). Alfred Nobel spent his summers at the Bjorkborn Manor from 1894 to 1896. He built a laboratory here which still exists today. It is now a museum where visitors can take a guided tour given by "Alfred Nobel himself" or someone else from his life. The exhibits cover Nobel's experimental activities, discuss how the prizes were set up and then concentrate on the man himself and what his life was like here in his last few years. Tours are given in Swedish and English. Café Alfred Nobel offers hot and cold drinks, sandwiches, and lighter lunches.

19.17 SUMMARY AND CONCLUDING REMARKS

The chapter starts with the contributions of three Swedish chemists to the history of the atom. Jakob Berzelius devised the modern symbols for the elements, determined many atomic weights and provided numerous examples of the Law of Definite Proportions but, like his contemporary John Dalton, rejected ideas that would have quickly advanced chemistry if they had been immediately accepted. Unfortunately, the Berzelius Museum is no longer available to visit in Stockholm, but Berzelius Square is one of the most beautiful settings to honor a chemist. Carl Wilhelm Scheele had a hand in discovering a number of elements including oxygen, for which he is often listed as a co-discoverer with Joseph Priestley. His statue in the Hops Garden and a replica of his pharmacy in Skansen Park are worthy scientific/historical sites. Svante Arrhenius's dissociation theory, presented in his doctoral thesis, was initially ridiculed but within two decades, it earned him the 1903 Nobel Prize in Chemistry.

The first several decades of awarding Nobel Prizes in chemistry and physics were characterized by extensive debate as to how to interpret Nobel's intent in selecting prizes, the proper role of the prizes during WWI, and the ramifications of strident interpersonal and international politics with particular emphasis on the out-sized role of Arrhenius, who was bent on revenge because of the early ridicule with which his dissociation theory had been received. Arrhenius had an unduly strong hand in denying awards to Dmitri Mendeleev and Ludwig Boltzmann and when and in what manner to honor Ernest Rutherford, Wilhelm Ostwald, Marie Curie, Max Planck, Albert Einstein, and many others. Determining the winners of other early awards involved: (1) an initial bias against theoretical physics, including the delay in recognizing quantum mechanics; (2) the Einstein dilemma characterized by virulent anti-Semitism; and (3) the Hahn/Meitner debacle that resulted in the controversial decision whereby Lise Meitner was unjustly denied a prize for her role in discovering fission. We visit the sites where the Nobel Prizes are awarded and then the Nobel Prize Museum. Travelers-with-the-atom will be well-informed as they stroll the museum, peer at the laureate banners parading far above them, take in the films, listen to the audio presentations, inspect the artifacts, hear the stories, and talk with fellow travelers.

Traveling with the Atom is a book about traveling in Europe and beyond to visit places related to one of the most carefully researched and enduring ideas in human history, the atomic concept. These places include homesteads, birthplaces, graveyards, squares, statues, mines, universities, laboratories, museums, libraries, lecture halls, apartments and individual rooms, estates, cathedrals and abbeys, and even castles in some of the most picturesque rural areas, charming small towns and villages, ordinary working-class municipalities, and some of the most elegant and romantic cities in Europe. During our travels, we have been reminded over and over that humankind has only slowly and haltingly unraveled these atomic insights. We have seen that the pathway to the modern atom has been characterized by ingenious experiments and clear-headed observations as well as serendipitous accidents and irrelevant or ill-conceived manipulations; by great insights as well as wrong-headed ideas; by persuasive arguments by humble men and women as well as pig-headed opinions driven by big egos; by logical discussions, papers, and meetings as well as by personal attacks, stinging diatribes, and heated debates. As scientists, students, friends, and companions, we have talked together about the many fascinating stories behind the development of the atomic concept so that when we have visited a site, we have been able to more fully appreciate exactly what we have in front of us. In summary, this book has been about two types of landmarks – the temporal landmarks of the history of the atomic concept and the physical landmarks that have been preserved all over the European continent and beyond to commemorate this achievement. What a great experience it is to "travel with the atom".

ADDITIONAL READING

- J. Wisniak, Jons Jakob Berzelius, A Guide to the Perplexed Chemist, *Chem. Educ.*, 2000, **5**, 343–350.
- P. Ball, How 2 Pro-Nazi Nobelists Attacked Einstein's 'Jewish Science', in *Serving the Reich: The Struggle for the Soul of Physics Under Hitler*, University of Chicago Press, Chicago, 2014.
- J. L. Marshall and V. R. Marshall, Rediscovery of the Elements, Carl Wilhelm Scheele, *The Hexagon*, 2005, pp. 8–13.

REFERENCES

1. C. Djerassi and R. Hoffmann, *Oxygen, A Play in 2 Acts*, Wiley-VCH Verlag GmbH Weinheim, Germany, 2001.
2. J. Marshall and V. Marshall, Northern Scandinavia: An Elemental Treasure Trove, in *Science History: A Traveler's Guide*, ed. M. V. Orna, American Chemical Society, Washington DC, 2014.
3. R. M. Friedman, *The Politics of Excellence*, Times Books, Henry Holt and Company, LLC, 2001.

Appendix

Traveler's Guide to the History of the Atom

Date	Scientist/Event	Short description	Ch	Country	Chart
430 BC	Leucippus Democritus	Greek a-tomic particles	2	Turkey	
1564	Galileo Galilei	Born in Pisa	9	Italy	9.1
1600	Giordano Bruno	Burned at the stake	9	Italy	9.1
1627	Robert Boyle	Born in Lismore Castle	2	Ireland	2.1
1635	Robert Hooke	Born on the Isle of Wight	2	England	3.1
1642	Isaac Newton	Born at Woolsthorpe Manor	3	England	3.1
1642	Galileo Galilei	Dies in Acetri, near Florence	9	Italy	9.1
1661	Robert Boyle	Publishes *The Sceptical Chymist*	2	England	3.1
1662	Robert Boyle	Boyle's Law relating gas pressure and volume	3	England	3.1
1666–1667	Isaac Newton	Year of miracles (*"annus mirabilis"*); Great Fire of London; Great Plague	3	England	3.1
1669	Hennig Brandt	Discovers phosphorus	2	Germany	
1679	Robert Boyle	Discovers method for producing high yields of phosphorus	2	England	3.2

Traveling with the Atom: A Scientific Guide to Europe and Beyond
By Glen E. Rodgers
© Glen E. Rodgers 2020
Published by the Royal Society of Chemistry, www.rsc.org

Appendix

Date	Scientist/Event	Short description	Ch	Country	Chart
1700	George Stahl	Advocated phlogiston theory	8	Germany	
1727	Isaac Newton	Buried in Westminster Abbey	3	England	3.2, 7.1
1756	Joseph Black	Discovers "fixed air" (CO_2)	3	Scotland	3.1, 3.4
1766	Henry Cavendish	Discovers "inflammable air" (H_2)	3	England	
1766	John Dalton	Born in Eaglesfield	4	England	4.1
1767	Joseph Priestley	Publishes *History and Present State of Electricity*	3	England	
1771	Antoine Lavoisier	Marries Marie-Anne Paulze, daughter of a "tax farmer"	8	France	
1772	Daniel Rutherford	Discovers "phlogisticated air" (N_2)	3	Scotland	
1774	Joseph Priestley	Discovers "dephlogisticated air" (O_2) Visits Antoine Lavoisier in Paris	3	England	3.5
1777	Carl Wilhelm Scheele	Independently discovers oxygen	19	Sweden	19.1
1778	Humphry Davy	Born at Penzance near Lands End	3	England	3.6
1789	Antoine Lavoisier	Publishes *Elementary Treatise on Chemistry*	8	France	8.1
1791	Joseph Priestley	Burned out of his home in Birmingham	3	England	3.5
1794	Joseph Priestley	Moves to Northumberland, PA (USA)	3	United States	
1794	Antoine Lavoisier	Guillotined at Place de la Révolution, now Place de la Concorde, Paris	8	France	8.1
1798	Humphry Davy	Appointed Director, Beddoes Institute, Bristol; calls N_2O "laughing gas"	3	England	3.6
1799	Joseph Proust	Discovers Law of Definite Proportion	4	France	
1799	Alessandro Volta	Announces invention of "voltaic pile", first battery	9	Italy	9.2
1801	Humphry Davy	Moves to the Royal Institution	3	England	3.2
1802	Joseph Louis Gay-Lussac	Gay-Lussac's Law (relating gas volume and temperature)	8	France	

(*continued*)

Date	Scientist/Event	Short description	Ch	Country	Chart
1803	John Dalton	Discovers Law of Multiple Proportions	4	England	4.2
1804	Napoleon Bonaparte	Appointed Emperor of France	8	France	
1807–1809	Humphry Davy	Discovers 6 elements in 3 years using Voltaic Pile	5	England	3.2
1808	John Dalton	Publishes *A New System of Chemical Philosophy* (atomic theory)	4	England	4.1, 4.2
1809	Joseph Louis Gay-Lussac	Law of Combining Volumes	8	France	
1811	Amadeo Avogadro	Avogadro's hypothesis (EVEN)	8	Italy	9.3
1812	Michael Faraday	Attends Davy's Farewell Lectures at the Royal Institution (Ri)	5	England	5.1
1813	Jöns Jacob Berzelius	Modern Symbols for Elements	19	Sweden	19.1
1814	Jöns Jacob Berzelius	Publishes table of atomic weights	19	Sweden	19.1
1814	Joseph v. Fraunhofer	Observed dark lines in solar spectrum, Still called "Fraunhofer Lines"	11	Germany	11.1
1814	André-Marie Ampère	Avogadro-Ampère gas theory (EVEN)	8	France	8.1
1820	Hans Christian Ørsted	Discovers electromagnetism	12	Denmark	12.1, 12.2
1825–1826	Michael Faraday	Establishes "Christmas Lectures" & "Friday Evening Discourses" at Ri	5	England	5.1
1829	Johann Wolfgang Döbereiner	Points out groups of three similar elements soon to be called "triads"	18	Germany	18.1, 18.2
1831	Michael Faraday	Publishes *Experimental Researches in Electricity* (lines of force; E and M fields)	5	England	7.1
1831	James Clerk Maxwell	Born in Edinburgh	5	Scotland	5.2

Appendix

Date	Scientist/Event	Short description	Ch	Country	Chart
1833	Michael Faraday	Rules of electrolysis; electrical current produces ions at electrodes	5	England	5.1
1843	Leopold Gmelin	Coins the term "triad" and extends list	18	Germany	
1848	Louis Pasteur	Discovers right- and left-handed molecules	8	France	
1851	London's "the Great Exhibition"	Finances "Exhibition of 1851 Scholarships" and the Science Museum	6	England	
1855–1862	James Clerk Maxwell	Publishes papers on lines of force and electromagnetic field	5	Scotland	5.3
1860	Robert Bunsen Gustav Kirchhoff	Spectroscopically discover the elements cesium & rubidium	11	Germany	11.2
1860	Stanislao Cannizzaro	Speaks at 1st Chemical Congress, Karlsruhe, clarifying atomic weights	9	Germany	
1860	James Clerk Maxwell	Distribution of Molecular Velocities; Kinetic-Molecular Theory of Gases	5	Scotland	5.4
1860	Alfred Nobel	First made nitroglycerine explode with success	19	Sweden	
1862	Anders Ångstrom	Identifies hydrogen in the Sun	11	Sweden	
1862	Alexandre-Émile Béguyer de Chancourtois	First orders elements by atomic weight	18	France	
1864	Julius Lothar Meyer	Proposes an early periodic table	18	Germany	18.1
1865	Josef Loschmidt	Calculates number of Molecules in a cm^3 of a gas at STP; molecular size	10	Austria	10.1
1865	John Newlands	Presents his "Law of Octaves" to the Chemical Society	18	England	18.1

(*continued*)

Date	Scientist/Event	Short description	Ch	Country	Chart
1867	Alfred Nobel	Invents dynamite by combining nitroglycerin with Kieselguhr	19	Germany	
1868	Pierre Janssen	Identifies the new element, helium, in the Sun	11	France	
1869	Dmitri Mendeleev	Publishes his first periodic table	18	Russia	18.5
1871	Ernest Rutherford	Born in Brightwater, New Zealand	6	New Zealand	6.1
1872	Ludwig Boltzmann	Maxwell–Boltzmann distribution of molecular energies in a gas (KMT)	10	Austria	10.1
1873	James Clerk Maxwell	Publishes *Treatise on Electricity and Magnetism* written at Glenlair House	5	Scotland	5.3
1874	James Clerk Maxwell	1st Professor of Physics at the new Cavendish Laboratory of Physics	3	England	5.4
1875	Lecoq de Boisbaudran	Discovers gallium (Mendeleev's eka-aluminum)	18	France	18.1
1877	Ludwig Boltzmann	Formulated his equation $S = k \log W$	10	Austria	10.1
1879	James Clerk Maxwell	Dies (age 48), buried at Parton Kirk	5	Scotland	5.3
1879	Lars Fredrik Nilson and Per Teodor Clève	Discover scandium (Mendeleev's eka-Boron)	18	Sweden	18.1
1884	John Joseph (J. J.) Thomson	Appointed (age 28) 3rd Cavendish Professor of Physics	5	England	5.4
1884	Svante Arrhenius	Proposes dissociation theory	19	Sweden	19.1
1885	Johann Balmer	Balmer equation that used integers to generate hydrogen wavelengths	11	Switzerland	
1886	Clemens Winkler	Discovers germanium (Mendeleev's eka-silicon)	18	Germany	18.3
1887	Heinrich Hertz	Discovers radio waves and the photoelectric effect	15	Germany	

Appendix

Date	Scientist/Event	Short description	Ch	Country	Chart
1895–1900	Ludwig Boltzmann Ernest Mach	Debate the existence of atoms	10	Austria	10.1
1890s	William Ramsay	Discovery of the "inert gases"	7, 18	England	7.1
1895	Wilhelm Röntgen	Discovers X-rays	13	Germany	13.1
	Ernest Rutherford	Emigrates from New Zealand to England	6	England	6.1
1896	Henri Becquerel	Discovers radioactivity	14	France	14.1
	Alfred Nobel	Estate sets up the Nobel Prizes	19	Sweden	19.2
1897	J. J. Thomson	Discovers the electron	5	England	5.4
1898	Pierre and Marie Curie	Announce the discovery of two elements, polonium and radium; coin the term "radioactivity"	14	Paris	14.1
1900	Max Planck	Introduces the quantum of energy in his explanation of blackbody radiation	15	Berlin	15.1
1901	Alfred Nobel	First Nobel Prizes awarded	19	Sweden	19.2
1902	Pierre and Marie Curie	Isolate nearly pure radium chloride and determine Ra's atomic weight	14	Paris	14.1
1902	Ernest Rutherford and Frederick Soddy	Characterize nuclear transformations and define half-life	6	Canada	6.2
1904	J. J. Thomson	Proposes "Plum-Pudding Model of the Atom"	5	England	5.4
1905	Albert Einstein	Publishes 3 seminal papers on (1) photoelectric effect (light quanta) wave-particle duality of light (2) Brownian motion due to atomic motions (3) Special theory of relativity	15	Switzerland	15.2, 15.3
1906	Pierre Curie	Killed by a horse-drawn wagon near Pont Neuf in Paris	14	Paris	14.2

(continued)

Date	Scientist/Event	Short description	Ch	Country	Chart
1906	Ludwig Boltzmann	Commits suicide	10	Italy	10.1
1909	Ernest Marsden Hans Geiger	Alpha-particle, gold-foil experiment in Rutherford's laboratory	6	England	6.3
1911	Ernest Rutherford	Announces discovery of the nucleus and the planetary atom	6	England	6.3
1911	Niels Bohr	Arrives in Cambridge to work with Thomson	12	England	6.4
1912	Niels Bohr	Moves to Manchester to work with Rutherford	12	England	6.3
1913	Niels Bohr	Presents the Bohr Atom with certain allowed orbits	12	Denmark	12.2
1913	Frederick Soddy	Formally proposes idea of isotopes	5	England	
1913	Henry Moseley	Discovers atomic number using X-rays	6	England	6.5
1914–1918	World War I	Henry Moseley killed at Gallipoli	6	Turkey	
1915	Ernest Rutherford	Performs first artificial nuclear transformation	6	England	6.3, 6.4
1919	Ernest Rutherford	Appointed 4th Cavendish Professor of Physics	6	England	5.4
1921	Marie Curie	First trip to America to receive 1 gram of radium from the "Women of America"	14	United States	
1923	Louis de Broglie	Links waves with electrons extending wave-particle duality to include matter	16	France	
1925	Wolfgang Pauli	Discovers the exclusion principle (No two electrons in an atom can have the same four quantum numbers)	—	Austria	
1925	Werner Heisenberg	Proposes "matrix mechanics"	16	Germany	

Date	Scientist/Event	Short description	Ch	Country	Chart
1926	Paul Dirac	Formulates his relativistic quantum mechanics	16	England	16.2
1926	Erwin Schrödinger	Constructs his wave equation and establishes "wave mechanics"	16	Switzerland	16.1
1927	Enrico Fermi	Appointed Chair of Physics at the University of Rome	17	Italy	17.1
1927	Werner Heisenberg	Discovers the uncertainty principle	16	Germany	
1929	Marie Curie	Second trip to America to receive money to purchase radium for Warsaw Institute	14	United States Poland	14.3
1932	John Cockcroft and Ernest Walton	Cockcroft-Walton Machine experiments verify Einstein's $E = mc^2$	6	England	6.6
1932	James Chadwick	Discovers the neutron	6	England	6.4
	Carl Anderson	Discovers the positron	14	United States	
	Irène Curie	Replaces her mother as director of the Radium Institute	14	France	14.1, 14.2
1934	Irène and Frédéric Joliot-Curie	Discover artificial radioactivity	14	France	14.2
1934	Enrico Fermi and his "Panisperna boys"	Start to bombard all elements with neutrons; discover "moderators"	17	Italy	17.1
1934	Otto Hahn and Lise Meitner	Bombard uranium with neutrons	17	Germany	17.2
1934	Ida Noddack	Maintained that Fermi had produced fission	17	Germany	
1938	Lise Meitner	Forced to emigrate to Sweden	17	Germany	17.2
1938	Otto Hahn and Fritz Strassman	Discover barium produced when uranium bombarded with neutrons	17	Germany	17.2
1938	Lise Meitner and Otto Frisch	Explain that Hahn and Strassman had discovered fission	17	Sweden	17.2

(*continued*)

Date	Scientist/Event	Short description	Ch	Country	Chart
1939	Werner Heisenberg	Appointed to lead German "uranium project"	17	Germany	17.3
1940	Erwin Schrödinger	Appointed Professor Dublin Institute of Advanced Studies	2	Ireland	2.2
1941	Werner Heisenberg and Niels Bohr	Meet at House of Honor in Copenhagen to discuss nuclear weapons	17	Denmark	17.2
1939–1945	World War II				
1944	Erwin Schrödinger	Publishes *What Is Life?*	2	Ireland	2.2
1953	James Watson Francis Crick	Discover DNA Double Helix	2	England	

Place Index

Entries and page numbers in **bold** indicate three atom (●●●), four atom (●●●●) and five atom (●●●●●) travel sites. Page numbers with suffix "*ch*" indicate a chart. Page numbers in *italics* refer to figures.

Abbeyside, County Waterford,
 Southern Ireland 172–173
Abdera, Greece 12
Aberdeen, Scotland 115
Alpbach, Austria 399–400
Annency, France 212
Arcueil, France 212
AUSTRIA *255ch*
 Alpbach 399–400
 **Graz (Physics Institut,
  University of Graz) 254**
 Vienna (*see* **Vienna**)

Bad Dürkheim, Germany 273
Bath, England 191
Benediktbeuren,
 Germany 267
Berlin, Germany 365–368,
 367ch, 383, 427–430,
 434–436, *435*, 440, 442
 **Archives of the Max
  Planck Society 368**
 Einstein plaques 383
 Fritz Haber
  Institute 435–436

Great Synagogue 383
Hahn Monument 436
1 Hessischestrasse
 (Meitner-Hahn
 Plaque) 434
**Humboldt University
 (Planck and Meitner
 statues) 366–367, 434**
Max Planck
 Residence 367–368
Thielallee
 (plaques in honor
 of the discovery of
 fission) 436
Bern, Switzerland 371–372,
 379–382, *380ch*
 Café Bollwerk 381
 **Einstein Exhibit at
  Historical Museum of
  Bern 382**
 **Einstein House and
  Plaque 379–381**
 Einstein Plaque at
  the Former Patent
  Office 381

519

Birmingham, England *57ch*, 62–64
Birstall, West Yorkshire, England *57ch*, 58
Bologna, Italy 231–232
Bowood House, Wiltshire, England *57ch*, **58–61**, 62
Bramley, Hampshire, England 437
Brandenberg, Germany 383–384
Brightwater, New Zealand (Lord Rutherford Memorial Reserve) 143–144
Bristol, England *65ch*, 68–72, 406–408, *407ch*
Bug, Weißdorf, Germany 453

Calne, Wiltshire, England *57ch*, 58–60
Cambridge, England *33ch*, 125–127, *126ch*, 147–148, 160–167, *164ch*, 292–293, 403, 408
    Cavendish Laboratory 52, 116–117, **125–126**, *126ch*, 127, 128, **135**, 160–165, **165–166**
    Cavendish Museum (at the New Cavendish Laboratory) **127**, 130, **135**, 164–165
    Crocodile Carving 165–166, *166*
    Eagle Public House 126–127, 167
    Newnham Cottage, Queen's Road 167
    Trinity College 44–46, *45ch*, 113
    Whipple Museum of the History of Science 127, 166–167
CANADA *see* **Montreal, Canada**
Caputh, Germany 383
Chatsworth House, Derbyshire, England 52
Chicago, Illinois, USA 426–427
Christchurch, New Zealand (Rutherford's Den) 147
Cognac, France 457–459
Como, Italy 228, 233–236, *233ch*
    Alessandro Volta Statue 235
    Camnago Volta cemetery (Volta tomb) 235–236
    Life Electric statue 235
    Tempio Voltiano (Volta Temple) 233–234, *234*
    Torre di Porta Nuovo or Torre Gattoni 235
    Volta Lighthouse 236
    Volta plaques 235
Copenhagen, Denmark *289*, 289–291, 304–309, *304ch*
    Assistens Kirkegård (Cemetery) 291, 307
    Carlsberg Honorary Residence 306–307, 430, 440–441
    Danish Historical Museum, Fredriksborg Castle 305
    Elephant Gate 306–307

Niels Bohr
 Birthplace 304
Niels Bohr Childhood
 Home 305
**Niels Bohr
 Institute 305–306
Ørsted Statue 289–290**
Rosenborg Castle 308
Round Tower
 (Rundetårn) 308–309
University of
 Copenhagen 305
Vor Frue Kirke (Church of
 Our Lady) 308
**Corsock, Dumfries and
 Galloway (James Clerk
 Maxwell Memorial
 Window)**, Scotland *122ch*,
124–125
CZECH REPUBLIC 12

Deansgrange Cemetery,
 County Dublin, Southern
 Ireland 173
DENMARK
 285–310, *295ch*
 Copenhagen (*see*
 Copenhagen)
 Kroppedal Museum 308
 Rudkøbing, Island
 of Langeland
 288–289
Derby Cathedral, England 53
Down House, Kent,
 England 191–192
Dublin, Southern
 Ireland 18–23, *20ch, 21,*
 25–28, *27*
Dungarvan, County
 Waterford, Southern
 Ireland 172–173

Eaglesfield, Cumbria,
 England 78, 79, *80ch*
Eastbourne, East Sussex,
 England 168
Edinburgh, Scotland 48–51,
49ch, 112, 113, 117–121,
118ch
 Edinburgh Academy 121
 Greyfriars Kirkyard *50,*
  50–51
 31 Heriot Row (Maxwell's
  Aunt Isabella
  Wedderburn's
  home) 121
 **James Clerk Maxwell
  Foundation 117–120**
 James Clerk Maxwell
  Statue *119,* 121
 Joseph Black
  Building, Edinburgh
  University 48–50
 National Museum of
  Scotland 48
 Scottish National Partrait
  Gallery 48
 Sylvan House 50
ENGLAND 30–31, 32–47, *33ch,*
 52–61, 62–64, 65–72, *65ch,*
 78–83, 89–96
 Bath 191
 Birmingham *57ch,* 62–64
 Birstall, West
  Yorkshire *57ch,* 58
 **Bowood House,
  Wiltshire** *57ch,* **58–61,**
  62
 Bramley, Hampshire 437
 Bristol *65ch,* 68–72,
  406–408, *407ch*
 Calne, Wiltshire *57ch,*
  58–60

ENGLAND *(continued)*
   Cambridge *(see*
     Cambridge, England)
   Chatsworth House,
     Derbyshire 52
   Derby Cathedral 53
   Down House,
     Kent 191–192
   Eaglesfield, Cumbria 78,
     79, *80ch*
   Eastbourne, East
     Sussex 168
   Freshwater, Isle of
     Wight *33ch,* 40–41
   Godmanchester,
     Cambridgeshire 433,
     441
   Grantham,
     Lincolnshire
     33ch, 43–44
   Heckmondwike, West
     Yorkshire *57ch,* 58
   Kendal, Cumbria 78,
     80ch, 81–83
   Lake District National
     Park, Cumbria *80ch*
   Leeds *(see* **Leeds,**
     England)
   London *(see* **London,**
     England)
   Oxford *(see* **Oxford,**
     England)
   Pardshaw Hall,
     Cumbria 78,
     79–81, *80ch*
   Penzance,
     Cornwall *65ch,* 66–68
   St. Michael's Mount,
     Cornwall *65ch,* 66
   Stalbridge, Dorset 33,
     33ch, 34
   Terling Place,
     Essex 128–129
   Warrington,
     Cheshire 55, 79
   Woolsthorpe,
     Lincolnshire 42
   Woolsthorpe Manor,
     Lincolnshire 43

Florence, Italy 23, 105,
   225–227, 425
   Basilica di Santa
     Croce 226, *227,* 425
   Foundation for Science
     and Technology 227
   Museo Galileo 225–226,
     425
Foxhill, New Zealand 144
FRANCE 196–220, 330–352,
   388–390
   Annency 212
   Arcueil 212
   Cognac 457–459
   Paris *(see* **Paris,**
     France)
   Ploubazlanec, l'Arcouest,
     Brittany 352–353
Freiberg, Germany 460–462,
   461ch
   Clemens Winkler
     Collection 460–461
   Clemens Winkler
     Memorial 460
   Hieronymus
     Theodor Richter
     and Ferdinand Reich
     Monument 462
   Terra Mineralia 462
   Winkler Birthplace 461
   Winkler Monument,
     Albert Park 462, *462*

Place Index 523

Freshwater, Isle of
  Wight 33ch, 40–41

Geneva, Switzerland 180
GERMANY 258, 260–261,
  266–269, 302, 320–325,
  321ch, 361–371, 376–379,
  378ch, 383–384, 400–402,
  404–406, 409–410, 427–430,
  434–436, 437–444, 442ch,
  453–454, 459–463, 469–470,
  496–497
  Bad Dürkheim 273
  Benediktbeuren 267
  Berlin (see **Berlin,**
    Germany)
  Brandenberg 383–384
  Bug, Weißdorf 453
  Caputh 383
  Freiberg (see **Freiberg,**
    Germany)
  Giessen 322–325
  Göttingen (see **Göttingen,**
    Germany)
  Grossbothen (Wilhelm
    Ostwald Park and
    Museum) 260–261,
    261
  Grunewald 365–366
  Haigerloch (Atomkeller-
    Museum) 441, *443*,
    443–444
  Heidelberg (see
    Heidelberg,
    Germany)
  Helgoland 409
  Jena (Döbereiner
    Hörsaal, Painting and
    Statue) 453ch, **454**
  Karlsruhe 239–240,
    241–242
  Leipzig 405, 440
  Munich (see **Munich,**
    Germany)
  Regensburg 255–256
  Remscheid Lennup
    (Deutsches
    Röntgen Birthplace
    and Museum)
    320–321
  Tübingen 463
  Ulm 377–379
  Varel 463
  Würzburg (Röntgen
    memorial) 322
Giessen, Germany 322–325
Glasgow, Scotland 168–169,
  188
Glenlair House, Dumfries and
  Galloway, Scotland
  121–124, *122ch*, *123*
Godmanchester,
  Cambridgeshire,
  England 433, 441
Göttingen,
  Germany 366, 369, 395,
  401–402, 405–406
  Bunsen residence
    plaque 273
  I. Physikalisches
    Institut 369, 406
  Science
    City 369, 406
  Stadtfriedhof
    Cemetery 369,
    405–406, **438**, *439*
Grantham, Lincolnshire,
  England 33ch, 43–44
Graz, Austria (Physics Institut,
  University of Graz) 254
GREECE 12
Gripsholm, Sweden 481

Grossbothen (Wilhelm Ostwald Park and Museum), Germany 260–261, *261*
Grunewald, Germany 365–366
Haigerloch (Atomkeller-Museum), Germany 441, *443*, 443–444
Havelock, New Zealand 144–146
Heckmondwike, West Yorkshire, England *57ch*, 58
Heidelberg, Germany 273–279, *274ch*
    August von Kekulé plaque 278
    Bergfriedhof (Bunsen grave) 276
    Bunsen and Kirchoff plaque 275
    Bunsen plaque, Plöck 276
    Bunsen statue, Anatomiegarten 273–275
    Bunsen's Laboratory and Residence 275
    Carl Bosch Museum 279
    Hörsaal Zentrum Chemie (Lecture Hall), Chemistry 276–277
    Kirchhoff-Institut für Physik 277
    Mendeleev Residence 278
    Pharmacy Museum (Apothekenmuseum) 278–279
    Philosopher's Way (Philosophenweg) 277–278
    Stadthalle (Town Hall) 276
    University Library 279
Helgoland, Germany 409
IRELAND (Southern Ireland) 10, 14–23, *14ch*, 25–28, 172–173, *172ch*
    Deansgrange Cemetery, County Dublin 173
    Dublin 18–23, *20ch*, 21, 25–28, *27*
    Lismore, County Waterford (Robert Boyle Room) *14ch*, 15
    Lismore Castle, County Waterford 15–18, 38
    Walton Causeway Park, Dungarvan, County Waterford 172–173
    Waterford *172ch*, 173
    Youghal, County Waterford *14ch*, 18, *19*
ITALY 105, 223–239, *225ch*, *238ch*, 240–241, *255ch*, 415–422
    Bologna 231–232
    Como (*see* **Como, Italy**)
    Florence (*see* **Florence, Italy**)
    Palermo, Sicily 241
    Pavia (Cabinet of Physics of Alessandro Volta) 232
    Pisa (*see* **Pisa, Italy**)
    Quarenga 238
    Rome (*see* **Rome, Italy**)
    Trieste 254–256, *256*
    Venice 249
    Vercelli 237

Place Index

Jena (Döbereiner Hörsaal, Painting and Statue), Germany *453ch*, 454

Karlskoga (Alfred Nobels Björkborn), Sweden 506
Karlsruhe, Germany 239–240, 241–242
Kendal, Cumbria, England 78, *80ch*, 81–83
Kroppedal Museum, Denmark 308
Kungälv, Sweden 436
Kutna Hora, Czech Republic 12

Lake District National Park, Cumbria, England *80ch*
LATVIA 261
Leeds, England 56–58, *57ch*
    Leeds Library 57–58
    Mill Hill Chapel 56
Leipzig, Germany 405, 440
Linköping, Sweden 482
Lismore, County Waterford, Southern Ireland 14–18, *14ch*
    Lismore Castle 15–18, 38
    Robert Boyle Room, Lismore Heritage Centre 15
Lomonosovo (formerly Oranienbaum), Russia 472
London, England 37–40, *39ch*, 46–47, 102–103, 107–111, *108ch*, 176–195, 454–455
    Benjamin Franklin House 54, 198
    Christopher Ingold Building, Gordon Street 193
    Corpus Christi Roman Catholic Church 39
    Cuming Museum 108–109
    Faraday House 110
    Faraday Memorial 107–108
    Faraday Plaque, Blandford Street 109
    Faraday Statue 109
    Highgate Cemetery 111
    Jermyn Street, St. James (Newton plaque) 46
    7 Kensington Park Gardens (Crookes plaque) 132
    Monument to the Great Fire of London 41
    Royal Institution **102–103, 109–110**
    Royal Observatory, Greenwich 191
    Royal Society, Carlton House Terrace 39–40
    Science Museum 96, 193–194
    Slade School of Art, Gower Street 194
    St. Helen's Church, Bishopsgate 41
    St. James's Church, Piccadilly 40
    St. Martin's Street (Newton marker) 47
    St. Paul's Cathedral 41, 157
    Universtiy College (UCL) 193
    West Norwood Cemetery 455

London, England (*continued*)
   West Square,
    Lambeth (Newlands
    RSC Plaque) 454
   Westminster Abbey 41,
    167–168, 176–195,
    189ch, 190ch

Manchester 83–84,
  91, 92–96, *93ch,*
  153–160, *159ch*
   John Dalton Building 94
   **John Rylands
   Library 93–94**
   Manchester Literary
    and Philosophical
    Society 83, 84, 90,
    93ch, 94–95
   Museum of Science
    and Industry 96
   New College 79
   Portico Library &
    Gallery 95
   Rutherford
    Building, University of
    Manchester 158–159
   Schuster
    Laboratory,
    University of
    Manchester 159–160
   Town Hall 92–93
Miletus, Turkey 11–12
Montreal, Canada 148–153,
  151ch
   McGill
    University 148–152
   **Rutherford Museum,
   McGill University
   150–152,** *152*
   Rutherford's
    Residence 152–153

   Schulich Library
    of Science and
    Engineering 152
Munich, Germany 267–269,
  268ch, 370, 409–410,
  437–438, *439*
   Alter Südfriedhof
    München 269
   **Deutsches Museum 269,
   437–438**
   Einstein memorial
    plaque 379
   Fraunhofer
    Glassmaker Relief,
    Thiereckstrasse 267
   Fraunhofer Statue,
    Maximilian-
    strasse 268–269
   Heisenberg sites
    409–410, 444
   The Hofbräuhaus 257,
    257
   Municipal Museum
    (Münchner
    Stadtmuseum) 268
   Planck family home 366
   Waldfriedhof
    (Heisenberg grave) 444
Museum of Modern Art, New
  York 12–13

Nelson, New Zealand 146
New York 12–13
NEW ZEALAND 142–146,
  145ch
   Historic Cape Light and
    Museum (Rutherford
    Gallery), Taranaki 146
   Lord Rutherford
    Memorial Hall,
    Foxhill 144

Place Index 527

Lord Rutherford Memorial Reserve, Brightwater 143-144
Nelson College, Nelson 146
Rutherford-Pickering Memorial, Havelock 144-146
Rutherford's Den, Christchurch 147
Northumberland, Pennsylvania (Priestley's house) 64-65
NORWAY 440
Rjukan (Vemork Power Plant) 440, 443

Oxford, England *33ch,* 34-37, 169-172, *171ch*
Clarendon Laboratory 37, 170
History of Science Museum 36-37, 170-171
Oxford University Museum of Natural History 37, 171-172
Trinity College 170
University College 34-36
48 Woodstock Road (Moseley Family Home) 170

Palermo, Sicily 241
Pardshaw Hall, Cumbria, England 78, 79-81, *80ch*
Paris, France 61, 104-105, 203-205, 206-210, *207ch,* 211-213, 215-216, 217, 218-219, 330-341, *335ch, 340ch,* 342-344, 346-352, 390, 494
Bibliothèque Nationale de France 351
Cimetière de Neuilly-sur-Seine (Ancien) 390
Cimetière du Père Lachaise 215
City Hall 210
College Marazin (now Academy of Sciences of the Institute of France) 206-208
Curie Artifacts and Tours 351-352
Curie graves 350-351, 352
Curie Graves 350-351, 352
Curie Museum (Musée Curie) 349-350
Curie residences 348, 351-352
Curie workplaces 349-350
Eiffel Tower 198, 210, 215, 217
Errancis Cemetery 209-210
Franklin statue 198
Gay-Lussac Café 216
La Concierge 208-209
La Petit Arsenal 208
Montmartre Cemetery 217
Musée des Arts et Métiers 208
Muséum d'Histoire Naturelle 333

Paris, France *(continued)*
　Panthéon
　　(Sorbonne) 350–351
　Parcours des
　　Sciences 351
　Pasteur Museum, Pasteur
　　Institute 218–219
　Place de la Concorde 209
　Rue Pacquet (Lavoisier
　　birthplace) 206
　Saint Merri Church 206
　Sceaux 351, 352
Parton (Near Dumfries),
　Scotland 122ch, 124
Pavia (Cabinet of Physics of
　Alessandro Volta), Italy 232
Penzance, Cornwall,
　England 65ch, 66–68
Pisa, Italy 224–225, 416,
　425–426
　　Ammannati
　　　House 224
　　Collegio Fermi 426
　　Domus Galilaeana
　　　224–225, 425–426
　　Palazzo della
　　　Carovana 426
Ploubazlanec, l'Arcouest,
　Brittany, France 352–353
POLAND 333–334, 353–357,
　354ch, 400 *see also* Warsaw,
　Poland

Quarenga, Italy 238

Regensburg,
　Germany 255–256
Remscheid Lennup
　(Deutsches Röntgen
　Birthplace and Museum),
　Germany 320–321
Riga, Latvia 261
Rjukan, Norway (Vemork
　Power Plant) 440, 443
Rome, Italy 224, 415–416,
　418–419, 423–425, 423ch
　Campo dei
　　Fiori 224, 424
　Centro Fermi 424
　Museum of Chemistry
　　"Primo Levi" 241
　Physics Museum of
　　the University of
　　Rome 424–425
　Via Gaeta (Fermi
　　birthplace) 423–424
Rudkøbing, Island of
　Langeland,
　Denmark 288–289
RUSSIA 449–450, 464–472,
　466ch
　Lomonosovo (formerly
　　Oranienbaum) 472
　St. Petersburg (*see* St.
　　Petersburg)
　Tobolsk, Siberia
　　466–467, 466ch

SCOTLAND 48–51, 112–113,
　117–125, 122ch, 168–169
　Aberdeen 115
　Corsock, Dumfries and
　　Galloway (James Clerk
　　Maxwell Memorial
　　Window) 122ch,
　　124–125
　Edinburgh (*see*
　　Edinburgh, Scotland)

Glasgow 168–169, 188
Glenlair House, Dumfries and Galloway, Scotland 121–124, *122ch,* 123
Parton (Near Dumfries) *122ch,* 124
Sicily 241
SOUTHERN IRELAND *see* IRELAND (Southern Ireland)
St. Michael's Mount, Cornwall, England *65ch,* 66
St. Petersburg 449–450, 464–465, 467–469, *468ch, 470,* 470–472
Alexander Nevsky Lavra cemeteries 472
Dmitry Mendeleev's Memorial Museum Apartment (D. I. Mendeleev Museum and Archives) 467–468
Lomonosov Collection 471
Mendeleev Statue and Three-Story-High Periodic Table 469
Monument to Mikhail Lomonosov 471
Rosstandart Metrology Museum at the D. I. Mendeleyev Institute of Metrology 468–469
Technologichesky Institut (Technology Institute) Metro Stop 469

Stalbridge, Dorset, England 33, *33ch,* 34
Stockholm, Sweden 429, 430, 433, **480–481,** *480ch,* **483–484,** 487, **503–506,** *504ch*
Alfred Nobel Grave 506
Arrenius Bust and Grave 487
Berzelius Grave, Solna cemetery 481
Berzelius Museum 480–481
Berzelius Statue in Berzelius Park 481
The Crown Pharmacy in Skansen Park 484
The Nobel Museum 505–506
Scheele Pharmacy and Exhibit 484
Scheele Statue 483–484
Stockholm City Hall 505
Stockholm Concert Hall 503–505
SWEDEN *295ch, 431ch,* 436, 459, 476, 480–482, 483–484, 487, 503–506
Gripsholm 481
Karlskoga (Alfred Nobels Björkborn) 506
Kungälv 436
Linköping 482
Stockholm (*see* **Stockholm, Sweden**)
Uppsala 481, 487
Ven island 307–308

SWITZERLAND 105, *378ch*,
    379–382
  Bern (*see* **Bern,
    Switzerland**)
  Geneva 180
  Zurich 24, 382, 391, 398

Tallahassee, Florida,
    USA 408
Taranaki, New Zealand 146
Terling Place, Essex,
    England 128–129
Tobolsk, Siberia, Russia
    466–467, *466ch*
Trieste, Italy 254–256, *256*
Tübingen, Germany 463
TURKEY 11–12

Ulm, Germany 377–379
UNITED KINGDOM *see*
    ENGLAND; SCOTLAND
UNITED STATES 64–65,
    422–423, 426–427, 433
  Chicago,
      Illinois 426–427
  New York 12–13
  **Northumberland,
    Pennsylvania
    (Priestley's
    house) 64–65**
  Tallahassee,
      Florida 408

Varel, Germany 463
Venice, Italy 249
Vercelli, Italy 237
Vienna 248–249, 252–253,
    256–257, 398–399,
    399ch, 434

**Arcades of the University
    of Vienna 248–249,
    399**
Café Landtmann
    256–257
Gravesite of Josef
    Loschmidt 249
Rathauspark (Bust of
    Ernst Mach) 254
Zentralfriedhof (Central
    Cemetery) 256

Warrington, Cheshire,
    England 55, 79
Warsaw, Poland 353–357,
    354ch, 400
  Central Agricultural
      Library 355
  First Radiological
      Laboratory in
      Poland 355
  **Maria Skłodowska-
    Curie Institute of
    Oncology 356**
  **Maria Skłodowska-Curie
    Museum 353**
  Maria Skłodowska-Curie
      Park and
      Statue 356
  Monument to Maria
      Skłodowska-
      Curie 353–354
  murals of Maria
      Skłodowska-
      Curie 354–355, 357
  Powązki
      Cemetery 356
  Schrödinger
      Equation 400

Place Index 531

Warsaw University of Technology 355
Waterford, Southern Ireland 173
Woolsthorpe, Lincolnshire, England 42
Woolsthorpe Manor, Lincolnshire, England 43

Würzburg (Röntgen-memorial), Germany 322

Youghal, County Waterford, Southern Ireland *14ch*, 18, *19*

Zurich, Switzerland 24, 382, 391, 398

Subject Index

Page numbers in *italics* refer to figures; the suffix "*ch*" indicates a chart, and "T" a table. Please note there is a separate index of places.

acids
    Arrhenius's dissociation theory 485, 486
    Lavoisier's oxygen-component theory 101, 104, 202
Adams, Douglas 111
Adams, John Couch 192
air *see* pneumatic chemistry
Albert, Prince 110
alchemy 12, 34, 36, 37, 44, 199–200, 203
Allegheny College, Pennsylvania 260
alpha rays (particles) 148, 149–150, 153–156, 157–158, 163, 330, 332, 346–347
aluminum 288, 347
Amaldi, Edoardo 418–419, *419,* 420
American Chemical Society 59, 64, 206–208, 324, *324*
ammonia, NH_3 ("alkaline air") 55, 71, 435 *see also* Haber–Bosch process

Ampère, André-Marie 105, 106, 216–217
Andersen, Hans Christian 288, 290, 291
Anderson, Carl 347
anesthetic, nitrous oxide as 70
Ångström, Anders 279
anti-atomists 250, 253, 259–260, 286, 372–373
antimatter 190, 347, 402
argon 128, 465
Arrhenius, Svante 258, *480ch,* 484–487, 491–492, 493–494, 497, 498
Astaire, Fred 17
Aston, Francis *164ch,* 165
Aston, William 499–500
atom, splitting the 157–159, 173, 500
atomic bombs 423, 433, 439–444, *442ch, 443*
atomic number 169, 330
atomic orbitals *396,* 396–397, *397*
atomic symbols 85–86, *87, 88,* 330, 476–477

atomic theory
  Berzelius's theory 477–478
  Bohr's model 294–299, *297*
  and the church 224
  Dalton's theory and symbols 77, 84–89, *87, 88,* 89–92, 477
  Gay-Lussac's Law of Combining Volumes 213–215
  hard sphere model 85–86
  modern theory 377, 389–390, 391–397
  physical *versus* chemical atoms 86, 246–248, 250, 252–253, 362–363, 372–373
  Rutherford's nuclear model 155–157
  Thomson's "plum-pudding" model 133, *133,* 154
atomic weights, relative 87–89, *88,* 90, 91–92, 239–240, 239T, 246T, 450, 477
Austen, Jane 52
autism 403
Avogadro, Amedeo 91–92, 216, 236–238, 239–240

Bacon, Roger 37
Balmer, Johann 280–281
Balmer series *297,* 297–299
barium 101, 430, 432
batteries 71–72, 229–231

Becher, Johann 200
Becquerel, Henri 210, 330–333, *335ch*
Becquerel's rays 148, 336
Beddoes, Thomas 67–68, 69
Beethoven, Ludwig van 249, 256
Berlioz, Hector 217
Bernhardt, Sarah 215
Berthollet, Claude 205, 210–212, 215
Berthollet, Pierre Eugene-Marcellin 350
Berthollet, Sophie 350–351
beryllium 419–420
Berzelius, Elizabeth Johanna Poppius 481
Berzelius, Jöns Jacob 89–90, 476–482, *479, 480ch,* 505
beta decay 417
beta rays 148, 149, 332
Biot, Jean-Baptiste 213
Black, Joseph 32, 48–51, *49ch,* 50
black-body radiation 273, 363–365
Blackburn, Jemima 120, 121
Blackett, Patrick 160
Bohr, Christian (father of Niels) 291, 292, 307
Bohr, Christian (son of Niels) 306, 307
Bohr, Ellen *290,* 291–292, 304, 307
Bohr, Harald 291, 307
Bohr, Jenny 291
Bohr, Margrethe (née Nørlund) 292, 294, *296,* 298, 300, 301, 307

Subject Index

Bohr, Niels 290, 291–307, *295ch, 296, 304ch*
    atomic model 294–299, *297*
    at Cambridge under J. J. Thomson 292–293
    at Copenhagen, Institute for Theoretical Physics 301–303, 305–306, 401, 402
    early life and PhD thesis 291–292
    fame enshrined in a poem 300–301
    and Otto Hahn 430, 432, 502
    and Heisenberg 440–441
    Knight of the Order of the Elephant 307
    at Manchester with Rutherford 158, 293–294, 301
    and Lise Meitner 429, 430, 432, 502
    Nobel Prize 305, 499
    Nobel Prize medals kept from Nazis 302–303, *303*
    and nuclear fission 432, 440–441
    travel sites *295ch,* 304–307, *304ch*
Boltzmann, Henriette 250–251, 253, 256
Boltzmann, Ludwig 112, 248, 250–257, *255ch, 256,* 428, 434, 496
bonds, chemical 477–478

book burning (Bücherverbrennung) 366, 367
Born, Max 300, 369, 395, 401, 402, 403–404, 405–406, 416, 501
Borodin, Alexander 472
boron 104, 213
Bosch, Carl 279
Boyle, Catherine 18–19, *21*
Boyle, Katherine (Countess of Ranelagh) 32–33, 34, 37, 38, 40
Boyle, Richard, 1st Earl of Cork 15–17, 18, *19,* 19–20
Boyle, Robert 13–23, *16, 20ch, 21,* 24, 31, 32–40, *33ch, 39ch*
    air experiments *35,* 35–36, 47
    hydrogen preparation 52
    phosphorus production 38
    travel sites 15–18, 18–23, *20ch, 33ch,* 34–35, 39–40, *39ch*
Boyle's Law 31, 36, 43
Bragg, Lawrence 117, 159–160
Brahms, Johannes 249
Braille, Louis 350
Brandt, Hennig 37
Britten, Benjamin 192
Broglie, Louis de *see* de Broglie, Louis
Bronowski, Jacob 111
Brothers Grimm 276
Brown, Ford Madox 92
Brownian motion 372–373, 500
Bruno, Giordano 224, 424
Buck, Pearl 422

Bunsen, Robert 240, 265, 269, 270–277, *272, 274ch,* 449
Bunsen burners 271, *272,* 276–277, *277*

calcium 101, 103
Calvin, John 180
calxes (oxides) 60–6161, 200, 201, 202
canal rays 134–135, 154
Cannizzaro, Stanislao 236, 238–242
carbon 86, *87, 88,* 105, 226, 246T, 477
carbon dioxide ("fixed air") 48, 55, 85
carbon monoxide 64, 71, 84
Carlsberg Foundation 292, 301
cathode rays 130–133, 134, 315, *316*
Cavendish, Henry 17, 32, 52–53, 116, 117, 203
Cavendish, William, 7th Duke of Devonshire 17, 52, 116–117, 127
Cavendish family (Dukes of Devonshire) 17
Cavendish Laboratory, Cambridge 52, 116–117, 125–126, 127, 128, 129–130, 135, 160–166
cerium 476, 479
cesium 271
Chadwick, James 158, 160, 163–164, 165, 346
Chantrey, Sir Francis 92, 194
Charles's Law 212–213

chemical bonding 477–478
chemical equations 206, 246–247
Chemical Landmarks *see* International Historic Chemical Landmarks; National Chemical Landmarks; National Historic Chemical Landmarks
chemistry equipment *see* equipment, laboratory
chlorine 478, 483
Chopin, Frédéric 215
chromium 349
chrysopoeia 34
Clerk Maxwell, James *see* Maxwell, James Clerk
Cleve, Per Teodor *452ch,* 459, 485, 486
Cockcroft, John 163–164, 165, 173
Coleridge, Samuel Taylor 69, 182
color-blindness 83, 91, 96, 114, 280
colors 113–114, 259, 260, 270–271
confined waves 391–393, *392*
consumption (tuberculosis), treatment of 67–68
Copernicus, Nicolaus 379, 437
copper *88,* 477
corpuscular hypothesis 12, 36, 43, 102
Courtois, Bernard 105
Crabtree, William 92
Crick, Francis 26, 126–127

Crookes, Sir William 131–132, 280
Crookes tubes 131, 132
Curie, Eugène 352
Curie, Ève 341, 343, 345, 347, 349
Curie, Irène *see* Joliot-Curie, Irène
Curie, Marie 333–357, *335ch, 336, 337, 340ch, 354ch*
    early life and education 333–334, 348, 355, 357
    and *l'Académie des Sciences* 341
    and Paul Langevin 341–342, 375, 494
    meets and marries Pierre 334–335, 348
    Nobel Prizes 339–340, 341–342, 494
    radiation poisoning 338–340, 343, 347
    discovers radium in pitchblende 336–338, 348, 349
    and the Radium Institute 343, 344, 346, 349–350, 353
    receives radium from women of America 344–345, *345*, 350
    at the Solvay Conference, Brussels 341, *342*, 375
    tragic and triumphal year (1911) 341–342, 348, 375
    travel sites *335ch, 340ch,* 348–357, *354ch*
    World War I *343,* 343–344
Curie, Pierre 333, 334–340, *335ch, 336, 337, 340ch,* 343, 348, 349, 350, 351, 352
Curtius, Theodor 276

D'Agostino, Oscar 418–419, *419*
Dalton, John 77–97, *78, 80ch, 81, 82, 93ch*
    atomic theory 71, 77, 84–89, *87, 88,* 96, 239, 477
    atomic theory, reactions to, 89–92
    as Clockman for Portico Library 95
    his closed mind in later life 91–92, 214, 215, 476–477
    his color-blindness 83, 91, 96
    honors 91
    Law of Multiple Proportions 85, 92
    Law of Partial Pressures 83
    rule of greatest simplicity 86–87
    travel sites 79–83, *80ch,* 92–96, *93ch*
Dante Alighieri 226
Darwin, Charles 191–192
Darwin, Erasmus 63
Davisson, Clinton 389, 390, 501

Davy, Edmund 100
Davy, Sir Humphry *39ch*,
   65–73, *65ch*, 100–103,
   102, 108ch, 407ch
   and Berzelius 478
   coal miner's safety lamp
      invention 105
   and Dalton's atomic
      theory 90–91, 102
   diamonds burned
      by 105, 226
   electricity
      experiments 71–72
   Faraday discovered
      by 103–105
   at the Pneumatic
      Institution (gas
      experiments) 68–71
   as poet 66, 70
   at the Royal
      Institution 72,
      101–104
   travel sites 37, *39ch*,
      65ch, 102–103, *108ch*
   Westminster Abbey
      memorial 180
de Boisbaudran, Paul Émile
   Lecoq 280, *452ch*, 457–459,
   465, 467
de Broglie, Louis 24, 388–390,
   500–501
de Broglie,
   Maurice *342*, 388
de Chancourtois, Alexandre-
   Emile Beguyer 451
de Hevesy, Georgy 302–303,
   303
Degas, Edgar 217
Delbrück, Max 436
Democritus 11–12

dephlogisticated air *see*
   oxygen
Devonshire, Dukes of *see*
   Cavendish family
Dirac, Paul 190, 402–403,
   406–409, *407ch*, 501
Dirac Equation 402–403
dissenters, English 54–55
dissociation theory 258,
   485–486, 492
DNA 26
Döbereiner, Johann
   Wolfgang 450, *452ch*,
   453–454, *453ch*
Doppler, Christian 249
Dostoyevsky,
   Fyodor 472
dot density
   diagrams 395–396, *396*
Dryden, John 42
Dumas, Alexandre 217
Duncan, Isadora 215
dynamic corpuscularity 43
dynamite 487–488
dysprosium 458

Eddington, Sir Arthur 81,
   82, *82*
Einstein, Albert *342, 367ch*,
   369–384, *371, 378ch, 380ch*
   explains Brownian
      motion 372–373
   and de Broglie 388, 395
   early life and
      education 369–371,
      379
   his blackboard at
      Oxford 171
   his chair at the Max
      Planck Society 368

Subject Index

marriage to Mileva
  Marić 370–372, 376,
  380
on Lise Meitner 429
Nobel prize 371, 373,
  374, 376, 497–499
in the patent office at
  Bern 371, 380–381
explains photoelectric
  effect with quantum
  theory 373–377
and Max Planck 360, 365,
  375, 376
relativity theory 82, 118,
  163, 171, 372, 381, 402,
  439, 498, 499
and Schrödinger 24, 388,
  395, 397
standing on Maxwell's
  shoulders 118, *120*,
  375
travel sites 269, *367ch*,
  377–384, *378ch, 380ch*
Einstein, Hans Albert 372, 380
Einstein, Mileva (née
  Marić) 370–372, 375, 376,
  382
eka-elements,
  Mendeleev's 456–460
electricity 54, 105–106,
  115–117, 228–231
electrochemistry 106, 107,
  485
electromagnetism 106,
  115–116, 217, 286–287, *287*
electrons
  as beta particles 332
  Bohr's atomic
    model 298–299,
    300–301
  discovery of 132–133
  Heisenberg's Uncertainty
    Principle 401–402
  in the photoelectric
    effect 373
  wave-particle duality 388,
    389–390, 393–397, 501
elements
  defined by Lavoisier 206
  discovered by
    Berzelius 476, 479
  discovered by Davy 66
  discovered by
    Scheele 482–483
  discovered by spectros-
    copy 271–272, 273,
    279–281
  Mendeleev's
    eka-elements 456–460
  named after atomic
    scientists 434
  symbols for 85–86, *87*,
    330, 476–477
  timeline of discovery *457*
  transuranic 421–422
  see also hydrogen; inert
    gases; oxygen;
    periodic table; radium;
    *and other individual
    elements*
Elgar, Sir Edward 192
entropy 251–252
equations, chemical 206,
  246–247
equipment, laboratory 59–60,
  60, 96, 151, *152*, 165, 226,
  227, 232, 254, 260, 322, 349
Erlenmeyer,
  Emil 240, 278
Euler, Leonhard 472

europium 458
Ewart, Peter 96

Faraday, Michael 103–110,
    108ch, 306
    as analytical chemist 105
    electricity studies
        105–106, 107, 131
    Maxwell and 115
    religious faith 111
    travel sites 50, 107–111,
        108ch
    Westminster Abbey
        memorial 183, 185
Fermi, Enrico 415–427, *418,
    419, 423ch*, 430, 431–432
    and Schrödinger 24
    travel sites 224, 225, 226,
        234, 423–427, *423ch*
Fermi, Laura (née Capon) 415,
    417, 421
fire and flames 200–201
fireworks 67
first world war *see* World War I
Fischer, Emil 368, 429,
    490–491
fixed air *see* carbon dioxide
flame tests 270–271
Fock, Vladimir A. 300–301
Foley, John 109
Franck, James 302
Franklin, Benjamin 54,
    197–198, *199, 207ch*
Franklin's kite
    experiment 54, 471
Franz Joseph, Emperor 253
Fraunhofer, Joseph von
    265–269, *268ch*
Fraunhofer lines 266, *267*,
    272–273, 280

French Revolution 63–64,
    203–205, 209, 217, 388
Freud, Sigmund 257
Frisch, Otto 430–431, 432, 433
Frisch, Robert 502

Galileo Galilei 23, 37,
    223–227, *225ch, 227*, 306,
    379, 425
gallium 280, 457–459, 458T
Galvani, Luigi 228–229,
    231–232
gases
    Boyle's air
        experiments *35*,
        35–36, 47
    inert 128, 193, 465
    kinetic molecular
        theory 43, 83–84, 112,
        113, 372
    Law of Partial
        Pressures 83
    pneumatic
        chemistry 47–48,
        51–54, 53T, 55–56, *60*,
        60–61, 62, 64, 67–72
Gay-Lussac,
    Geneviève 213
Gay-Lussac, Joseph Louis 91,
    210, 212–216
Gay-Lussac's Law 212–213
Geiger, Hans 153–155, 157,
    158, 160
Geissler, Heinrich 131
genetic studies 26
germanium 459–461
Germer, Lester 389, 501
Gibbons, Orlando 192
Glinka, Mikhail 472
Gmelin, Leopold 450–451

Subject Index

Goethe, Johann Wolfgang 454
gold 34, *88,* 302–303, 477
gold-making (chrysopoeia) 34, 37
Goldstein, Eugen 134
gravity 43
Gray, Henry 111
Great Fire of London (1666) 41, 42
Green, George 188
Grimm brothers 276
Guillaume, Charles 498
Guinness, Sir Benjamin Lee 21

Haber, Clara Immerwahr 427, 435
Haber, Fritz 427, 435–436, 483, 497
Haber–Bosch process 258, 279, 435, 436
Hahn, Otto 269, 365, 369, 427–436, *428,* 437, 438, *439,* 502–503
Hales, Stephen 47
half-life of radioactive elements 150
Halley, Edmond 191
Hanckwitz, Ambrose Godfrey 38
Happiness Formula (Ostwald) 259
hard sphere model 85–86
Hawking, Stephen 43, 403, 408
Heaviside, Oliver 116, 119
heavy water 440, 441, 443, 444
Heisenberg, Werner 269, 368, 400–405, 409–410, 439–444, *442ch,* 501
Heisenberg's Uncertainty Principle 401–402
helium 193, *272,* 279–280, 465
Helmholtz, Hermann von 276
Herschel, Caroline 190–191
Herschel, Sir John Frederick 191
Herschel, Sir William 183, 190–191
Hertz, Gustav 302
Hertz, Heinrich 116, 373
Himmler, Heinrich 439
Hisinger, Wilhelm 476, 479, 481
Hitler, Adolf 365–366, 422
Hooke, Robert 31, *33ch,* 35, 37–38, 39, *39ch,* 40–41, 181–182
Hutton, James 51
Huygens, Christiaan 375
hydrogen
    atomic symbol 86, *87, 88,* 477
    atomic weight *88,* 246T
    emission spectrum *272,* 280–281, *281,* 297, 297–299, 401
    wave mechanics of 393–395, *397*
hydrogen, H_2, ("inflammable air") 52, 71, 203, 478
hydrogen chloride, HCl ("marine acid air") 55, 202
hydrogen cyanide 482

indium 280
inert gases 128, 193, 465
inflammable air *see* hydrogen
International Chemical Congress, Karlsruhe 239–240, 449, 463
International Historic Chemical Landmarks 59, 206–208
Invisible College (of natural philosophers) 32–33, 37
iodine 105, 347–348
ions 107, 485, 486
iron 273
isotopes 134–135, 168–169, 330, 347

Janssen, Pierre 279
Joliot-Curie, Frédéric 163, *335ch, 340ch,* 346–348, 349, 352–353, 417–418
Joliot-Curie, Irène 163, *335ch,* 336, *340ch,* 342, 343–344, 345, 346–347, 348, 349, 351, 352–353, 417–418
Jordan, Pascual 401, 402, 403–404
Joule, James Prescott 92–93, 96, 192–193

Kaiser Wilhelm Institute (KWI) for Chemistry 427, 428–429, 430, 436
Kaiser Wilhelm Institute (KWI) for Physical Chemistry and Electrochemistry 435
Kaiser Wilhelm Institute (KWI) for Physics 440, 441
Kant, Immanuel 286

Kapitza, Peter (Pyotr) 46, 161, 166, 306
Karlsruhe Congress 239–240, 449, 463
Kekulé, August 239–240, 278, 455–456
Kelvin, William Thomson 115, 188
Kepler, Johannes 308, 379
Kierkegaard, Søren 291
kinetic molecular theory of gases 43, 83–84, 112, 113, 372
Kirchhoff, Gustav 269, 270–273, 274, *274ch,* 275, 276, 277, 449
Kölliker, Rudolf Albert von 318

laboratories 12, 33–34, 35, 38, 39, 41, 44–46, 58–60, *60,* 64, 109, *110, 204, 208,* 322, 323, 325, 349
Lagrange, Joseph-Louis 209
Landmarks, Historic Chemical *see* International Historic Chemical Landmarks; National Chemical Landmarks; National Historic Chemical Landmarks
Langevin, Paul 341–342, *342,* 350, 388
Laplace, Pierre 215
latent heat 50
Laud, William, Archbishop of Canterbury 19
Laue, Max von 369
laughing gas *see* nitrous oxide, N_2O

Lavoisier, Antoine 198–210, *204, 207ch*
    combustion and oxidation experiments 201–203, 482–483
    death 205, 209
    Elementary Treatise on Chemistry 67, 205, 206
    Law of Conservation of Mass 201
    in Paris 203–205
    Priestley and 55, 61
    travel sites 206–210, *207ch, 208*
Lavoisier, Marie Anne Paulze 198, *199, 204,* 204–205
Law of Combining Volumes 91, 213–215
Law of Conservation of Mass 201, 470–471
Law of Definite Proportion (or Constant Composition) 84–85
Law of Multiple Proportions 85, 92
Law of Partial Pressures 83
Le Bel, Joseph 218
lead *88,* 316–317, 319
Lémery, Nicholas 52
Lenard, Phillip 498
Leucippus 11–12
Levi, Primo 241
Lewis, C. S. 172
Liebig, Justus von 269, 323–325
light 42–43, 116, 218, 363–365, 373–375 *see also* spectroscopy
Lincoln, Abraham 93

Linnaeus, Carl 483
Lister, Joseph, 1st Baron Lister 192
Lit & Phil society *see* Manchester Literary and Philosophical Society
Litvinenko, Aleksandr 348
Lomonosov, Mikhail *468ch,* 469–472
Lonsdale, Dame Kathleen *82,* 82–83
Loschmidt, Josef 248–249, *255ch,* 256
Louis XVI, King of France 209
Łukaszczyk, Franciszek 356
Lunar Society 63, 67–68
Lyman, Theodore 281

Macdonald, Sir William 148
Mach, Ernest 249–250, 252–253, 254, *255ch,* 496
Machiavelli, Niccolò 226, 425
magnesium 101, 270, 273
Manchester Literary and Philosophical Society (Lit & Phil) 83, 84, 90, 94–95
Manhattan Project 422, 432
Marcet, Jane 103, 105, 106
March, Hilde 25, 400
Marie Antoinette, Queen of France 209
Marsden, Ernest 154–156, 157, 158, 160
Marx, Karl 111
matrix mechanics 401–402
Max Planck Society 366, 368
Maximilian Joseph I, King of Bavaria 266, 268–269

Maxwell, James Clerk 107,
    111–127, *114, 118ch, 119,
    120, 122ch, 126ch*
  electricity and
    magnetism
    studies 115–117
  light (and color)
    experiments 113–114,
    115, 116, 375
  travel sites 50, 117–127,
    118ch, 122ch, 126ch
  Westminster Abbey
    memorial 183,
    185–186
Maxwell, John Clerk 112–113,
    124
Maxwell–Boltzmann
    distribution 251
Maxwell's equations 115–116,
    294, *296,* 375
Meitner, Lise 427–435, *428,*
    431ch, 435, 436–437
  and Boltzmann 252
  denied a Nobel Prize 433,
    502–503
  discovery of nuclear
    fission 427–435
  and the Plancks 365, 428
  travel sites 367, 434–435,
    436–437, *438*
Mendeleev, Dmitri 449–450,
    455, 455–457, 458–460, 463,
    464–469, *466ch*
  his later years 464–465,
    486
  Nobel nomination
    blocked by
    Arrhenius 486, 492
  and the periodic
    table 449–450, *452ch,*
    455–457, 458–460, 463,
    465, *470*
  travel sites 278, 466–469,
    466ch, 468ch
mephitic air *see* nitrogen
mercury *88, 272*
metal oxides (calxes) 60–61,
    200, 201, 202
methane 92, 228
Meyer, Julius Lothar 240,
    452ch, 463–464
Meyer, Viktor 276
Michelangelo Buonarroti 226,
    227, 425
moderators, neutron 421,
    424, 440
Moissan, Henri 215, 465, 492
mole, coining of term 258
molecules 83–89, *87,* 214–215,
    218, 240, 246–247, 477–478
Molière 215
Morrison, Jim 215
Moseley, Henry (Harry) 37,
    158, 160, 169–172, *171ch*
Moussorgsky, Modest 472
Mozart, Wolfgang
    Amadeus 249
Mussolini, Benito 422

Napoleon Bonaparte
    (Napoleon I) 104,
    211–212, 231, *232*
National Chemical Landmarks
    (RSC) 79, 170, 194
National Historic Chemical
    Landmarks (ACS) 64
Nazis 24–25, 302–303, 356,
    365, 404, 429–430, 439, 440
neon 134, 193, *272,* 499
neptunium 421

Subject Index

Nernst, Walther 342, 369
neutron bombardment 418–423, 429, 430
neutron moderators 421
neutrons 162, 163–164, 330, 346
Newlands, John 451–452, 451T, 452ch, 454–455
Newton, Sir Isaac 31, 33ch, 37, 39ch, 41–47, 45ch, 118, 119, 182, 183–184, 379
Newton-John, Olivia 402
Nicholas II, Tsar 467
Nilson, Lars Fredrik 452ch, 459, 467
nitric acid 71
nitric oxide, NO ("nitrous air") 55, 71, 85, 489
nitrogen ("mephitic" or "phlogisticated air") 51–52, 86, 87, 246T
nitrogen dioxide, NO_2 ("phlogisticated nitrous air") 55
nitroglycerin 487, 489
nitrous acid 71
nitrous oxide, N_2O ("diminished or dephlogisticated nitrous air") 55, 65, 68–69, 70–71
Nobel, Alfred 486, 487–490, 489, 504, 506
Nobel Prizes 487–506, 504
    biases and prejudices 339–340, 492, 494–497, 498–499, 502–503
    for Chemistry 128, 153, 169, 215, 260, 341–342, 347, 433, 436, 486, 490–494, 491T, 493, 499–500
    early days of 490–497, 491T, 495T
    Einstein and 371, 373, 374, 376, 497–499
    established by Alfred Nobel 487–490
    the Hahn/Meitner debacle 433, 501–503
    Liebig's training influence 324
    medals hidden from Nazis 302–303, 303
    Mendeleev's nomination blocked by Arrhenius 486, 492
    nuclear physics and 501–503
    for Physics 24, 128, 133, 160, 163, 164, 166, 173, 190, 302, 319, 346, 347, 365, 374, 389–390, 394, 395, 403–404, 422, 494–497, 495T, 500, 501–503
    quantum mechanics finally honored 500–501
    travel sites 503–506
    winners buried in Stadtfriedhof, Göttingen 369
noble gases 128, 193, 465
Noddack, Ida 422
nuclear chemistry nomenclature 330
nuclear fission 422–423, 427–434, 436, 440–441, 443, 502–503

nuclear model of
   atom 155–157
nuclear physics 414–415,
   417–423, 501–503
nuclear weapons *see* atomic
   bombs
nucleons 330 *see also*
   neutrons; protons

Odling, William 452
Ohm, Georg Simon 269
orbitals, atomic *396,*
   396–397, *397*
Ørsted, Anders 288–289
Ørsted, Hans 37
Ørsted, Hans
   Christian 105, 285–291,
   289, 295ch, 304ch
Ørsted, Sophie 288
Ostwald, Wilhelm 253, 254,
   258–261, 371, 373, 485–486,
   493–494, 496–497
Ostwald process 258, 435
oxidation 200–201,
   202–203
oxides (calxes) 60–61, 200,
   201, 202
oxygen ("dephlogisticated
   air")
     atomic symbol 86,
        87, 88, 477
     atomic weight *88,* 246T
     as component of
        acids 101, 104, 202
     discovery of 53–54,
        55–56, 60–61, 62, 478,
        482–483
     named by
        Lavoisier 202–203

Pasteur, Louis 216, 218–219
periodic table 449–452, *452ch,*
   455–460, *456,* 463, 464, 465,
   469
Perrin, Jean 131–132, *342,*
   350, 500
Petty, William,
   2nd Earl of Shelburne 58,
   61, 62
phlogisticated air *see* nitrogen
phlogiston 200–203, 470
phosphorus 37, 38, *88,* 246T,
   347, 348
photoelectric effect 373–374,
   374
photography 270, *271*
photons 373–374, 376–377
photosynthesis 56
Pickering, William
   Hayward 145–146
Planck, Erwin 365
Planck,
   Hermann 368, 369
Planck, Marga (née von
   Hoesslin) 368, 369
Planck, Marie (née
   Merck) 365, 367
Planck, Max 269, 306, *342,*
   360–369, *362, 367ch,* 376,
   428, 496, 497
Planck's constant 295, 300,
   364, 369, 394
Planck's equation 295
platinum *88,* 454
plutonium 421–422, 423
pneumatic chemistry 35,
   35–36, 47–48, 51–54, 53T,
   55–56, *60,* 60–61, 62, 64,
   67–72

Subject Index 547

Pneumatic Institution, Bristol 65*ch*, 68–72
Poincaré, Henri 342, 494
pointillistic paintings 12–13, *13*
polonium 337, 346, 348, 420
Popova, Anna Ivanova 464
positrons 190, 346–347, 402
potassium 101, 288, 477
Potter, Beatrix 277, *277*
Priestley, Joseph 53–65, 57*ch*, 63
    discovery of oxygen and other gases 55–56, *60*, 60–61, 62, 64, 203, 482–483
    electricity experiments 54
    on phlogiston 63, 200
    religious beliefs 54–55, 61, 63–64
    soda water ("windy water") preparation 55
    travel sites 37, 56–61, 57*ch*, 64–65
Priestley, Mary 62, 64–65
Priestley Medal 64
Priestley Society 57–58
protactinium 428
protons 154, 157–158, 163, 330
Proust, Joseph 84
Proust, Marcel 215
Prout, William 89
Prout's hypothesis 89
Purcell, Henry 192

Quaker scientists 81–83, *82*, 91
Quaker sites 79–81, *80ch*
quantum mechanics (wave mechanics) 24, 190, 391–397, 395, 401–405, 500–501
quantum numbers 393, 394
quantum theory 295–299, 360–361, 363–365

radioactivity
    artificial 347–348, 417–420
    Cockcroft and Walton's experiments 163–164
    discovery by Becquerel and the Curies 328–333, *332*, 336–340, 341–348, 349–350, 351
    Rutherford's experiments 148–150, 153–156
    Soddy and the concept of isotopes 168–169
    see also X-rays
radium 337–339, *339*, 343, 344–345, 350, 419–420
radon 419–420
Raleigh, Sir Walter 16–17, 18
Ramsay, Sir William 128, 181, 192, 193–194, 328–329, 427, 452*ch*, 465, 492
Rasetti, Franco 418–419, *419*, 429
Rayleigh, John William Strutt, 3rd Baron 128–129, 180–181, 452*ch*, 465
Rayleigh radiation 128

Reich, Ferdinand 280, 462
relativity, theory of 82, 118, 171, 372, 381, 402, 439, 498, 499
religious beliefs
    Joseph Black and Scottish Covenanters 50–51
    Robert Boyle 22–23, 24
    Faraday and Sandemanians 111
    Galileo and the church 224, 226
    Priestley and rational dissenters 54–55, 61, 63–64
    Quaker scientists 81–83, *82*, 91
Richter, Hieronymus Theodor 280, 462
Rimsky-Korsakov, Nikolai 472
Roget, Peter Mark 70
Rømer, Ole *295ch, 304ch,* 308–309
Röntgen, Bertha (née Ludwig) 314, 315–316, 317, *317,* 322
Röntgen, Wilhelm 312–322, *321ch,* 496–497
    discovers X-rays 315–320, *317*
    early life and education 313–314
    travel sites 269, 320–322, *321ch*
Röntgen rays *see* X-rays
Roscoe, Henry Enfield 276–277
Rossetti, Christina 111
Rossini, Gioachino 226, 425

Royal Institution 72, 90, 101–104, 106–107, 109–110, 115, 133
Royal Society 39–40
    Dalton and 90–91, 92
    Darwin's funeral 192
    Davy Medal recipients 454, 464, 486
    Lonsdale as first woman Fellow 82–83
    Newton's funeral 183
    origins as the Invisible College 32–33, 37
    Priestley and 55, 58
Royal Society of Chemistry 59, 79, 170, 194
rubidium 272
rule of greatest simplicity (Dalton) 86–87
rusting of metals 200
Rutherford, Daniel 32, 50, 51–52
Rutherford, Eileen 162
Rutherford, Mary ("May") 143, 146, 147, 149, 161, 167
Rutherford, Sir Ernest, 1st Baron Rutherford of Nelson *126ch,* 142–168, *144, 151ch, 159ch, 164ch,* 342
    alpha particle–gold foil experiments 154–156
    anti-submarine detection (sonar) 157
    and Bohr 293–294, 299, 301, 302, 306
    at Cambridge 147–148, 160–163, 164–167

Subject Index

created 1st Baron
   Rutherford of
   Nelson 146, 162–163
Crocodile nick-
   name 165–166, *166*
and the Curies 338
death 167
early life in New
   Zealand 142–147
on Faraday 103
and Fermi 418
in Manchester 153–160,
   293–294
at McGill University,
   Montreal 148–152
and Moseley 169
Nobel Prize for
   Chemistry 153,
   492–493
radiation studies
   148–150, 153–156
splitting the atom
   157–159, 500
travel sites 46, *126ch*,
   143–147, *145ch*,
   150–153, *151ch*,
   158–160, *159ch*,
   164–168, *164ch*
Westminster Abbey
   memorial 186–188
Rydberg constant 280,
   297–298, 299

Sacks, Oliver 319–320
samarium 458
Sax, Adolph 217
scandium 459
Scheele, Carl Wilhelm *480ch*,
   481, 482–484

Schoenberg, Arnold 249
Schrödinger, Anny 24, 25, 400
Schrödinger, Erwin 11, *20ch*,
   23–28, 388, 390–405, *399ch*,
   401, 402, 403, 434, 500–501
Schrödinger equation
   393–395, 399–400
Schrödinger's Cat 398, *404*,
   404–405
Schubert, Franz 249
scurvy, treatment of 55
second world war *see* World
   War II
Segrè, Emilio 418–419, *419*
selenium 479, 482
Seurat, Georges 12–13, *13*
Shelburne, William Petty, 2nd
   Earl of 58, 61, 62
Shelley, Percy Bysshe 35
Siegbahn, Manne 429, 430,
   433, 502, 503
silver 12, *88*
Skłodowska-Curie, Marie *see*
   Curie, Marie
Soddy, Frederick 150, *151ch*,
   168–169, 499–500
sodium 101, *272*, 273, 477
Solvay, Ernest *342*
Solvay Conference 1911,
   Brussels 341, *342*, 375, 496
Sommerfeld, Arnold *342*, 401
Southey, Robert 69–70, 182
spectroscopy 266, 270–273,
   272, 279–281, 297–299, 337,
   401, 425
splitting the atom 157–159,
   173, 500
Stahl, Georg 200
Stein, Gertrude 215

Stendahl 217
Stokes, Sir George 192
Stoney, George 132
Strassman, Fritz 429, 430, 431–432, 433, 436, 502
Strauss, Johann II 249
strontium 101, 450T
Strutt, John William, 3rd Baron Rayleigh 128–129, 180–181, *452ch,* 465
sulfur dioxide, SO_2 ("vitriolic acid air") 55
Sullivan, Arthur 109
Swift, Jonathan 20
symbols, atomic 85–86, *87,* 330, 476–477

Tait, Peter Guthrie 113, 117, 121
Tchaikovsky, Pyotr 472
Tennyson, Alfred Lord 40–41
Tesla, Nikola 318–319
thallium 280
Thatcher, Margaret 44
Thomson, George Paget 389–390, 501
Thomson, Sir John Joseph ("J. J.") *126ch,* 128, 129–130, 132–135, 306
    and Niels Bohr 292–293, 299
    canal rays and discovery of isotopes 134–135
    cathode rays and the discovery of the electron 132–133, 389–390
    at the Cavendish Laboratory 129–130, 132–135, 148–149, 162, 292–293
    on Ernest Rutherford 148–149, 160
    travel sites 46, *126ch,* 135
    Westminster Abbey memorial 186
Thomson, Rose 186
Thomson, Thomas 89
Thomson, William, 1st Baron Kelvin 115, 188
thorium 168, 336, 479
Thynne, Lord John 192
tin 66, 477
Tolkien, J. R. R. 172
Torricelli, Evangelista 131
Travers, Morris 465
tuberculosis (consumption), treatment of 67–68
Tycho Brahe *295ch, 304ch,* 307–308, 309

ultraviolet catastrophe 363–365
Uncertainty Principle 401–402
uranium 150, 168, 330, 331–333, *332,* 421, 422, 429, 430–432, 441, 444
urine, phosphorous from 37, 38

vacuum technology 131
valences 455–456
van Helmont, Jan Baptista 47
van't Hoff, Jacobus 218, 486, 490, 491
Vaughan Williams, Ralph 192
Vauquelin, Louis Nicolas 349
Victoria, Queen 111
Volta, Alessandro 71–72, 101, 105, 212, 227–231, *232,* 232–236, *233*

Subject Index 551

voltaic piles 71–72, 101, 212, 227, 229–231, *230, 232*
von Laue, Max 302

Wallace, Alfred Russel 192
Wallach, Otto 369
Walton, Ernest 163–164, 165, 172–173, *172ch*
Watson, James 26, 126–127
Watt, Gregory 67, 68
Watt, James 56, 63, 67, 194
Watts, Isaac 103, 104
wave equations 393–395, 399–400
wave mechanics (quantum mechanics) 24, 190, 391–397, 395, 401–405, 500–501
wave-particle duality 375, 388, 389–390, 501
Wedgwood, Josiah 49, 63, 67
Wedgwood, Thomas 67
Whewell, William 54
Wilde, Oscar 27, 215
Wilhelm II, German Emperor 318

Wilson, Charles Thomson Rees 500
Windaus, Adolf 369
"windy water" (soda water) preparation 55
Winkler, Clemens *452ch,* 459–462, *462,* 467
Wollaston, William 89
Wordsworth, William 46, 69, 182
World War I 157, 160, 170, 301, 343–344, 388, 427, 428, 435–436, 483, 496–497
World War II 302–303, 356, 365, 366, 405, 433, 440–441, 443–444, 472
Wren, Sir Christopher 41, 110
Wurtz, Adolphe 210

X-rays (Röntgen rays) 148, 312, 313, 315–320, *317,* 321, 322, 343

Young, Thomas 181

Zola, Émile 217
Zsigmondy, Richard 369